Precision Agriculture for Sustainability and Enviro Protection

CW01432593

Precision agriculture (PA) involves the application of technologies and agronomic principles to manage spatial and temporal variation associated with all aspects of agricultural production in order to improve crop performance and environmental quality. The focus of this book is to introduce for a non-specialist audience the role of PA in food security, environmental protection, and sustainable use of natural resources, as well as its economic benefits.

The technologies covered include yield monitors and remote sensing, and the key agronomic principles addressed are the optimal delivery of fertilizers, water and pesticides to crops only when and where these are required. As a result, it is shown that both food production and resource efficiency can be maximized, without waste or damage to the environment, such as can occur from excessive fertilizer or pesticide applications. The authors of necessity describe some technicalities about PA, but the overall aim is to introduce readers who are unfamiliar with PA to this very broad subject and to demonstrate the potential impact of PA on the environment and economy.

The book shows how farmers can place sustainability of the environment at the centre of their operations and that this is improved with the application of PA. The range of topics described includes sampling and mapping, weed and pest control, proximal and remote sensing, spatio-temporal analysis for improving management, management zones and water management. These are illustrated with case studies on sampling and mapping, biofuels from sugarcane and maize, paddy rice cultivation, and cotton production.

Margaret A. Oliver is a Visiting Professor in the Department of Soil Science at the University of Reading, UK, and was for several years editor of the *Journal of Precision Agriculture*.

Thomas F. A. Bishop is a Senior Lecturer in the Faculty of Agriculture and Environment at the University of Sydney, Australia.

Ben P. Marchant is an Environmental Scientist at the British Geological Survey and was previously at Rothamsted Research, Harpenden, UK.

Other books in the Earthscan Food and Agriculture Series

Precision Agriculture for Sustainability and Environmental Protection

Edited by
Margaret A. Oliver,
Thomas F. A. Bishop and
Ben P. Marchant

earthscan
from Routledge

First published 2013 by Routledge

2 Park Square, Milton Park, Abingdon, Oxfordshire OX14 4RN
711 Third Avenue, New York, NY 10017

Routledge is an imprint of the Taylor & Francis Group, an informa business

First issued in paperback 2018

British Library Cataloguing-in-Publication Data
A catalogue record for this book is available from the British Library

Library of Congress Cataloging-in-Publication Data
Precision agriculture for sustainability and environmental protection /
edited by Margaret A. Oliver, Thomas F. A. Bishop, and Ben P. Marchant.
  pages cm. -- (Earthscan food and agriculture series)
  Includes bibliographical references and index.
  ISBN 978-0-415-50440-9 (hbk : alk. paper) -- ISBN 978-0-203-12832-9
  (ebk : alk. paper) 1. Precision farming. 2. Sustainable agriculture.
  I. Oliver, M. A. (Margaret A.) editor of compilation. II. Bishop, Thomas
  F. A., editor of compilation. III. Marchant, Ben P., editor of compilation.
  S494.5.P73P725 2013   631.5--dc23
  2013017040

ISBN: 978-0-415-50440-9 (hbk)
ISBN: 978-1-138-36415-8 (pbk)

Typeset in ITC Galliard
by Bookcraft Ltd, Stroud, Gloucestershire

Contents

PART 3
Management   **133**

8   Site-specific management and delineating management zones   135
DENNIS L. CORWIN

9   Precision weed management   158
ROLAND GERHARDS

10 Site-specific irrigation water management   172
ROBERT G. EVANS AND E. JOHN SADLER

11 The economics of precision agriculture   191
BEN P. MARCHANT, MARGARET A. OLIVER, THOMAS F. A. BISHOP
AND BRETT M. WHELAN

12 Spatially distributed experimentation: tools for the
optimization of targeted management   205
ROGER G. V. BRAMLEY, ROGER A. LAWES AND SIMON E. COOK

PART 4
Case studies   **219**

Case study 1   Sampling and mapping in precision agriculture   221
MARGARET A. OLIVER

Case study 2   Precision agriculture in sugarcane production   233
JOSÉ P. MOLIN, GUSTAVO PORTZ AND LUCAS RIOS DO AMARAL

Case study 3   Precision rice farming for small-scale paddy
fields in Asia   243
BYUN-WOO LEE AND KYU-JONG LEE

Case study 4   Farmer perceptions of precision agriculture for
fertilizer management of cotton   252
ROLAND K. ROBERTS, JAMES A. LARSON, BURTON C. ENGLISH AND
J. COLBY TORBETT

Future prospects   265
MARGARET A. OLIVER, THOMAS F. A. BISHOP AND BEN P. MARCHANT

Index of authors   269
Subject index   273

Figures

Case studies

Tables

Case studies

Contributors

Viacheslav I. Adamchuk, Department of Bioresource Engineering, McGill University, Canada

Lucas Rios do Amaral, Biosystems Engineering Department, University of São Paulo, Brazil

Thomas F. A. Bishop, Department of Environmental Sciences, Faculty of Agriculture and Environment, University of Sydney, Australia

Roger G. V. Bramley, CSIRO Ecosystem Sciences, Glen Osmond, Australia

Paul G. Carter, Washington State University, USA

Ric Coe, World Agroforestry Centre, Nairobi, Kenya

Simon E. Cook, Centre for Water Research, University of Western Australia, Crawley, Australia

Dennis L. Corwin, USDA-ARS, US Salinity Laboratory, Riverside, CA, USA

A. Gordon Dailey, Rothamsted Research, Harpenden, UK

Burton C. English, Department of Agricultural and Resource Economics, University of Tennessee, USA

Robert G. Evans, Benton City, WA, USA

Anja Gassner, World Agroforestry Centre, Nairobi, Kenya

Roland Gerhards, Department of Weed Science, University of Hohenheim, Germany

Jerry L. Hatfield, National Laboratory for Agriculture and the Environment, Ames, IA, USA

Ana Horta, Centre for Environmental and Marine Studies, University of Aveiro, Portugal

Newell R. Kitchen, USDA-ARS Cropping Systems and Water Quality Research Unit, Columbia, MO, USA

James A. Larson, Department of Agricultural and Resource Economics, University of Tennessee, USA

Roger A. Lawes, CSIRO Ecosystem Sciences, Wembley, Australia

Byun-Woo Lee, Department of Plant Science, Seoul National University, Republic of Korea

Kyu-Jong Lee, Department of Plant Science, Seoul National University, Republic of Korea

Ben P. Marchant, Rothamsted Research, Harpenden, UK; British Geological Survey, Nottingham, UK

José P. Molin, Biosystems Engineering Department, University of São Paulo, Brazil

Margaret A. Oliver, Soil Research Centre, Department of Geography and Environmental Science, University of Reading, UK

Gustavo Portz, Biosystems Engineering Department, University of São Paulo, Brazil

Roland K. Roberts, Department of Agricultural and Resource Economics, University of Tennessee, USA

E. John Sadler, USDA-ARS Cropping Systems and Water Quality Research Unit, Columbia, MO, USA

Fergus Sinclair, World Agroforestry Centre, Nairobi, Kenya

J. Colby Torbett, Louisiana Field Office, National Agricultural Statistics Service, Baton Rouge, LA, USA

Raphael A. Viscarra Rossel, CSIRO Land and Water, Bruce E. Butler Laboratory, Canberra, Australia

Richard Webster, Rothamsted Research, Harpenden, UK

Brett M. Whelan, Department of Environmental Sciences, Faculty of Agriculture and Environment, University of Sydney, Australia

Stephen L. Young, University of Nebraska-Lincoln, West Central Research and Extension Center, North Platte, NE, USA

Preface

The driving force behind putting this book together was twofold. First, the world's population has reached the 7 billion mark and is set to increase to around 9 billion by 2050, pressure on the land and other resources continues to increase and food security is likely to be an increasing problem. Precision agriculture (PA) provides a methodology and approach that can help to manage land optimally and to preserve the environment for a sustainable agriculture. The basis of PA tends to be known to specialists in agriculture and associated sciences, but it is not widely dispersed in the public domain. Therefore, the second purpose was to provide the background to PA by a group of established experts in their field. The book aims to be didactic and to point the way to readers for more in-depth study. Many chapters also include a case study that provides greater depth on the topic. The book has four sections: the background to PA and food security, the specialist techniques involved in PA, management based on PA, and a set of case studies.

Chapter 1 sets the scene with an overview of what PA is about in the general sense. Crops remove nutrients from the soil in a field, and conventional practice in agriculture has been to apply fertilizers uniformly to the soil to remedy the loss. This approach takes no account of the spatial variation in the crop's growing conditions and the spatial variation in yield. Some parts of the field are likely to be more depleted in nutrients than others, and PA aims to take this into account by spatially variable management. A major difficulty with PA is to identify where fields need more or less of the relevant nutrients or irrigation water. (As we see in Part 2, there are many techniques to assist with this and these make PA a broad subject that involves specialists in all of these areas.) Chapter 2 examines the causes of spatial variation in agricultural production (such as the availability of water to the crop, soil condition, temperature, pests, diseases, weeds and so on) and the role of PA in food production, quality and security. Precision agriculture is generally associated with large modern farms in Australia, Europe and the USA that use the latest technologies to monitor and interpret variation in their fields, to prescribe variable management strategies and then to act upon these prescriptions. Such technologies might not be available to farmers in the developing world, but PA is urgently required in these regions to address threats to food security. Chapter 3 describes various initiatives aimed at using more low-cost PA solutions to increase agricultural production, sustainability and food security of farmers with smallholdings in the developing world. The authors emphasize that in these circumstances socio-economic factors such as poverty and availability of labour are as likely to limit yield as the biophysical factors which often limit production in Europe and the USA. Therefore, these socio-economic factors must be considered when PA initiatives are implemented in the developing world.

Part 2 comprises four chapters that describe some of the techniques applied in PA. There are many examples in this book where factors such as soil properties, crop yields and weed populations are measured across fields. These measurements must be interpreted to produce maps of the key factors that affect the crop. The maps are then used to decide upon variable-rate treatments that might include nutrients, lime and pesticides, and seeding rate. For example, in Chapter 4 we see how geostatistical analyses of potassium concentrations in the soil can be combined with mathematical models of crop growth to select variable-rate treatments of potassium fertilizer. It is inevitable that some degree of uncertainty will be associated with the maps, and the chapter describes how it can be quantified and accounted for when making decisions. The remote sensing of land has provided valuable insight into the inherent variation present within fields that is often evident in images of the crops as well. Chapter 5 describes the background to remote sensing and the different forms and resolutions that are now available. Several indices based on the soil and crops are available that can be used to assist management in PA. Proximal sensing is allied to remote sensing as both approaches are non-destructive and non-invasive. Chapter 6 gives an overview of the types of proximal sensors that are of value in PA. Electrical conductivity measurement has been widely applied to assess the degree of variation in the soil within fields. The case study uses a sensor to determine pH and to develop a spatially variable lime-application map. Chapter 7 outlines the importance of space-time modelling for PA as crop growth varies not only in space but is highly dynamic in time. In the case study a space-time model for soil moisture is developed based on a combination of spatial and temporal variables and space-time geostatistics.

Part 3 comprises five chapters that relate to management based on PA. Chapter 8 uses proximal sensors to delineate zones that are as homogeneous as possible for management. Uniform and site-specific management are compared, the components of the latter are considered and also the reason why there is a need for a site-specific approach. The case study illustrates how this can be achieved through the delineation of management zones that are fairly homogeneous. Weeds and pests cause serious losses of yield. It is crucial for food security that they are controlled by herbicides and pesticides. Chapter 9 describes how weeds have been controlled uniformly over fields both chemically and mechanically, but with modern techniques to map weeds and the machinery now available, they can be managed site-specifically. This can involve financial savings to the farmer, but of great importance are the environmental benefits that accrue from this. Chapter 10 examines a crucial aspect of PA if more marginal land has to be brought into use to ensure food security. Irrigation has a very long history in making marginal areas productive, but it requires careful management to avoid problems of salinity and structural decline in the soil. In addition, competition for water will increase and agriculture, which is the largest consumer of water, requires large amounts for irrigation. A precise approach to irrigation will be needed to meet both economic and environmental constraints. The case study illustrates how variable the requirements for water are within a centre pivot irrigation system and how the sprinkler system can respond to the varying needs. The economics of PA is a thread that runs through the whole subject. Precision agriculture can be costly in its sophisticated form, but need not necessarily be so. Chapter 11 examines the economics of PA and the difficulties involved in assessing the benefits. Other chapters show how information on within-field variation and the crop response to this variability can be quantified. However, such information is inherently uncertain, and the responses are affected by factors such as the weather or disease rates that can be difficult to predict. The chapter emphasizes that economic assessments of PA must account for this uncertainty and variation. To gain insight into the economics of PA, field experimentation is required to estimate models of the responses of crops to PA. Chapter 12 describes how such experiments can be conducted efficiently and outlines the development of the ideas related to within-field experimentation with examples from grain cropping and viticulture. In particular, it shows that crop response to changed input levels varies dramatically across the fields and that there is great potential to optimize inputs spatially.

In Part 4 we present four case studies that discuss the challenges of particular aspects of PA and of using PA for particular crops in more detail. Case study 1 shows how geostatistics can be used with a range of data to guide future soil and crop sampling for PA. It shows the effects of sampling intensity on the accuracy of predictions of extractable K in the soil for precise management. The second case study shows how autoguidance (using GPS) of machinery can reduce overall inputs of fertilizers and pesticides to the field and how the use of optical crop sensors can help with the management of N applications. In Asia, paddy rice fields and farms are generally small, and high-technology PA systems are impractical and expensive. However, Case study 3 describes how a simple system based upon digital photographs of the crop can ensure that nitrogen fertilizer is applied efficiently. Case study 4 discusses cotton production and revisits the problem of assessing the benefits of PA. The authors describe how assessments based upon questionnaires might be more efficient than the economic assessment exercises described in Chapter 11.

Margaret Oliver, Thomas Bishop and Ben Marchant

Part 1

Precision agriculture and food security

1 An overview of precision agriculture

Margaret A. Oliver

Introduction

The world population is growing and is forecast to reach around 9 billion by 2050 and more than 10 billion by 2100 according to a United Nations press release of May 2011. The competition for land by a myriad of users means that land available for agriculture is likely to decrease despite technological improvements that might extend the boundaries of land that can support agriculture. Therefore, with the population continuing to increase farmers must try to produce more from less land while at the same time protecting the environment and ensuring food security. Recent research has suggested that the global demand for cereals will increase by 75 per cent between 2000 and 2050 (IAASTD, 2008). A larger world population increases the challenges we face to manage land in such a way so as to sustain it and the wider environment in a healthy condition for future generations. Much of the progress in increasing yields to meet the growing population and also higher standards of living in many parts of the world in the middle part of the twentieth century stemmed from improvements in crop varieties bred to give larger yields and produce less waste, for example straw from cereals (Gale and Youssefian, 1985).

In 2009 38 per cent of the Earth's land area was agricultural (land occupied by arable crops, under permanent crops and permanent pastures) (FAOSTAT, 2011), and of this less than a third is used for arable crops (11 per cent of the world's land area). The arable land provides us not only with food, but also with fibre for clothing, household and industrial goods, and fuel. The FAO (2010) estimates that >75 per cent of the Earth's land area is unsuitable for rainfed agriculture and that only 3.5 per cent is suitable for agriculture without any physical constraints. In developed countries land is being withdrawn from cultivation for building, manufacturing industry, roads and so on, and consequently agriculture must be intensified to maintain production. In many developing countries there is scope to bring new land into production with technological advances and new crop varieties. However, this is not true for parts of Asia, where almost all cultivable land is already in use. Where it is possible to extend the cultivable area, such as in Brazil, this is at the expense of native forest, which has an important role in the global ecosystem. Intensification of agriculture and changes in land use can have serious consequences for the environment by increasing soil

4   *Oliver*

erosion, desertification, salinization and flooding, all of which are increasing in many parts of the world. I discuss these issues briefly later in this chapter.

Crops of all kinds remove nutrients from the soil, but not evenly – their effects vary from place to place. The nutrients removed need to be replaced by the application of fertilizers and or manures to the soil. Crop yields increased substantially in the second half of the twentieth century: between 1961 and 1999 global production of the major cereal crops doubled (Wiebe and Gollehon, 2006). This increase has been mainly in developed countries and without additional land; it is the result of several factors such as improved crop varieties, greater use of fertilizers, herbicides and pesticides, increased use of irrigation, improved understanding of the processes involved in crop production and technological advances. The latter have enabled a better understanding of processes by monitoring crop development, weeds and pests. Research and plant breeding have led to genetic improvement of crops, which has probably contributed most to the increase in crop yields of the last half century (Eggli, 2008). Plant breeding has modified plants so that the harvest index (HI), that is the ratio of harvested biomass to the total crop weight, is greater. A good example of plant breeding to achieve a larger HI was the breeding of new short-stemmed cereal varieties in the 1960s in which the amount of straw was greatly reduced and more energy was used for the grain (Gale and Youssefian, 1985). Improvements from plant breeding have also changed the crop-growing seasons, improved the use of nitrogen and resistance to certain diseases, contributed to better storage and transport and so on. The disadvantages, however, are that some of the newer cultivars require more fertilizer, pesticides and water, and these increase input costs to the farmer and create problems for the environment. Water is becoming increasingly scarce because of the many competing uses for it. Agriculture consumes large volumes of water for irrigation, and as the shortages and costs of water increase, irrigation must be managed more carefully (see Chapter 10). Genetic engineering of crops is also playing a major part in increasing yields and resistance to diseases and other stresses (Chivian and Bernstein, 2008).

The intensification of agriculture to produce larger yields and poor agricultural management have tended to degrade the environment resulting in losses of soil and pesticides from fields. In addition, over-application of nitrogen and phosphorus has led to losses of these nutrients into surface and groundwaters and also on to land where they are not required. This has made us aware of a need to manage land in a sustainable way such that farmers can provide the food and other raw materials that we require, but at the same time ensure that the land remains in a condition suitable to continue farming it for future generations. There is a general consensus that farmers need to do their job so that we can all eat and clothe ourselves, but at the same time farmers need to maintain the soil and water in a healthy condition. This means maintaining a good soil structure, organic matter and nutrient status, pH and biodiversity, and limiting the losses of soil itself, nitrogen, phosphorus and pesticides into the environment elsewhere. Therefore, farmers must understand how their actions might lead to soil erosion and compaction, and losses of plant nutrients and other agrochemicals from their land by surface, vertical and lateral flows so as to sustain the quality of the soil and water (Hatfield, 2000). Although organic farming achieves some of these goals, it is not without drawbacks, and it cannot feed a growing world population. More promising solutions are the return to a more integrated approach to farming (Marsh, 2000; Tinker, 2000) and precision agriculture (PA) or site-specific management (SSM). Integrated farming aims to minimize inputs to achieve good yields, and furthermore to apply them only when necessary (Spedding, 2003). The basis of this approach is to integrate beneficial natural processes into modern farming practices and to minimize pollution (Tinker, 2000). Crop rotation is essential in an integrated system; it ensures a better nutrient balance than with monoculture and also some resistance to diseases. The other solution, PA, is to some extent linked with the intentions of the integrated approach because it aims to apply inputs (fertilizers, seeds, pesticides, water and so on) at the rate at which they are required rather than uniformly. The concepts of PA and sustainability are inextricably linked (Bongiovanni and Lowenberg-DeBoer, 2004). With uniform management a single application rate of fertilizers, seeds, pesticides, lime, water, etc. is used for the entire field, with the result that some parts are likely to receive too much and others too little. This could lead to increased pollution of ground- and surface waters, and greater pressure from weeds, pests or diseases (Froment *et al.*, 1995). Precision agriculture in arable farming is the subject of this book, and in the next section I shall describe the background to it and its many components

Precision agriculture

In looking for specific definitions of precision agriculture, I came across this statement by Rick Heard (2006): 'If you ask 10 people what is their definition of precision agriculture you would probably get that many different answers in return.' The term 'precision agriculture' appears to have been used first in 1990 as the title of a workshop held in Great Falls, Montana, sponsored by Montana State University. Before this, the terms 'site-specific crop management' or 'site-specific agriculture' were used. The first two international conferences on what we now regard as 'precision agriculture' referred to site-specific management in the title, but by the third conference in 1996 the term 'precision agriculture' was used. By the mid 1990s, 'precision agriculture' became the preferred term for what we know as modern precision agriculture, a concept that first emerged in the USA in the 1980s.

There is a tendency to consider PA as a modern concept related to agricultural systems in parts of the developed world; however, it is not (see Chapter 3). Precision agriculture has been practised by farmers since the early days of agriculture. Farmers divided their land into smaller areas, the characteristics of which they knew well, and they grew crops where the conditions were most suitable. The early farmers managed their land precisely to ensure that they produced enough food for their families as do subsistence farmers of today; it was a matter of life or death. In Britain there is evidence of small fields that were relatively uniform, each of which could be managed as a unit, and that have since been joined to form much larger fields that are consequently more variable (see Frogbrook *et al.*, 2002, for an example). Modern PA, however, may help both farmers of the developed world with large farms and farmers in Africa and Asia with smallholdings to achieve greater yields and to manage their land better (McBratney *et al.*, 2005).

The National Research Council of the USA (1997, p. 17) gave a clear definition of modern PA as follows: 'Precision agriculture is a management strategy that uses information technologies to bring data from multiple sources to bear on decisions associated with crop production.' It suggested that PA has three components: obtaining data at an appropriate scale, interpretation and analyses of the data, and implementation of a management response at an appropriate scale and time. The intensity and resolution of some of the spatial information involved in PA mean that the revolution to modern PA is essentially about a change in the scale of operation and management. In addition, individuals, governments, non-governmental organizations (NGOs) and land managers have become increasingly aware of the effects of land use on environmental systems or ecosystems. The ability to determine within-field variation and to manage it to improve the economy of agricultural activities and to mitigate their effects on the environment is central to the concept of PA. The data used in PA are often at a fine spatial resolution, for example data from yield monitors, proximal sensor data, remotely sensed data, digital elevation models and so on. There are also many data on soil and crop properties at much coarser scales.

At the heart of PA is the fact that the soil, weather and microclimate vary both spatially and temporally. As there is no simple definition of modern precision agriculture, I shall give its aims. The need for precision agriculture, both old and new, arises because soil, drainage, insolation and topography are rarely uniform over farms or within fields. Heterogeneity is a feature of land (and of the broader environment), so one should not expect to manage the land uniformly. Precision agriculture aims to apply nutrients, soil ameliorants such as lime, water, pesticides and herbicides only where and when they are needed. The purpose is to optimize agricultural production, that is to improve productivity, crop quality and food safety, to improve farm economy and food traceability (Peets *et al.*, 2009) and at the same time reduce undesirable impacts on the environment and improve sustainability. This can be achieved by varying inputs and types of tillage across fields. Farmers in many parts of the world are required by law to manage their land so that groundwater is not polluted. Precision agriculture can help to fine-tune existing management to reduce the leaching of N, for example (McBratney *et al.*, 2005). Precise management entails an understanding of crop requirements for profitable yield, how efficiently the crop uses nutrients, how much the soil can provide and the temporal patterns of nutrient uptake. This requires soil information, but there is also much ongoing research on the use of the plant to indicate its nutrient status and requirements. For example, multispectral sensors such as Quickbird (Bausch and Khosla, 2010) and hand-held sensors such as Greenseeker (Inman *et al.*, 2007) can provide information that will help farmers to optimize N fertilizer inputs. Most pesticides are used for weed control, and with a better understanding of weed dynamics their use could be much better controlled (see Chapter 9). Variation in time and space must be characterized properly (see Chapter 7), and this is inextricably linked with the technological advancements I describe below.

History of modern precision agriculture

Spatial variation in the soil and the environment more generally has been known about since mankind started to manage the land. The soil is remarkably variable, as a result of variation in the soil's parent material, microclimate, drainage, landscape, previous uses and so on. Consequently crop development, yields, quality, pests, weeds and diseases vary spatially and temporally. It is this variation that PA aims to understand and manage in a local or site-specific way, essentially the variation within fields.

The work of Gilbert and Lawes in the latter part of the nineteenth century and their successors at Rothamsted Research, Harpenden, UK, could also be considered as precision agriculture because they wanted to assess the benefits of various combinations and amounts of crop nutrients and also of different crop varieties. The aim was to increase yields, something which applications of cheap fertilizers could achieve, and there were no concerns about their impact on the environment until the last quarter of the twentieth century. Until the 1980s precise or site-specific management was at the farm level, and the management unit was the farm. The soil of any one field was sampled to determine the mean value of crop nutrients and pH, and these were amended uniformly over the field. Similarly, crop yield was based on the total weight taken from the field.

In the mid 1970s to early 1980s farmers became increasingly aware of the potential benefits of better record-keeping and understanding of soil and crop input requirements (Robert, 1999). Robert also described the outcome of a study in the late 1970s by CENEX (Farmers Union Central Exchange, Inc. in the USA) and the Control Data Corporation that showed that farmers were beginning to realize the magnitude of variation in the soil and in crop performance within fields and of the potential benefits of management within fields by zones. This reflected a change in the scale of operation from management at the farm level to recognizing and managing the variation within fields. This change in emphasis resulted from the increase in size of farm machinery in the developed countries and rapid technological change (Schueller, 1997). Farmers removed hedges, fences and ditches and merged fields into increasingly larger units to accommodate the new machines (Frogbrook *et al.*, 2002). The original fields, which could be managed reasonably as a unit, were now parts of larger fields with their inherent variation added together and increasing the within-field variability.

It was during the 1980s that the quiet revolution in agriculture based on information technology began (Cochrane, 1993). One of the most significant steps was the introduction of a yield meter by the Massey Ferguson Company in 1982, which meant that yields could be recorded continuously for the first time. In 1984 Massey Ferguson's field trial in the United Kingdom investigated whether yield varied within fields and whether the yield meter could be used to measure it (Oliver, 2010). Global positioning systems (GPS) were not available at the time, and yield was recorded manually as machines passed over grid squares marked by poles. Farmers found that yields of wheat varied within fields by 10 t ha^{-1}. Schueller and Bae (1987) also did field trials with a combine harvester fitted with input sensors and a grain flowmeter, a portable computer to record data and location measurement equipment. The latter consisted of a microwave transponder comprising a digital distance measuring unit (DDMU), a master transponder connected to the DDMU and two remote transponders placed outside the field. These early trials of yield monitors had no GPS, but in 1991 global positioning systems were available on tractors, and yield mapping became fairly routine. The early GPSs had an accuracy of only 100 m, which was not good enough for mapping, but by the mid 1990s accuracy improved to 5–10 m with differential GPS (DGPS). This improved further after 2000 when the US Department of Defense made access to its global positioning satellites generally available. With differential GPS (DGPS), which requires an additional receiver fixed at a known location nearby, sub-metre accuracy can now be attained.

Cereal yields were the first to be recorded automatically with GPS recorders onboard tractors. Schueller (1997) described how one major manufacturer of grain harvesters claimed that a third of its new combines were equipped with yield monitors. Considerable effort has gone into methods to record yield reliably; the first were for cereals, but there are now monitors for other crops including maize, potatoes, cotton, sugar beet, grapes and other fruits (Griffin, 2010). Yield monitors for grain crops are the most widely available and used. There are now four main techniques for measuring yield (Johnson, 1996): mass flow, weight, optical and nuclear. The first two are common in North America and the latter two in Europe. The data from monitors are often 'noisy' and require filtering and calibration before they can be used for mapping.

The aim of recording yield is to enable one to map the spatial variation in the crop to try to gain insight into the factors that affect yield such as drainage, landscape effects, soil structure, texture, bulk density, nutrient status, pH and soil moisture. This means that intense data are also required on several soil properties, some of which are permanent, such as texture, and others more transient, such as nutrient status and pH. If relations between soil and other environmental factors can be identified, the farmer can consider precise local or variable management of the field. Even before yield monitors had become more widely established, SoilTeq (Luellan, 1985) created a spreader that could change the blend and rate of fertilizer on-the-go, that is what we now know as variable-rate application (VRT). The first VRT machines were used in 1985 by CENEX (Robert, 1999). There has been a mushrooming of technology and services in response to the needs for variable-rate applications of fertilizers and pesticides. However, VRT application depends critically on the availability of accurate digital maps of the relevant properties for precise management (see Chapter 4 and Case study (CS) 1). A major stumbling-block to the wider spread and adoption of PA is the sparsity of soil and crop information, although there are examples of on-the-go measurement of pH (Viscarra Rossel and McBratney, 1997) that result in detailed maps for management.

Before the 1990s maps, other than of the soil and possibly landscape, played little part in agricultural management. Schafer *et al.* (1984) stated that maps of soil type and topography could be used to control fertilizer and pesticide applications and tillage operations. The first yield map of Searcy *et al.* (1989) showed the effect of compaction from farm machinery on yield (see Oliver, 2010). The question then arises as to how to map the data from yield monitors, remote and proximal sensors and more conventional surveys of the soil, crop and landscape. The data from yield monitors and various sensors are usually intense and not a challenge to the many conventional interpolators available for predicting values at unsampled places. The generally sparse data from conventional surveys of the soil and crops are a challenge, however, and the question arose as to how to map such data. Most people who work with spatial data eventually want to map properties from sparse sample data because scattered values are difficult to interpret. In addition, there is a desire to use this discrete information to say something about the surrounding area where there is no information. The aim is to use the sample values to predict at unsampled sites, and this can usually be achieved by some method of interpolation (see Webster and Oliver, 2007, for examples). There are several drawbacks to obtaining estimates from conventional methods of interpolation to produce contour maps for management: the estimates are usually more or less biased, the errors are not determined and there is no way of knowing whether the data are spatially correlated. Unless data are spatially correlated, no method of interpolation should be used. It was to deal with the challenge of sparse data that an interest in geostatistics arose early in the development of PA. To manage land in a site-specific way with variable-rate applications of fertilizers, pesticides, herbicides and seeds requires accurate digitized maps; otherwise site-specific management (SSM) might be no better than uniform application. The challenge is to quantify spatial variation in the soil, crop development, pests, weeds and yield to improve management. This aspect of precision agriculture is developed in Chapter 4.

Environmental impacts of agriculture and how PA can help

The type of agriculture that now dominates the developed world is often termed 'industrial agriculture'. It is highly mechanized, often monoculture, and based on large farms and fields with large applications of fertilizers, water, pesticides and other chemicals. It is and has been having wide-ranging effects on the environment, and increasingly it seems that it is not sustainable. Nevertheless, the agricultural story has been one of success in terms of feeding the increasing population of the world, and the number of malnourished people declined over the 40 years to 1997 (Tilman, 1999), but has since increased (FAO, 2010). Changes within ecosystems, however, mean that industrial agriculture is not sustainable, and different approaches are needed to reverse this trend and to ensure food security (i.e. 'when all people at all times have access to sufficient, safe, nutritious food to maintain a healthy and active life', FAO, 1996). The latter and sustainability are inextricably linked, just as are the concepts of PA and sustainability (Bongiovanni and Lowenberg-DeBoer, 2004). Tilman (1999) stated: 'A "more of the same approach" to the doubling of agricultural production will have significant environmental costs, costs that could be lowered by processes that increase the efficiency of fertilizer use, such as precision agriculture and by incentives for their use.' Maintaining the health of ecosystems will make agriculture more sustainable and also ensure good water quality, biodiversity in the soil, maintenance of soil organic carbon and reduce losses from the land. I discuss the effects of agriculture on the environment below and indicate how PA may reduce them. All forms of agriculture have effects on the environment, but they are magnified by modern agricultural methods.

Fertilizer and pesticide applications

The principal fertilizers are nitrogen, phosphorus and potassium, and of these the first two have the greatest effect on the environment. In conventional farming, fertilizers and pesticides are applied uniformly over fields at a given time. This leads to over-application in some places and under-application in others. Both have cost implications to the farmer, with a waste of materials on the one hand and a potential loss of yield on the other. The environmental costs, however, stem from over-application, which allows N, P and pesticides to move out of the field into the ground- and surface waters and to other areas of land where they are not required. With greater knowledge about the soil and understanding of crop requirements and the crop's condition, fertilizers and pesticides can be applied in more precise amounts, and when and where they are needed. This is a vital role that is embodied in the concept of precision agriculture. For example, remote and proximal sensors (see Chapters 5 and 6) are now available that can monitor a crop's N requirements, and also identify weeds (see Chapter 9) and certain crop diseases (see Chapter 5).

Nitrogen, especially in the form of nitrates, and phosphorus have a profound effect on water quality. In many parts of the world there are strict limits on how much nitrate is allowable in drinking water supplies; in Europe this is 50 mg l^{-1} and in the USA it is 10 mg l^{-1}. In the United Kingdom and more generally in the European Union, farmers with land designated in a nitrate-vulnerable zone (NVZ) have to follow strict guidelines (see DEFRA, n.d.) on when and how much N fertilizer they may apply. Nitrogen moves readily in the environment because of denitrification and in solution as nitrate. Phosphorus is much less mobile because it is bound more tightly to soil particles, but it can be removed in solution after years of heavy applications of fertilizer. Both N and P can be removed from land by soil erosion in the solid phase.

Nitrogen and phosphorus from fertilizer applications accumulate in surface water (rivers, lakes and oceans) leading to eutrophication. This results in greater growth of plants and algae that can eventually lead to oxygen depletion (hypoxia) and the loss of life for the animals in the water, and a reduction in biodiversity. Goolsby *et al.* (2000) report an example of a large hypoxic zone where the Mississippi river discharges into the Gulf of Mexico that they consider to be the result of agriculture. Bongiovanni and Lowenberg-DeBoer (2004) summarize research on the economic and environmental benefits of variable-rate (VRT) applications of N in the USA and Europe; more of the studies showed a benefit than not. Maine *et al.* (2010) showed that overall there were benefits from VRT applications of N for maize in South Africa. The timing of nutrient applications is also important to ensure their efficient use by the crop. The aims of PA are to match the spatial and temporal demands of the crop for inputs of nutrients and also water and pesticides. A precision agriculture approach has the potential to make modern agriculture more environmentally friendly.

Characteristics of the soil and spatial and temporal variation in its properties affect the incidence of pests (animal pests, pathogens and weeds) and also the efficacy and fate of pesticides (Patzold *et al.*, 2008). Over-application of pesticides results in their accumulation in ground- and surface waters and can lead to resistance in both weed and animal pest species.

Pesticides are used to deal with weeds, insects and diseases, but most are applied for weed control (Hatfield, 2000). Sensors and other forms of mapping can help with more precise control of weeds and diseases so as to avoid blanket applications of herbicides. In addition, equipment is available to control weeds, insects and disease site-specifically. Much research has been done on the movement of pesticides through soil and on the soil properties that affect the movement (Price *et al.*, 2009). Where the soil is sandy and very permeable, farmers should apply no more pesticide than necessary, whereas on clay soil the pesticides are far less mobile and excesses are unlikely to leach away before they degrade. Furthermore, integrated pest management should allow farmers to reduce the use of pesticides and minimize resistance to the chemicals. It involves the use of crop hybrids that have resistance to common diseases and pests, that is host resistance, crop diversification and biological control. These are aims that PA also embraces.

Soil erosion

Erosion has been serious in arable agriculture for millennia, and farmers still need to do more to limit soil loss from their land. The topsoil is the most fertile part of the soil, and its loss leaves in place an impoverished soil that requires ever more applications of nutrients to maintain yields. Soil forms so slowly that we should regard it as a non-renewable resource. In addition, the material lost from the soil can end up in reservoirs and water courses resulting in costly operations to remove it and also contributing to an increase in flooding. Soil erosion is spatially variable because it depends on soil type, topography, drainage and management (see Chapter 10 for more detail on these factors). The precision agriculture approach can help to mitigate the effects of soil erosion by guiding farmers on their cultivation practices. If the soil is erodible because of its texture (large sand and or silt contents) or landscape position (on a steep slope) or both, farmers should leave the stubble for as long as possible before ploughing or drill directly into it, that is, practise conservation tillage.

Tilled land is prone to soil erosion because it is bare for some time, and farmers aim to produce a fine tilth that provides suitable conditions for the crop to germinate and establish. Conservation tillage is increasing; it involves leaving at least 30 per cent of plant residues on the surface and the use of minimal tillage or direct drilling. According to Jones *et al.* (2006), 46 per cent of arable land in the UK was under conservation tillage at that time (2006). Conservation tillage can reduce soil erosion and runoff, and increase the soil's organic matter and so stabilize soil aggregates and improve soil structure. It aims to minimize soil erosion and also to reduce the use of fuel in tractors. Tillage can be varied in a PA approach, especially one based on management zones (see Chapter 8). There are drawbacks to conservation tillage, such as the need to use more herbicides to control weeds and increasing soil compaction, but overall the effects have been beneficial.

Soil organic matter

Soil organic matter plays a vital role in the soil in maintaining soil structure, nutrient status, water-holding capacity, cation exchange capacity (CEC) and biodiversity. Biodiversity is important for sustainable agriculture because the many kinds of organisms mediate soil processes such as nutrient cycling, organic matter decomposition, hydrology, structural stability and gaseous exchanges. Agricultural operations, in particular ploughing and monoculture, have led to the depletion of organic matter by oxidation and direct losses in erosion. Price *et al.* (2009) note that organic matter is a major factor in determining herbicide adsorption and hence its availability in the soil solution.

Irrigation

Irrigation is practised over large parts of the semi-arid world, but its use needs to be carefully controlled. Irrigation removes water from aquifers over very large areas and results in many problems where management is poor. Large areas of irrigated land have been affected by salinization and sodicity because the water contains salts that accumulate in the soil when the water evaporates or is transpired. Saline groundwater can also rise to the surface when too much irrigation water has been applied and drainage is inadequate. Sodium salts are the main cause of both salinity and sodicity; the latter occurs when exchangeable Na^+ (ESP) accounts for >15 per cent (note that this is an approximate percentage given by US Salinity Laboratory, but there is no widely accepted critical ESP) of the exchangeable cations. Excess sodium causes clay in the soil to disperse and soil structure to deteriorate leading to waterlogging and reduced crop productivity because of nutrient imbalances. Irrigation should be targeted to sites where drainage is suitable or where drainage can be improved sufficiently. Precision agriculture embraces precise management of irrigation by aiming to apply water only where and when it is required (see Chapter 10).

Precision agriculture and other types of sustainable agriculture

As with other types of agriculture that promote themselves as sustainable and environmentally friendly, these are also benefits that should accrue from PA. Organic farming is claimed to be the most sustainable form of agriculture. As for PA, there are many definitions of what organic farming means. Essentially it avoids the use of synthetic chemicals on crops and soil, and relies on the management of various forms of organic matter and legumes to enhance the chemical, biological and physical properties of the soil, crop rotation and natural biological cycles to optimize crop production, and encourages natural predators for crop protection. Nutrient applications in the form of rock phosphates are allowed. Although there are advantages to the environment from this approach, it has drawbacks such as large nitrate flushes from applications of manures to the soil in adjacent water bodies and groundwater, and problems from weeds. It is also likely that phosphorus and potassium are declining in the soil under organic farming (Tinker, 2000). Benefits to the consumer, in terms of the end product from cereal and vegetable crops, have not been proved (Tinker, 2000; Smith-Spangler *et al.*, 2012). Yet the cost of organic food is at a premium because yields are small and the production costs are large. From the point of view of food security, organic farming cannot meet the increasing demands for food. Precision farmers probably meet more of the requirements for mitigating the effects of nitrate in ecosystems than do organic farmers, and they are also aware of the need to maintain their land in good condition. In a PA approach farmers would apply manures to areas where leaching and surface losses will be least. A modern approach to agriculture does not preclude the applications of organic matter to the soil, and this is being done increasingly through conservation tillage in PA.

Integrated cropping systems are a 'half-way house' with the same aims as PA, but they aim to integrate natural processes into modern farming practices. This includes crop rotations and leys to improve the organic matter status of the soil (see above).

Conservation agriculture is gaining support worldwide and could be regarded as an aspect of both conservation tillage and precision agriculture (Dumanski *et al.*, 2006). The aims are to optimize inputs and yields, but at the same time protect and improve the condition of the land. Conservation agriculture integrates ecological management with modern agricultural production. It minimizes mechanical disturbance of the soil, retains adequate amounts of organic residues on the soil surface and includes crop rotation. As in precision agriculture it advocates the precise placement of inputs such as seeds, fertilizers, pesticides and other agricultural chemicals. Conservation agriculture uses integrated pest management to reduce the use of pesticides. Zero or minimum tillage is important in conservation agriculture, and 105 million hectares worldwide are now under zero tillage (Derpsch and Friedrich, 2009). Zero tillage can involve a range of practices that include direct drilling to place seeds and other inputs, or creating narrow slots or trenches for applying seeds and fertilizer. Derpsch and Friedrich also state that all crops can be grown under this system of tillage. Zero tillage reduces losses from the soil, runoff and oxidation of organic matter and improves the quality of soil and water. The biomass retained at the soil surface maintains the soil's organic matter content, structure, hydraulic properties and biodiversity and diminishes the consumption of fuel in machines.

Dryland farming is practised over large parts of the world that are semi-arid such as the prairies of North America, the steppes of Russia, the Middle East, India, Australia, large parts of Africa and South America. Yields are generally small, and the farming is limited to certain crops, but careful management does enable such marginal areas to be used successfully for agriculture. In the southern Canadian prairies, the Great Plains of North America, the Russian steppes and Australia the main crop is winter wheat, but the choice of crop depends on when the main rain falls. There is no irrigation, and this type of agriculture depends on the retention of rainfall. Runoff is minimized, often by terracing and or contour ploughing, by leaving crop residues on the soil surface and cropping every other year in some places. Dryland farming has links with both conservation agriculture and PA because its aim is to conserve soil moisture with mulches and crop residues, minimize cultivation to avoid evaporation, minimize nutrient losses, use crop varieties that are drought resistant and remove weeds to minimize evapotranspiration. In fact, Twomlow *et al.* (2008) refer to precision conservation agriculture that embraces the principles of both PA and conservation agriculture in dry areas.

Geostatistics

It is clear from the above that precision agriculture and other endeavours to manage land in a site-specific way depend on knowing how soil and crop properties vary spatially, often from sparse sample information. Geostatistics developed to analyse data on the metal content of ore bodies in mines (Matheron, 1963). It has proved equally successful in other fields such as petroleum engineering, hydrology, meteorology, soil science, the environment generally and precision agriculture specifically. Geostatistics provides the tools, variography and kriging, to predict attributes accurately for digital maps that can then be used for precise management (for more information on theory see CS1 and Webster and Oliver, 2007).

Geostatistics in precision agriculture

The first paper to apply geostatistics in precision agriculture was by Mulla and Hammond (1988). Its aim was to introduce geostatistics for mapping P and K in the soil and to determine the nature and extent of the spatial variation in these nutrients and the sampling intensity necessary to identify the major patterns in the soil. The authors stated that variable-rate management needed appropriate sampling to provide accurate maps of the crop nutrients, and used geostatistics to interpolate between the measured values. Geostatistics was already well established in soil science from the work of Richard Webster and his team in the early 1990s (Burgess and Webster, 1980; Burgess *et al.*, 1981). Mulla's work during the late 1980s and early 1990s laid the foundations for the adoption of geostatistics in PA (Mulla, 1989, 1993; Bhatti *et al.*, 1991). During the same period, Miller *et al.* (1988) and Webster and Oliver (1989) were applying geostatistics in an agricultural context; they were on the track of PA but not in an explicit way. In 1988 at the International Geostatistics Congress held in Avignon, I was told that tractors had onboard computers that would be able to use kriged maps to guide fertilizer and other applications. At the time this seemed far-fetched and fanciful as few people were even familiar with microprocessors, which were essential for the types of portable operations associated with agriculture.

The growing enthusiasm for geostatistics in the PA community was supported by the National Research Council (1997, p. 4). The council made the point that 'current mapping techniques are limited by a lack of understanding of the geostatistics necessary for displaying spatial variability of crops and soils' and 'an increased knowledge base in geostatistical methods should improve interpretation of precision agriculture data' (p. 59). By the Fourth International Conference on PA in 1998 there were 19 papers with geostatistical applications and in 2008 there were 23.

The variogram is the central tool of geostatistics; it summarizes the way that properties vary from place to place, and information from the variogram is essential for geostatistical prediction, kriging. For many environmental applications kriging is most likely to be used for interpolation and mapping. The values of properties are usually estimated at the nodes of fine grids, and the variation can then be displayed by isarithms or by layer shading. The kriging variances or standard errors can be mapped similarly: they are guides to the reliability of the estimates. Where sampling is irregular, such a map may indicate if there are parts of a region where sampling should be increased to improve the estimates. Accurate kriged predictions, and therefore the resulting maps, depend on accurate variograms and spatially correlated or dependent data from which to predict. Soil and crop properties can vary at disparate spatial scales both within and between fields. Therefore, the spatial scales of variation in the properties of most importance for site-specific management should be used to guide sampling. To sample to provide spatially dependent data requires a sampling intensity that relates to the spatial scale; the data should be at an interval that is well within the correlation range of the spatial variation (see CS1).

Kitchen *et al.* (2002) state that PA management involves much more than simply creating 'pretty maps'. Precision agriculture is interdisciplinary and requires agronomic and computer skills and an understanding of the causes of variation in yield to know what can and cannot be improved. Case study 1 helps to show how variation can be quantified to produce accurate maps and how such maps can be used.

Conclusions

To maintain food availability at levels sufficient to feed the increasing world population while at the same time protecting soil and environmental health, agriculture must be carried out with a greater awareness of the crop, water, nutrient and environment interfaces to ensure sustainability and food security. Crop nutrients added as fertilizers to the soil will become more costly and they can affect the environment adversely when over-applied. The supply of water for crop irrigation is likely to become more scarce and increasing competition for water will result in increased prices – it is not a free commodity, as is so often assumed. Precision agriculture provides a means of managing such inputs to cropping systems in an environmentally sustainable way through site-specific management. Other inputs such as seeds and pesticides can be applied site-specifically, which can help with the farm's economy and also reduce adverse environmental impacts. The following chapters illustrate that PA is not just for large-scale farming, as in the USA and Australia, but also for operations on much smaller scales such as in rice cultivation (CS3). Many tools are used in PA such as remote (Chapter 5) and proximal (Chapter 6) sensing, geostatistics to obtain precise predictions of values at unsampled places (Chapter 4) and from spatio-temporal data (Chapter 7). The four case studies provide in-depth examples of sampling and prediction in PA (CS1) and the application of PA in sugarcane (CS2), rice (CS3) and cotton cultivation (CS4).

References

Bausch, W. C. and Khosla, R. (2010) 'QuickBird satellite versus ground-based multi-spectral data for estimating nitrogen status of irrigated maize', *Precision Agriculture*, 11, 274–90.

Bhatti, A. U., Mulla, D. J. and Frazier, B. E. (1991) 'Estimation of soil properties and wheat yields on complex eroded hills using geostatistics and thematic mapper images', *Remote Sensing of the Environment*, 37, 181–91.

Bongiovanni, R. and Lowenberg-DeBoer, J. (2004) 'Precision agriculture and sustainability', *Precision Agriculture*, 5, 359–87.

Burgess, T. M. and Webster, R. (1980) 'Optimal interpolation and isarithmic mapping of soil properties. I. The semi-variogram and punctual kriging', *Journal of Soil Science*, 31, 315–31.

Burgess, T. M., Webster, R. and McBratney, A. B. (1981) 'Optimal interpolation and isarithmic mapping of soil properties. IV. Sampling strategy', *Journal of Soil Science*, 32, 643–54.

Chivian, E and Bernstein, A. (2008) *Sustaining Life: How Human Health Depends on Biodiversity*, New York: Oxford University Press.

Cochrane, W. W. (1993). *The Development of American Agriculture: A Historical Analysis*, Minneapolis, MN: University of Minnesota Press.

DEFRA (n.d.) 'Nitrate vulnerable zones'. Available online at <http://www.defra.gov.uk/food-farm/land-manage/nitrates-watercourses/nitrates/> (accessed February 2013).

Derpsch, R. and Friedrich, T. (2009) 'Development and current status of no-till adoption in the world, in *Proceedings of 18th Triennial Conference of the International Soil Tillage Research Organization (ISTRO)*, 15–19 June 2009, Izmir, Turkey (on CD).

Dumanski, J., Peiretti, R., Benites, J. R., McGarry, D. and Pieri, C. (2006) 'The paradigm of conservation agriculture', in *Proceedings of World Association of Soil and Water Conservation*, paper no. P1–7, pp. 58–64.

Eggli, D. B. (2008) 'Comparison of corn and soybean yields in the United States: historical trends and future prospects', *Agronomy Journal*, 100, S79–S88.

FAO (1996) *Report of the World Food Summit*, Rome: FAO.

FAO (2010) *Statistics Year Book*. Online. Available online at <http://www.fao.org/fileadmin/templates/ess/ess_test_folder/Publications/yearbook_2010/a04.xls> (accessed February 2013).

FAOSTAT (2011) 'FAOSTAT: Agriculture'. Available online at <http://www.fao.org> (accessed February 2013).

Frogbrook, Z. L., Oliver, M. A., Salahi, M. and Ellis, R. H. (2002) 'Exploring the spatial relations between cereal yield and soil chemical properties and the implications for sampling', *Soil Use and Management*, 18, 1–9.

Froment. M., Dampney. P., Goodlass. G., Dawson. C. and Clarke. J. (1995) *A Review of Spatial Variation of Nutrients in the Soil*. Report of Project CE0139 to MAFF, Ministry of Agriculture, Fisheries and Food, London.

Gale, M. D. and Youssefian, S. (1985) 'Dwarfing genes in wheat', in G. E. Russell (ed.) *Progress in Plant Breeding*, London: Butterworths, pp. 1–35.

Goolsby, D. A., Battaglin, W. A., Lawrence, G. B., Artz, R. S., Aulenbach, B. T., Hooper, R. P., Keeney, D. R. and Stensland, G. J. (2000) *Fluxes and Sources of Nutrients in the Mississippi–Atchafalaya River Basin, Topic 3*. Report for the Integrated Assessment on Hypoxia in the Gulf of Mexico. NOAA Costal Ocean Program Decision Analysis Series, no. 17, Silver Spring, MD.

Griffin, T. W. (2010) 'The spatial analysis of yield data', in M. A. Oliver (ed.) *Geostatistical Applications for Precision Agriculture*, Dordrecht: Springer, pp. 89–116.

Hatfield, J. L. (2000) *Precision Agriculture and Environmental Quality: Challenges for Research and Education*. Available online at <http://www.arborday.org/programs/papers/PrecisionAg.html> (accessed February 2013).

Heard, R. (2006) 'Perspective: defining precision agriculture', PrecisionAg Special Reports. Available online at <http://www.precisionag.com/works> (accessed February 2013).

IAASTD (2008) *Agriculture at a Crossroads: Global Summary for Decision Makers*. Available online at <http://www.agassessment.org/> (accessed February 2013).

Inman, D., Khosla, R., Reich, R. M. and Westfall, D. G. (2007) 'Active remote sensing and grain yield in irrigated maize', *Precision Agriculture*, 8, 241–52.

Johnson, R. C. (1996) *Target Farming: A Practical Guide to Precision Agriculture*, 2nd edn, Saskatoon, Saskatchewan, Canada: R. C. Johnson.

Jones, C. A., Basch, G., Baylis, A. D., Bazzoni, D., Bigs, J., Bradbury, R. B., Chaney, K., Deeks, L. K., Field, R., Gomez, J. A., Jones, R. J. A., Jordan, V., Lane, M. C. G., Leake, A., Livermore, M., Owens, P. N., Ritz, K., Sturny, W. G. and Thomas, F. (2006) *Conservation Agriculture in Europe: An Approach to Sustainable Crop Production by Protecting Soil and Water?* Jealott's Hill, Bracknell, UK: SOWAP.

Kitchen, N. R., Snyder, C. J., Franzen, D. W. and Wiebold, W. J. (2002) 'Educational needs of precision agriculture', *Precision Agriculture*, 3, 341–51.

Luellan, W. R. (1985) 'Fine-tuned fertility. Tomorrow's technology here today', *Crops and Soils Magazine*, 38, 18–22.

McBratney, A. B., Whelan, B., Ancev, T. and Bouma, J. (2005) 'Future directions in precision agriculture', *Precision Agriculture*, 6, 7–23.

Maine, N., Lowenberg-DeBoer, J., Nell, W. T and Alemu, Z. G. (2010) 'Impact of variable-rate application of nitrogen on yield and profit: a case study from South Africa', *Precision Agriculture*, 11, 448–63.

Marsh, J. (2000) *Integrated Farm Management: A Farm Strategy for the 21st Century*, Stoneleigh, UK: LEAF, Royal Agricultural Society of England.

Matheron, G. (1963) 'Principles of geostatistics', *Economic Geology*, 58, 1246–66.

Miller, M. P., Singer, M. J. and Nielsen, D. R. (1988) 'Spatial variability of wheat yield and soil properties on complex hills', *Soil Science Society of America Journal*, 52, 1133–41.

Mulla, D. J. (1989) 'Using geostatistics to manage spatial variability in soil fertility', in C. M. Renard, R. J. Van den Beldt and J. F. Parr (eds) *Soil, Crop and Water Management in the Sudano–Sahelian Zone*, Pantcheru, India: ICRISAT.

Mulla, D. J. (1993) 'Mapping and managing spatial patterns in soil fertility and crop yield', in P. C. Robert, W. E. Larson and R. H. Rust (eds) *Proceedings of Site Specific Crop Management: A Workshop on Research and Development Issues*, Madison, WI: American Society of Agronomy; Crop Science Society of America; Soil Science Society of America, pp. 15–26.

Mulla, D. J. and Hammond, M. W. (1988) 'Mapping soil test results from large irrigation circles', in J. S. Jacobsen (ed.) *Proceedings of the 39th Annual Far West Regional Fertilizer Conference*, Pasco, WA: Agricultural Experimental Station Technical Paper 8597, pp. 169–71.

National Research Council (1997) *Precision Agriculture in the 21st Century*, Washington, DC: National Academy Press.

Oliver, M. A. (2010) 'An overview of geostatistics and precision agriculture', in M. A. Oliver (ed.) *Geostatistical Applications for Precision Agriculture*, Dordrecht: Springer, pp. 1–34.

Patzold, S., Mertens, F. M., Bornemann, L., Koleczek, B., Franke, J., Feilhauer, H. and Welp, G. (2008) 'Soil heterogeneity at the field scale: a challenge for precision crop protection', *Precision Agriculture*, 9, 367–90.

Peets, S., Gasparin, C. P., Blackburn, D. W. K. and Godwin, R. J. (2009) 'RFID tags for identifying and verifying agrochemicals in food traceability systems', *Precision Agriculture*, 10, 382–94.

Price, O. R., Oliver, M. A., Walker, A. and Wood, M. (2009) 'Estimating the spatial scale of herbicide and soil interactions by nested sampling, hierarchical analysis of variance and residual maximum likelihood', *Environmental Pollution*, 157, 1689–96.

Robert, P. C. (1999) 'Precision agriculture: research needs and status in the USA', in J. V. Stafford (ed.) *Precision Agriculture '99*, Sheffield, UK: Sheffield Academic Press, pp. 19–33.

Schafer, R. L., Young, S. C., Hendrick, J. G. and Johnson, C. E. (1984) 'Control concepts for tillage systems', *Soil & Tillage Research*, 4, 313–20.

Schueller, J. K. (1997) 'Technology for precision agriculture', in J. V. Stafford (ed.) *Precision Agriculture '97*, Oxford, UK: BIOS Scientific Publishers, pp. 19–33.

Schueller, J. K. and Bae, Y. H. (1987) 'Spatially-attributed automatic combine data acquisition', *Computers and Electronics in Agriculture*, 2, 119–27.

Searcy, S. W., Schueller, J. K., Bae, Y. H., Borgelt, S. C. and Stout, B. A. (1989) 'Mapping of spatially-variable yield during grain combining', *Transactions of the ASAE*, 32, 826–9.

Smith-Spangler, C., Brandeau, M. L., Hunter, G. E., Bavinger, J. C., Pearson, M., Eschbach, P. J., Sundaram, V., Liu, H., Schirmer, P., Stave, C., Olkin, I. and Bravata, D. M. (2012) 'Are organic foods safer or healthier than conventional alternatives? A systematic review', *Annals of Internal Medicine*, 157, 348–66.

Spedding, A. (2003) *Integrated Farm Management.* Easy Guide 52, RuSource Briefings Collection. Available online at <http://ofi.openfields.org.uk> (accessed February 2013).

Tilman, D. (1999) 'Global environmental impacts of agricultural expansion: the need for sustainable and efficient practices', *Proceedings of the National Academy of Science*, 96, 5995–6000.

Tinker, P. B. (ed.) (2000) *Shades of Green: A Review of UK Farming Systems*, Stoneleigh, UK: Royal Agricultural Society of England.

Twomlow, S., Urolov, J. C., Jenrich, M. and Oldrieve, B. (2008) 'Lessons from the field: Zimbabwe's conservation agriculture task force', *Journal of SAT Agricultural Research*, 6, 1–9.

Viscarra Rossel, R. A. and McBratney, A. B. (1997) 'Preliminary experiments towards the evaluation of a suitable soil sensor for continuous "on-the-go" field pH measurements', in J. V. Stafford (ed.) *Precision Agriculture '97*, Oxford, UK: BIOS Scientific Publishers, pp. 493–501.

Webster, R. and Oliver, M. A. (1989) 'Disjunctive kriging in agriculture', in M. Armstrong (ed.) *Geostatistics*, vol. 1, Dordrecht: Kluwer Academic Publishers, pp. 421–32.

Webster, R. and Oliver, M. A. (2007) *Geostatistics for Environmental Scientists*, 2nd edn, Chichester, UK: J. Wiley & Sons.

Wiebe, K. and Gollehon, N. (eds) (2006) *Agricultural Resources and Environmental Indicators, 2006 Edition /EIB–16*, Washington, DC: Economic Research Service, USDA.

2 The role of precision agriculture in food production and security

Jerry L. Hatfield and Newell R. Kitchen

Introduction

The increasing world population will continue to place demands on the agricultural production system to produce increasing amounts of food for the foreseeable future to feed a population that is likely to exceed 9 billion people by 2050 (United States Census Bureau, n.d.). The growth in population towards 9 billion will be accompanied by a decrease in the amount of land available for agricultural production because of the increased space required by people for cities and associated infrastructure, for example roads, reservoirs, water treatment plants and so on. A relatively simple analysis of these two facts would lead to the conclusion that food production will have to become more efficient (i.e. more food produced on similar or less land resource) to meet the expected population demand. To add to the complexity of this problem, there is the possibility of climate change associated with a potential increase in the variability of precipitation, which could dramatically alter the reliability of critical components in crop production (Hatfield *et al.*, 2011). If we couple the increasing human population with a declining available soil resource (Lal, 1997) and more variable precipitation (Karl *et al.*, 2009), then the urgency to be able to address this problem and provide solutions for agriculture intensifies. One of the questions to be asked is what role precision agriculture could potentially have in contributing to increased and improved food production to achieve food security. Food security as defined by FAO is 'a situation that exists when all people, at all times, have physical, social and economic access to sufficient, safe and nutritious food that meets their dietary needs and food preferences for an active and healthy life' (FAO, 2002). If we link this definition with agriculture then understanding the limitations to crop production and the role of precision agriculture to address these problems, we begin to develop potential solutions to ensure food security for future generations.

Variability in production represents a multi-dimensional problem. It occurs among years induced by variation in the weather, within fields induced by variation in the soil, drainage conditions, physiography, aspect, salinity, nutrient management, and across years and within fields induced by the legacy of management decisions and their interactions with the weather during the growing season. The multi-dimensionality of this problem is rarely considered in how we view the overall agricultural production system. In this chapter, we will concentrate on the management decisions and the within-season weather because these offer the greatest possibility for the realization of the potential of precision agriculture.

Causes of variability in production

Soil water is a critical factor in plant growth and any limitation in the soil to supplying optimum amounts of water at critical growth stages limits plant growth. Although this basic agronomic tenet has been thoroughly studied and is well understood at a single-plant level, it has not been investigated extensively at landscape and field scales to achieve the management objectives of precision agriculture. Water availability is critical to food production, and understanding the role of soil water will be central to soil resource assessment and soil management opportunities that precision agriculture offers in the future. We propose that a better understanding of this principle when scaled up may provide a foundation for improved management of agricultural fields. Hatfield (2012) showed that variability in maize (*Zea mays* L.) yields across different fields was related more to water availability than to nitrogen application rates. The foundation for this observation was based on the differences in soil water use by the plants on different soil types within these fields (Hatfield *et al.*, 2007; Hatfield and Prueger, 2011). These differences were as large as 300 mm of cumulative evapotranspiration during the growing season of a maize crop and as much as 200 mm of transpiration. The primary factor causing these differences was the soil water-holding capacity related to differences in soil organic matter content and texture of the different soil types within the field. Elsewhere, within-field variation in other important soil properties has been found to suppress soil water-holding capacity and diminish yield. Examples include soil with large sand and small clay content (James and Godwin, 2003; Sandras *et al.*, 2003; Jiang and Thelen, 2004) and soil with a large clay content and or an impervious argillic horizon because of the reduced rooting depth (Kitchen *et al.*, 1999; Kitchen *et al.*, 2003). Jiang *et al.* (2008) showed that soil water-holding capacity on an eroded slope where the soil was an alfisol (alisol) could be 50 to 60 per cent less than on the summit and footslope landscape positions. Distributions of soil water availability across a field are related to position in the landscape, soil texture and soil organic matter content.

Throughout the course of the growing season, separation among soil types within fields was most pronounced during the grain-filling period because of a shortage of available soil water, except in the years with more than adequate rainfall during this period. The shortage of water in some years with less than adequate rainfall during grain-filling led to a reduction in water use efficiency because of the decreased grain yield induced by water stress (Hatfield and Prueger, 2011). This effect can be explained by reduction in grain yield relative to the total seasonal water use. Yield patterns within fields were observed to depend upon the total rainfall during the growing season, with soil on the upper slope producing the largest yields in the years with above-normal rainfall and the lowest yields in years with less than normal rainfall (Jaynes and Colvin, 1997; James and Godwin, 2003). Sadler *et al.* (2000a,b) conducted a field-scale study of drought-stressed maize and found a relationship between soil map units and grain yield; however, these relationships alone did not explain the variation in yield within the field. Variation among sites within a given soil type was significant, and they suggested that improved understanding of the variation in yield would require more detailed observations of crop water stress within the growing season and greater resolution of the soil properties. Kaspar *et al.* (2003) evaluated maize yields across a field in central Iowa and showed that in growing seasons with less than normal rainfall there was a negative relationship between grain yield and relative elevation, slope and curvature, whereas in years with above-normal rainfall there was a positive relationship with these terrain attributes. Kumhálová *et al.* (2011) expanded on this concept by demonstrating that yield and crop nutrient concentrations of wheat (*Triticum aestivum* L.) were spatially related to topography in which more soil water was available in the areas of the landscape near the bottom of the slope. The correlation between water-flow accumulation from the higher areas of the slope to the lower areas and yield were strongest in the dry years and were weak for the wetter years, which was similar to observations made by other researchers (Kitchen *et al.*, 1999; Kaspar *et al.*, 2003). Thus, the observations by Hatfield and Prueger (2011) on differences in water use within fields explain differences observed in the other studies because few other studies have measured crop water use directly but inferred water use patterns through seasonal rainfall patterns. The implications for food security from water availability would suggest that a combination of seasonal rainfall patterns, soil water-holding capacity and water redistribution as driven by landscape features greatly affect food production and therefore provide valuable information on how fields should be managed for maximum production.

Bouman (2007) addressed this problem in an analysis of crop-production systems and the potential to increase crop water productivity or water use efficiency with the goal of increasing food production and saving water. Bouman's (2007) approach was based on four principles that can apply to precision agriculture. These principles were: '(1) increase transpirational crop water productivity; (2) increase the storage size for water in time and space; (3) increase the proportion of non-irrigation water inflows to the storage pool; and (4) decrease the non-transpirational water outflows of the storage pool'. His second principle describes the reason for the observed variation in yield across fields in central Iowa, and efforts to increase crop productivity per unit amount of water will have to begin with understanding how to increase the size of the storage pool in soil for water. In other soil types, for example claypan soil in central Missouri, emphasis will have to be on increasing the movement of water into the soil for storage to increase water availability. The results obtained to date across a number of locations would suggest that there will be no single solution to increasing soil water availability for crop production.

Machado *et al.* (2000) showed that the variation in maize yield across a field was due to a combination of biotic (insects, diseases and weeds) and abiotic (water and temperature) factors, and additional soil nitrate-N levels only affected yields when there was adequate soil water. They also observed that the presence of biotic stresses linked with insects, diseases and weeds across the field and the effect on maize yield were more unpredictable than abiotic stresses (Machado *et al.*, 2000). Efforts to quantify the role of precision agriculture practices will have to account for soil water availability if we are to understand the potential that precision agriculture may have to increase food production. Once soil water availability is quantified spatially and temporally, precision soil and water management practices can be embraced to improve crop water use efficiency.

One way to draw particular attention to the impact of variable within-field soil water availability has been to transform yield map data into monetary metrics. Massey *et al.* (2008) used 10 years of site-specific yield data for maize, soya bean (*Glycine max* (L.) Merr.) and grain sorghum (*Sorghum bicolor* L.) across a 36.5-ha field with claypan soil in central Missouri to quantify temporal changes in profit and loss response. While some field areas were profitable almost every year, other large areas of the field had negative profit most years. The latter occurred in areas of the field with considerable topsoil erosion. Brock *et al.* (2005) also found that high-yielding management zones in a maize–soya bean rotation were associated with poorly drained level soil types, whereas low-yielding zones were associated with eroded soil or soil on more sloping areas. Over the past several decades, the impact of landscape-dependent factors causing differences in water availability for the growing crop and yield have been well documented through research in a number of different locations, for instance Iowa (Spomer and Prest, 1982), Nebraska (Jones *et al.*, 1989), Colorado (Wood *et al.*, 1991; McGee *et al.*, 1997), Saskatchewan (McConkey *et al.*, 1997) and New York (Timlin *et al.*, 1998).

Lal (2009) stated that soil degradation was the cause of reduced crop yield through the increase in drought-prone areas of the field or imbalance in the elements important for crop growth. Lal (1993, 1997) observed that the processes causing soil degradation were loss of soil structure, compaction, crusting, erosion related to the physical attributes of soil, nutrient depletion, acidification, salinization, element imbalance related to the soil chemistry, depletion of soil organic matter and alteration of the diversity and activity of soil micro-organisms representing the biological component of soil. Degradation can perpetuate where areas of a field have limited productivity, primarily related to the availability of water and nutrients across a field caused by a reduction in soil organic matter content. Although soil degradation occurs within fields in site-specific ways, it is also a global issue (Lal, 1997). Therefore, soil resource management that affects future food security is also a global issue (Lal, 2010). The results cited above primarily demonstrate the impact of soil water in the USA; however, there are many examples throughout the world of yield variation within fields and of where soil degradation has changed water availability in certain landscape positions. Pena-Barragan *et al.* (2010) showed that sunflower (*Helianthus annus* L.) yields varied across fields in Spain with landscape position and the presence of competing *Ridolfia segetum* within the field. These concepts apply to other crops; for example Arno *et al.* (2011) observed that the yield of grapes varied in Spain, and they attributed the variation to differences in the soil condition, topography and mineral contents of the petiole. They suggested that yield variation would have to be overcome by improved nutrient management coupled with an understanding of the role of soil and topography on soil water availability to the crop. In Portugal, maize yield was shown to be strongly related to the distance to water flow accumulation lines, with yield decreasing as the distance to flow lines increased (Marques da Silva and Silva, 2006). Similarly, cereal yield was strongly correlated with flow accumulation in the Czech Republic, but was strongest for drier years (Kumhálová *et al.*, 2008). Soil and terrain attributes that modify site-specific soil water within fields are often found to be important when accounting for the variation in yield, with the following as a few examples of the many studies carried out in recent years: Australia (Robinson *et al.*, 2009), Italy (Basso *et al.*, 2009; Casa and Castrignanò, 2008), Belgium (Vitharana *et al.*, 2008) and Canada (Chi *et al.*, 2009).

Precision agriculture has focused on nutrient management with applications of nitrogen, phosphorus, potassium and magnesium, and also on problems of soil acidity with applications of lime to the soil. However, this might not be the most fruitful aspect of precision agriculture to focus upon to provide a long-term solution to food production. Further, Lal (2009) indicated that poor plant growth and limited nutrient availability in degraded soil was the cause of poor human nutrition in many areas of the world. In conclusion, the spatial distribution of soil water availability has both direct and indirect effects on almost all factors limiting plant growth, production and food quality. To increase food production and provide greater food security, we propose that more and enhanced precision agriculture science and technologies are needed to support soil management decisions that will improve soil water availability.

Role of precision agriculture in food production

Prior to the modern era of mechanized agriculture, fields were often much smaller and the subdivision of land to these parcels was often based on soil variation. Therefore, *precision* management was achieved naturally because each soil and crop management was manual and specific to the field. This type of precision management still characterizes food production in developing areas of the world. But for much of the world over the past century mechanization has resulted in the aggregation of small fields and parcels into large fields. These fields now encompass multiple and complex soil and landscape variation. Thus today, the principles of precision agriculture are based on the integration of technology for the efficient temporal and spatial management of large fields. Although there has been a promise of enhanced management decisions leading to improvements in production efficiency and reductions in inputs by site-specific management, precision agriculture still primarily remains a set of diagnostic tools in which the analysis is conducted to determine why yield varies across the field as compared to a prescriptive approach to reduce yield variation. For example, the results obtained by Arno *et al.* (2011) and Pena-Barragan (2010) for different crops in Spain identify the causes of yield variation and suggest approaches with potential to reduce field variability; however, they did not demonstrate whether these approaches reduced yield variation. This approach offers one application of precision agriculture techniques, whereas other approaches focus on understanding the variation in the field to create the optimal yield and quality that can lead to an increase in within-field variation in yield.

One of the primary applications of precision agriculture technology has been to improve the understanding of spatial variation within fields. There are a number of different tools and approaches that can help to achieve this. In some cases, precision agriculture tools have been revolutionary for measuring and analysing the impact of soil and landscape properties on crop productivity. For example, multiple sensing platforms have been commercialized to measure soil profile apparent electrical conductivity (EC_a). Since these sensors have been developed, studies have consistently shown how soil EC_a can be an effective surrogate measurement for important soil properties that affect yield, such as salinity, clay content, cation exchange capacity (CEC), clay mineralogy, soil pore size and distribution, soil moisture content and temperature (McNeill, 1992; Rhoades *et al.*, 1999). When mobilized and linked with GPS, it allows for rapid and high-resolution acquisition of soil EC_a at a scale similar to the recording of crop yield. Subsequently, linking soil EC_a and yield data has been used successfully to determine soil and landscape factors that control yield over a wide range of different crops and climate zones, for example maize (Jaynes *et al.*, 1993; Kitchen *et al.*, 1999), maize, soya bean, sorghum, winter wheat (*Triticum aestivum* L.) (Kitchen *et al.*, 2003), cotton (*Gossypium hirsutum* L.) (Corwin *et al.*, 2003) and maize–soya bean rotation (Jiang and Thelen, 2004). Understanding how these factors link together will provide the foundation for the application of precision agriculture methods.

The use of satellite imagery to quantify spatial variation within fields has been extensive and quantifying field variation is necessary to determine how to improve field management to achieve the goal of food security. One example was reported by Begue *et al.* (2008) for sugarcane (*Saccharum officinarum* L.), where the normalized difference vegetation index (NDVI) from SPOT images during the growing season was used to quantify patterns within fields. They concluded from these observations that it was necessary to know the stage of crop development in order to interpret the spatial pattern across the field. Single images were not sufficient to provide the information required for diagnosis of the crop's condition. The pattern of vigour, as determined by NDVI, can be used to diagnose differences in growth. There are a number of different indices that could be applied to satellite imagery to estimate many different crop characteristics, for example leaf-area index, ground cover, leaf chlorophyll, biomass and light interception. These have been summarized by Hatfield *et al.* (2008). There are also many different indices that could be applied to assess the variation in crop vigour across the field. Wang *et al.* (2012) used a combination of different indices to quantify variation in rice yields within and among rice (*Orzya sativa* L.) paddy fields in Taiwan. They used the spatio-temporal patterns to classify paddy fields into different yield categories for the potential management of nitrogen application practices. These techniques can be applied to fields to help ensure optimum food production for a given land area and these methods will have to be adopted to facilitate these applications.

One of the most common applications of precision agriculture technology has been in the detection of crop nitrogen stress within fields. Bongiovanni and Lowenberg-DeBoer (2004) summarized the available information relating precision agriculture to sustainability and environmental quality, and concluded that reductions in nitrogen inputs could be achieved through enhanced management decisions and applying nutrients spatially to fields. They cited examples of reductions in nitrogen fertilizer applications that increased farm profitability; however, no major improvement in crop yields was obtained through these changes. These findings could be extended to encompass the positive environmental impact, because of the reduced nitrogen inputs and more efficient use of nitrogen by the crop. Numerous field studies have been conducted to assess the ability to detect nitrogen stress and then apply different amounts of nitrogen fertilizer to different parts of a field according to the crop requirements and the soil condition. Raun *et al.* (2002) showed that improvements in nitrogen use efficiency (NUE) were achievable through the combination of optical sensing and variable-rate application. They found NUE improved by more than 15 per cent when nitrogen fertilizer was applied; however, the improvements were a result of more efficient use of the nitrogen applied because the rates were near the optimal rates required by the plant compared only to increases in crop yield. Kitchen *et al.* (2010) found that optical reflectance sensing worked effectively for determining within-season nitrogen fertilizer rates for maize in relation to soil type, fertilizer cost and the price of maize, which all affected the analysis. Profit to farmers using these sensors would range from US$25 to US$50 ha^{-1}. Subsequent field-scale studies over 55 fields using this same canopy sensor technology resulted in an average profit increase of US$42 ha^{-1} (Scharf *et al.*, 2011). In about 25 per cent of these fields, reflectance sensing prescribed more nitrogen fertilizer than the producer-based rate, but higher yields were achieved. Cao *et al.* (2011) showed that it was possible to apply this technology to small-scale fields (such as those in densely populated developing countries) and to achieve an improvement in NUE through a reduction in nitrogen application rates. They were also able to show increases in wheat yields with improved assessments of nitrogen status in the fields prior to grain-filling. In a more recent analysis Raun *et al.* (2011) showed, with regard to nitrogen management in maize and wheat, that yield and nitrogen responsiveness were independent of one another. The implications of these findings would suggest that, for a more effective implementation of precision agriculture technologies, other factors that affect crop yield, such as soil water availability, need to be included. This was confirmed in a study by Christensen *et al.* (2005) in which they observed, for maize canopies, the inability to detect nutrient deficiency effectively without an understanding of the plant water status. Jain *et al.* (2007) applied hyperspectral imagery to potato (*Solanum tuberosum* L.) fields and found that it was possible to detect leaf nitrogen content. This knowledge could be applied to nitrogen management of the crop. These findings have been questioned by Schepers and Holland (2012), who related crop vigour to yield and suggested that the uncertainties in weather and variation in the soil contribute to yield variation, and attempts to relate within-season management to crop yield have to be more inclusive than has been suggested in previous studies.

For irrigated agricultural lands, within-field variation in soil water-holding capacity, which was most often caused by the variation in soil texture, reduces the effectiveness of conventional irrigation practices. Over the past decade variable-rate automated irrigation systems have been developed (Perry *et al.*, 2002; see Chapter 10). Hedley and Yule (2009) applied the concepts of spatial variation to evaluate soil water availability and irrigation management in North Island, New Zealand. They were able to show, through the implementation of available soil water maps for a field, an improvement in irrigation management resulting in a saving of over 20 per cent of the applied water. In many areas of the world with scarce water resources, these savings would be substantial. Variable-rate systems for centre pivots have been commercialized, such as that offered by Valley Irrigation Co. (Valley Irrigation, 2011).

Some precision agriculture techniques for understanding the variation in yield that work well in some settings have been ineffective in others. For example, the spatial analysis of changes at the field scale has recently been evaluated by Inman *et al.* (2008) using a combination of reflectance, soil colour and crop yield. In this analysis, they coupled early-season NDVI, soil colour-based management zones and relative maize yields, and found that linking NDVI with the colour-based soil zones did not increase the ability to explain yield within the field. However, there are other studies that show positive relationships among different remote sensing indices and yield variation (Arno *et al.*, 2011; Pena-Barragan *et al.*, 2010; Wang *et al.*, 2012). These techniques offer considerable potential, and the application needs to be developed to contribute to improved food security through precision agriculture.

Role of precision agriculture in food quality

Management of nitrogen to enhance grain quality could increase the potential to provide more protein for the world's population. Engel *et al.* (1999) showed that it was possible to relate nitrogen management to protein in wheat during natural water stress conditions at the grain-filling stage. With this information it was possible to maintain the protein levels close to the critical values for high-quality wheat grain. Reyns *et al.* (2000) evaluated wheat yield, grain moisture and grain protein content across a field in Belgium and observed that they were not uniform across the field. Stewart *et al.* (2002) evaluated yield and protein content in durum wheat (*Triticum turgidum* subsp. *durum* (Desf.) Husn.). They found areas of the field with lower available soil water capacity leading to water stress during grain-fill where there was little soil organic matter and the soil texture was coarse; these effects reduced grain yield, but interacted with soil nitrogen content to increase protein levels in the grain. Miao *et al.* (2006) observed that some maize grain quality properties showed moderate or strong spatial dependence. For example, maize oil did not show any spatial dependence compared to starch content or vitreousness, which showed strong spatial dependence. They also observed that the variation in the quality properties of maize grain and in yield were less than the variation in soil properties. They did observe, however, significant differences between the two hybrids evaluated in this study for both grain quality and yield. In a related study, Pettersson *et al.* (2006) were able to develop a relationship for barley (*Hordeum vulgare* L.) yield and grain protein using reflectance values at the early vegetative growth stage, an index for elevated daily maximum temperatures during grain-filling and normalized electrical conductivity (EC_a) of the soil. These indices relate the early-season vigour with potential weather stress on the plant during grain-filling and are a surrogate for water availability from the soil. Shoji *et al.* (2005) found in a rice paddy field that spatial variation in grain yield and protein content was related to the availability of water and nutrient uptake at different micro-elevations within the field. Their conclusion was that micro-elevation and micro-climatic differences were sufficient to cause differences in yield and protein content. These types of combination models may be critical to understanding the dynamics of crop response across fields more fully. Attention directed towards grain quality could potentially produce large dividends in food production and food security because of the enhanced nutrient content of the food.

Food security

Food security is a complex problem extending beyond food production and quality. Flood (2010) pointed out that food security would require production systems with healthy plants which can be achieved through improved nutrient or pest management. This is an appropriate place for precision agriculture to play a role in terms of potentially identifying areas of fields or landscapes with insect or disease pressures or nutrient deficiencies. Lal (2009) suggested that soil degradation and the impacts on nutrient availability leading to poor plant growth contribute to the lack of food security. Li *et al.* (2011) examined the constraints on four major food crops – wheat, rice (*Oryza sativa* L.), sorghum (*Sorghum vulgare* (L.) Monech.) and chickpea (*Acer arietinum* L.) – in South Asia and concluded that drought-induced water stress contributed to 20 to 30 per cent of the reduction in yield from potential yield. They suggested that low soil fertility and poor fertilizer and pest management also contributed significantly to yield reductions. Garrity *et al.* (2010) integrated soil restoration with nutrient management under a tree inter-cropping system to demonstrate that improvements in crop production were possible across large areas in Africa. These examples have not been linked with precision agriculture directly, but the concepts of precision agriculture could be coupled with crop diversification across the landscape in order to reduce the potential risk arising from uncertainties in the weather. Such diversification with the goal of increasing plant water availability and use across the landscape will create major advancements in food production and food security because of the reduced risk in crop failure.

The recent analysis by Wirsenius *et al.* (2010) of the land changes under different growth scenarios that will be required to meet the food consumption needs of a growing population indicates that an increase in production per unit of land will be necessary. Principles of precision agriculture offer the potential to begin increasing productivity and potentially food quality. However, there are no easy solutions to the problem of being able to continue to increase our food supply until we focus on understanding the underlying reasons for the variability within fields and the potential impact of continued degradation of the soil resource on our ability to supply the crop with water and nutrients.

Challenges

Precision agriculture offers the potential to improve food production and food security. However, the challenge will be to move beyond the diagnostic phase into a more prescriptive phase of application of these technologies. Nitrogen management has been one of the areas in which precision agriculture technology has been applied successfully. However, one of the emerging challenges would be to relate this information to soil water-holding capacity in order to achieve the maximum benefit to both production and grain quality from N applications. Food security can be enhanced through integration of the spatial information at the field scale combined with information about the most effective management practices to be implemented within the field. There is evidence that combining remote sensing with soil maps and agronomic assessments will provide new insights into improved management practices. Feeding an ever-increasing population on the limited resources available will require integration of information from a variety of sources into improved methods for managing each individual field. This is the goal of precision agriculture. To meet the challenge of feeding the world requires that we begin to implement a much more integrated approach to managing our production systems.

References

Arno, J., Rosell, J. R., Blanco, R., Ramos, M. C. and Martinez-Casasnovas, J. A. (2011) 'Spatial variability in grape yield and quality influenced by soil and crop nutrition characteristics', *Precision Agriculture*, 13, 393–410.

Basso, B., Cammarano, D., Chen, D., Cafiero, G., Amato, M., Bitella, G., Rossi, R. and Basso, F. (2009) 'Landscape position and precipitation effects on spatial variability of wheat yield and grain protein in southern Italy', *Journal of Agronomy and Crop Science*, 195, 301–12.

Begue, A., Todoroff, P. and Pater, J. (2008) 'Multi-time scale analysis of sugarcane within-field variability: improved crop diagnosis using satellite time series', *Precision Agriculture*, 9, 161–71.

Bongiovanni, R. and Lowenberg-DeBoer, J. (2004) 'Precision agriculture and sustainability', *Precision Agriculture*, 5, 359–87.

Bouman, B. A. M. (2007) 'A conceptual framework for the improvement of crop water productivity at different spatial scales', *Agricultural Systems*, 93, 43–60.

Brock, A., Brouder, S. M., Blumhoff, G. and Hoffman, B. S. (2005) 'Defining yield-based management zones for corn–soybean rotations', *Agronomy Journal*, 97, 1115–28.

Cao, Q., Cui, Z., Chen, X., Khosla, R., Dao, T. H. and Miao, Y. (2011) 'Quantifying spatial variability of indigenous nitrogen supply for precision nitrogen management in small-scale farming', *Precision Agriculture*, 13, 45–61.

Casa, R. and Castrignanò, A. (2008) 'Analysis of spatial relationships between soil and crop variables in a durum wheat field using a multivariate geostatistical approach', *European Journal of Agronomy*, 28, 331–42.

Chi, B.-L., Bing, C.-S., Walley, F. and Yates, T. (2009) 'Topographic indices and yield variability in a rolling landscape of western Canada', *Pedosphere*, 19, 362–70.

Christensen, L. K., Upadhyaya, S. K., Jahn, B., Slaughter, D. C., Tan, E. and Hills, D. (2005) 'Determining the influence of water deficiency on NPK stress discrimination in maize using spectral and spatial information', *Precision Agriculture*, 6, 539–50.

Corwin, D. L., Lesch, S. M., Shouse, P. J., Soppe, R. and Ayars, J. E. (2003) 'Identifying soil properties that influence cotton yield using soil sampling directed by apparent soil electrical conductivity', *Agronomy Journal*, 95, 352–64.

Engel, R. E., Long, D. S., Carlson, G. R. and Meier, C. (1999) 'Method for precision nitrogen management in spring wheat: I. Fundamental relationships', *Precision Agriculture*, 1, 327–38.

Flood, J. (2010) 'The importance of plant health to food security', *Food Security*, 2, 215–31.

Food and Agriculture Organization (FAO) (2002) *The State of Food Insecurity in the World 2001*, Rome: Food and Agricultural Organization of the United Nations.

Garrity, D. P., Akinnifesi, F. K., Ajayi, O. C., Weldesemayat, S. G., Mowo, J. G., Kalinganire, A., Larwanou, M. and Bayala, J. (2010) 'Evergreen Africa: a robust approach to sustainable food security in Africa', *Food Security*, 2 197–214.

Hatfield, J. L. (2012) 'Spatial patterns of water and nitrogen response within corn production fields', in G. Aflakpui (ed.) *Agricultural Science*, Intech. Available online at <www.intechopen.com>, pp. 73–96.

Hatfield, J. L. and Prueger, J. H. (2011) 'Spatial and temporal variation in evapotranspiration', in G. Gerosa (ed.) *Evapotranspiration: From Measurements to Agriculture and Environmental Applications*, Intech. Available online at <www.intechopen.com>, pp. 3–16.

Hatfield, J. L., Prueger, J. H. and Kustas, W. P. (2007) 'Spatial and temporal variation of energy and carbon dioxide fluxes in corn and soybean fields in central Iowa', *Agronomy Journal*, 99, 285–96.

Hatfield, J. L., Gitelson, A. A., Schepers, J. S. and Walthall, C. L. (2008) 'Application of spectral remote sensing for agronomic decisions', *Agronomy Journal*, 100, S117–S131.

Hatfield, J. L., Boote, K. J., Kimball, B. A., Ziska, L. H., Izaurralde, R. C., Ort, D., Thomson, A. M. and Wolfe, D. W. (2011) 'Climate impacts on agriculture: implications for crop production', *Agronomy Journal*, 103, 351–70.

Hedley, C. B. and Yule, I. J. (2009) 'Soil water status mapping and two variable-rate irrigation scenarios', *Precision Agriculture*, 10, 342–55.

Inman, D., Khosla, R., Reich, R. and Westfall, D. G. (2008) 'Normalized difference vegetation index and soil color-based management zones in irrigated maize', *Agronomy Journal*, 100, 60–6.

Jain, N., Ray, S. S., Singh, J. P. and Panigrahy, S. (2007) 'Use of hyperspectral data to assess the effects of different nitrogen applications on a potato crop', *Precision Agriculture*, 8, 225–9.

James, I. T. and Godwin, R. J. (2003) 'Soil, water and yield relationships in developing strategies for the precision application of nitrogen fertiliser to winter barley', *Biosystems Engineering*, 84, 467–80.

Jaynes, D. B. and Colvin, T. S. (1997) 'Spatiotemporal variability of corn and soybean yield', *Agronomy Journal*, 89, 30–7.

Jaynes, D. B., Colvin, T. S. and Ambuel, J. (1993) 'Soil type and crop yield determinations from ground conductivity surveys', ASAE Paper 933552, St. Joseph MI: ASAE.

Jiang, P. and Thelen, K. D. (2004) 'Effect of soil and topographic properties on crop yield in a north-central corn–soybean cropping system', *Agronomy Journal*, 96, 252–8.

Jiang, P., Kitchen, N. R., Anderson, S. H., Sudduth, K. A. and Sadler, E. J. (2008) 'Estimating plant-available water using the simple inverse yield model for claypan landscapes', *Agronomy Journal*, 100, 830–6.

Jones, A. J., Mielke, L. N., Bartles, C. A. and Miller, C. A. (1989) 'Relationship of landscape position and properties to crop production', *Journal of Soil and Water Conservation*, 44, 328–32.

Karl, T. R., Melillo, J. M. and Peterson, T. C. (eds) (2009) *Global Climate Change Impacts in the United States*, Cambridge: Cambridge University Press.

Kaspar, T. C., Colvin, T. S., Jaynes, D. B., Karlen, D. L., James, D. E., Meek, D. W., Pulido, D. and Butler, H. (2003) 'Relationship between six years of corn yields and terrain attributes', *Precision Agriculture*, 4, 87–101.

Kitchen, N. R., Sudduth, K. A. and Drummond, S. T. (1999) 'Soil electrical conductivity as a crop productivity measure for claypan soils', *Journal of Production Agriculture*, 12, 607–17.

Kitchen, N. R., Drummond, S. T., Lund, E. D., Sudduth, K. A. and Buchleiter, G. W. (2003) 'Soil electrical conductivity and topography related to yield for three contrasting soil-crop systems', *Agronomy Journal*, 95, 483–95.

Kitchen, N. R., Sudduth, K. A., Drummond, S. T., Scharf, P. C., Palm, H. L., Roberts, D. F. and Vories, E. D. (2010) 'Ground-based canopy reflectance sensing for variable-rate nitrogen corn fertilization', *Agronomy Journal*, 102, 71–84.

Kumhálová, J., Matějková, Š., Fifernová, M., Lipavský, J. and Kumhála, F. (2008) 'Topography impact on nutrition content in soil and yield', *Plant, Soil and Environment*, 54, 255–61.

Kumhálová, J., Kumhála, F., Kroulík, M. and Matějková, Š. (2011) 'The impact of topography on soil properties and yield and the effects of weather conditions', *Precision Agriculture*, 12, 813–30.

Lal, R. (1993) 'Tillage effects on soil degradation, soil resilience, soil quality and sustainability', *Soil Tillage Research*, 21, 1–7.

Lal, R. (1997) 'Soil degradation and resilience of soils', *Philosophical Transactions of the Royal Society (B)*, 352, 997–1010.

Lal, R. (2009) 'Soil degradation as a reason for inadequate human nutrition', *Food Security*, 1, 45–57.

Lal, R. (2010) 'Beyond Copenhagen: mitigating climate change and achieving food security through soil carbon sequestration', *Food Security*, 2, 169–77.

Li, X., Waddington, S. R., Dixon, J., Joshi, A. K. and de Vicente, C. (2011) 'The relative importance of drought and other water-related constraints for major food crops in South Asian farming systems', *Food Security*, 3, 19–23.

McConkey, B. G., Ullrich, D. J. and Dyck, F. B. (1997) 'Slope position and subsoiling effects on soil water and spring wheat yield', *Canadian Journal of Soil Science*, 77, 83–90.

McGee, E. A., Peterson, G. A. and Westfall, D. G. (1997) 'Water storage efficiency in no-till dryland cropping systems', *Journal of Soil and Water Conservation*, 52, 131–6.

McNeill, J. D. (1992) 'Rapid, accurate mapping of soil salinity by electromagnetic ground conductivity meters', in G. C. Topp, W. D. Reynolds and R. E. Green (eds) *Advances in Measurements of Soil Physics Properties: Bringing Theory into Practice*, SSSA Special Publication 30, Madison, WI: ASA, CSSA, and SSSA, pp. 201–29.

Machado, S., Bynum, E. D., Jr., Archer, T. L., Lascano, R. J., Wilson, L. T., Bordovsky, J., Segarra, E., Bronson, K., Nesmith, D. M. and Xu, W. (2000) 'Spatial and temporal variability of corn grain yield: site-specific relationships of biotic and abiotic factors', *Precision Agriculture*, 2, 359–76.

Marques da Silva, J. R. and Silva, L. L. (2006) 'Evaluation of maize yield spatial variability based on field flow density', *Biosystems Engineering*, 95, 339–47.

Massey, R. E., Myers, D. B., Kitchen, N. R. and Sudduth, K. A. (2008) 'Profitability maps as an input for site-specific management decision-making', *Agronomy Journal*, 100, 52–9.

Miao, Y., Mulla, D. J. and Robert, P. C. (2006) 'Spatial variability of soil properties, corn quality and yield in two Illinois, USA fields: implications for precision corn management', *Precision Agriculture*, 7, 5–20.

Pena-Barragan, J. M., Lopez-Granados, F., Jurado-Exposito, M. and Garcia-Torres, L. (2010) 'Sunflower yield related to multi-temporal aerial photography, land elevation and weed infestation', *Precision Agriculture*, 11, 568–85.

Perry, C., Pocknee, S., Hansen, O., Kvien, C., Vellidis, G. and Hart, E. (2002) 'Development and testing of a variable-rate pivot irrigation control system', ASAE Technical Paper 02–2290, St. Joseph, MI: ASAE.

Pettersson, C.-G., Söderström, M. and Eckersten, H. (2006) 'Canopy reflectance, thermal stress, and apparent soil electrical conductivity as predictors of within-field variability in grain yield and grain protein of malting barley', *Precision Agriculture*, 7, 343–59.

Raun, W. R., Solie, J. B., Johnson, G. V., Stone, M. L., Mullen, R. W., Freeman, K. W., Thomason, W. E. and Lukina, E. V. (2002) 'Improving nitrogen use efficiency in cereal grain production with optical sensing and variable rate application', *Agronomy Journal*, 94, 815–20.

Raun, W. R., Solie, J. B. and Stone, M. L. (2011) 'Independence of yield potential and crop nitrogen response', *Precision Agriculture*, 12, 508–18.

Reyns, P., Spaepen, P. and de Baerdemaeker, J. (2000) 'Site-specific relationship between grain quality and yield', *Precision Agriculture*, 2, 231–46.

Rhoades, J. D., Chanduvi, F. and Lesch, S. M. (1999) 'Soil salinity assessment: methods and interpretation of electrical conductivity measurements', FAO Irrigation and Drainage Paper 57, Rome: Food and Agricultural Organization of the United Nations.

Robinson, N. J., Rampant, P. C., Callinan, A. P. L., Rab, M. A. and Fisher, P. D. (2009) 'Advances in precision agriculture in south-eastern Australia. II. Spatio-temporal prediction of crop yield using terrain derivatives and proximally sensed data', *Crop and Pasture Science*, 60, 859–69.

Sadler, E. J., Bauer, P. J. and Busscher, W. J. (2000a) 'Site-specific analysis of a droughted corn crop: I. Growth and grain yield', *Agronomy Journal*, 92, 395–402.

Sadler, E. J., Bauer, P. J. Busscher, W. J. and Miller, J. A. (2000b) 'Site-specific analysis of a droughted corn crop: II. Water use and stress', *Agronomy Journal*, 92, 403–10.

Sandras, V., Baldock, J., Roget, D. and Rodriguez, D. (2003) 'Measuring and modelling yield and water budget components of wheat crops in coarse-textured soils with chemical constraints', *Field Crops Research*, 84, 241–60.

Scharf, P. C., Shannon, D. K., Palm, H. L., Sudduth, K. A., Drummond, S. T., Kitchen, N. R., Mueller, L. J., Hubbard, V. C. and Oliveira, L. F. (2011) 'Sensor-based nitrogen application out-performed producer-chosen rates for corn in on-farm demonstrations', *Agronomy Journal*, 103, 1683–91.

Schepers, J. S. and Holland, K. H. (2012) 'Evidence of dependence between crop vigor and yield', *Precision Agriculture*, 13, 276–84.

Shoji, K., Kawamura, T., Horio, H., Nakayama, K. and Kobayashi, K. (2005) 'Variability of micro-elevation, yield, and protein content within a transplanted paddy field', *Precision Agriculture*, 6, 73–86.

Spomer, R. G. and Prest, R. F. (1982) 'Soil productivity and erosion of Iowa loess soils', *Transactions of the ASAE*, 25, 1295–9.

Stewart, C. M., McBratney, A. B. and Skerritt, J. H. (2002) 'Site-specific durum wheat quality and its relationship to soil properties in a single field in northern New South Wales', *Precision Agriculture*, 3, 155–68.

Timlin, D. J., Pachepsky, Y., Snyder, V. A. and Bryant, R. B. (1998) 'Spatial and temporal variability of corn grain yield on a hillslope', *Soil Science Society of America Journal*, 62, 764–73.

United States Census Bureau (n.d.) 'Total mid-year population for the world, 1950–2050'. Available online at <www.census.gov/population/international/data/idb/worldpopulation.phb> (accessed 18 April 2012).

Valley Irrigation (2011) *Variable Rate Irrigation (VRI)*, Valley, NE: Valmont Industries. Available online at <www.valleyirrigation.com/page.aspx?id=2342&pid=42> (accessed 28 December 2011).

Vitharana, U. W. A., Van Meirvenne, M., Simpson, D., Cockx, L. and de Baerdemaeker, J. (2008) 'Key soil and topographic properties to delineate potential management classes for precision agriculture in the European loess area', *Geoderma*, 143, 206–15.

Wang, Y.-P., Chen, S.-H., Chang, K.-W. and Shen, Y. (2012) 'Identifying and characterizing yield limiting factors in paddy rice using remote sensing yield maps', *Precision Agriculture*, 13, 553–67.

Wirsenius, S., Azar, C. and Berndes, G. (2010) 'How much land is needed for global food production under scenarios of dietary changes and livestock productivity increases in 2030?', *Agricultural Systems*, 103, 621–38.

Wood, C. W., Peterson, G. A., Westfall, D. G., Cole, C. V. and Willis, W. O. (1991) 'Nitrogen balance and biomass production of newly established no-till dryland agroecosystems', *Agronomy Journal*, 83, 519–26.

3 Improving food security through increasing the precision of agricultural development

Anja Gassner, Ric Coe and Fergus Sinclair

Introduction

Although long practised intuitively by subsistence farmers (Vanlauwe *et al.*, 2010; Heijting *et al.*, 2011), precision agriculture (PA) came of age in the 1990s as a technically driven means to improve industrialized agriculture. It promised benefits to both farmers and society through increasing production efficiency while improving stewardship of the environment (Srinivasan, 2006). These principles are central to the recent resurgence of interest in eco-efficiency (Keating *et al.*, 2010), driven by global food price spikes overlying progressive concern about the degradation of agroecosystems worldwide. These drivers have refocused global concern on the dual aims of improving food security while protecting the environment (Godfray *et al.*, 2010; Mueller *et al.*, 2012). This makes it a particularly appropriate time to take stock of the relevance of the principles of PA for agricultural improvement in developing countries.

Industrial agriculture, based on inorganic fertilizers, pesticides and other inputs, generally uses the risk-averse premise that given uncertainty in space and time, uniform within-field treatment is the best strategy (McBratney and Whelan, 1999). In contrast, PA recognizes the fine-scale heterogeneity of agricultural fields, as many subsistence farmers have traditionally done. Whereas subsistence farmers often have to concentrate inputs in fertile microsites as a risk-minimizing strategy (Vanlauwe *et al.*, 2010), in industrialized agriculture PA focuses on optimizing farm inputs by translating site-specific crop demands into variable management practices (Srinivasan, 2006; Mueller *et al.*, 2012).

In the early days of PA the ultimate goal was to understand within-field variation in plants and soil and then to tailor management to address this variability (e.g. Bouma, 1997). Precision agriculture was very much driven by technological advance, both in global navigation satellite systems as well as microcomputers and farm machinery. After 15 years of research and implementation of PA mainly in Europe and the USA, scientists came to the conclusion that it was more fruitful to identify the main processes that limit yield rather than to address all of the fine-scale variability (e.g. Dobermann *et al.*, 2002). This shifted the focus from a technically driven to a more results orientated approach. Still, the overall impression people have of PA is of an intensive crop management system, served by high-end technology (Cook *et al.*, 2003; Gebbers and Adamchuk, 2010).

Precision agriculture has been demonstrably successful in large-scale, mechanized, commercial or what can be termed industrial farming (Lowenberg-DeBoer, 2001) and especially with high-value cash crops that receive large amounts of agrochemical inputs and enter markets with strong differentiation based on quality, such as viticulture and horticulture (Srinivasan, 2006; Gebbers and Adamchuk, 2010). There are few documented examples of PA being applied to smallholder farms in developing countries. This is consistent with the predominance of high input industrial farming in North America and Europe (referred to as the North), with concomitant environmental impacts, whereas smallholders in developing countries (referred to as the South) often struggle to apply sufficient agrochemicals or irrigation water to maintain reasonable yields (Mueller *et al.*, 2012). Precision agriculture technologies in the North were developed largely to reduce wastage and leakage of agrochemicals, whereas in the South they are aimed at maximizing the effects of small quantities of agronomic inputs (Donovan and Casey, 1998; Srinivasan, 2006). It has been argued that high investment costs and associated increases in risk make PA unsuitable for smallholder farmers in developing countries (Cook *et al.*, 2003; Mohd Noor *et al.*, 2005).

While PA has not delivered the technological revolution in the agricultural sector that was predicted (Tey and Brindal, 2012), it has succeeded in reintroducing the concept of locally adapted interventions to both agricultural practitioners and scientists and in highlighting the need for information about the spatial and temporal variation of factors affecting yield. In the light of the current food crisis, especially in sub-Saharan Africa, the region with the highest demographic growth in the world, development agencies and governments are debating strategies to achieve a 'new green revolution for Africa' (Godfray *et al.*, 2010; Mueller *et al.*, 2012). Proponents of large-scale agricultural commodification argue that the millions of dollars of foreign investment involved will develop local infrastructure, facilitate transfer of skills and technology, create jobs, alleviate poverty and help to ensure food security in host countries. Others emphasize that tailor-made solutions that are inclusive, responsive to the needs of the poor and mindful of existing knowledge and local realities are more likely to bring about success in the fight against hunger in Africa (Nord and Luckscheiter, 2010). Whichever view prevails, it is clear that small-scale agriculture, dominated by fine-scale variation in yield determining factors, will continue to be a major source of livelihood for millions of rural people in Africa for a long time to come. In this chapter we take stock of the relevance of PA for addressing the needs of smallholder farmers and suggest extensions of the concepts for application in the smallholder context.

PA technologies for subsistence farmers

In a classical sense PA addresses mainly agronomic factors that influence crop yields, but social factors are mainly seen as drivers of these. The standard recipe for PA is a stepwise process that entails:

1   defining the yield-limiting factor or factors at a given time;
2   mapping these factors across the region of interest;

3   designing a variable-rate management strategy that addresses the spatial variability of these factors and

4   assessing and monitoring environmental and economic benefits of implementing variable-rate management strategies.

The gap between average yields presently achieved by farmers and yield potential is determined by the yielding ability of available crop varieties or hybrids and the degree to which crop and soil management practices allow expression of this genetic potential (Cassman, 1999; Mueller *et al.*, 2012). Supporting the expression of this genetic potential while increasing the efficiency of use of farm resources by adjusting crop management according to field variability and site-specific conditions is intuitively appealing to most agricultural practitioners. At face value, PA technologies seem to be especially appropriate tools for agriculture in developing countries, where policies that promote management of land and water resources for sustainable intensification have remained elusive (Gebbers and Adamchuk, 2010; Godfray *et al.*, 2010).

For classical PA to improve yields successfully, the yield-limiting factors need to be clearly defined and responsive to agronomic practices. The yield potential of a crop variety grown by a farmer does not only depend on its genetic makeup, but also on the inherent agronomic potential of the site. No increase in yield can be expected if the field chosen for planting is not suitable for the growth of the crop.

Cassman (1999) points out that most of the achievements of the green revolution in Asia were made on irrigated fields. He argues that the success of ecological intensification of cropping systems in unfavourable rainfed environments will be relatively small because present yields are very small and the primary constraint is lack of water. Approaches to management of soil fertility are often overly simplistic and the complex interaction between soil and plant interactions poorly understood. For example, long-term fertilizer trials have shown that nutrient imbalance, rather than a simple deficiency, can have more severe effects on yields (Cassman, 1999). There is other evidence that mineral fertilizers alone cannot address soil fertility as the yield-limiting factor because the underlying soil biology accounts for much variation in responsiveness of crop yield (Barrios *et al.*, 2012). Soil degradation is a major cause of small yields in Africa, Asia, and South and Central America. Inappropriate farming methods, deforestation and overgrazing were identified as the primary causes, leaving substantial areas unsuitable for intensive agriculture (Godfray *et al.*, 2010). Although the production practices and physical processes that cause erosion are well understood, technical solutions to prevent this kind of degradation are rarely adopted (Mueller *et al.*, 2012).

Before we discuss in detail why, in a smallholder context, a more holistic view of PA is needed, we look at how development practitioners have integrated PA technologies into their natural resource management programmes. These are selected examples that illustrate how the application of PA principles has played out in the developing world rather than an exhaustive catalogue.

Site-specific nutrient management of cereal in Asia

Site-specific nutrient management (SSNM) aims to record and predict the spatial variation of the nutrient supply in fields and to address this with variable fertilizer rates (Srinivasan, 2006). The principles of SSNM were developed for rice through more than a decade of research beginning in the mid 1990s and involving countries across Asia and in Africa. The experiences with rice were subsequently used to develop SSNM principles for maize and wheat, which were ready for delivery by 2010. Delivery of SSNM for rice from 2002 to 2008 focused on developing and promoting printed guidelines for large rice-growing regions. Uptake by farmers was limited because of the amount and sophistication of knowledge required to use the printed materials to develop field-specific guidelines for individual farms (Timsina *et al.*, 2010; Global Rice Science Partnership, 2010).

Micro-dosing fertilizer application in millet production systems in Niger

Micro-dosing technology has been developed by the International Crop Research Institute for the Semi-Arid Tropics (ICRISAT) in an attempt to increase the affordability of mineral fertilizer while giving plants enough nutrients for optimal growth. This micro-dosing technology consists of applying relatively small quantities of fertilizer (from 2 to 6 g hill^{-1}) at sowing time, thus substantially reducing the recommended amount of fertilizer that subsistence farmers need to apply, for example 10-fold for di-ammonium phosphate from 200 to 20 kg ha^{-1}. The implementation of this technology has resulted in greater nutrient use efficiency (Twomlow *et al.*, 2010).

Precision conservation agriculture

Precision conservation agriculture (PCA) assists farmers to be successful when applying conservation agriculture by tailoring practices to local circumstances (Jerich, 2011). Conservation agriculture (CA) is defined by three simple principles: (1) minimizing soil disturbance, (2) using crop rotations and or associations and (3) keeping soil covered with crop residue (Giller *et al.*, 2011). An example of PCA land preparation could include hand-dug planting holes, precise lime application around the root zone of each plant and precise spatial positioning of plants (Jerich, 2011). The universal applicability of CA principles for smallholder farmers in Africa has been questioned. More tailored approaches that adapt some or all three principles for different circumstances are seen as critical to their appropriateness (Giller *et al.*, 2011).

Precision manuring

Results indicate that farmers can improve management of manure applied to cropped areas simply by rotating the night-time tethering sites of their animals.

Through this strategy of precision manuring, they can concentrate manure application on the 'bad spots' or 'tired soils' that are most in need of nutrients and organic matter. Deliberate application of manure, compost and other fertilizers to low-yielding parts of fields is a common strategy employed by smallholder farmers (Vanlauwe *et al.*, 2010). The practice is especially useful for poor farmers, since they do not have enough land to ignore areas of declining soil fertility. Village-level management of precision manuring shows promise for enabling dryland communities to fine-tune the management of agro-pastoral systems across whole landscapes, resulting in larger and more sustainable yields (Taddesse *et al.*, 2003).

Supplementary irrigation

Supplementary irrigation (SI) is the addition of water to essentially rainfed crops during times of serious rainfall deficits. The combined use of rainfall and irrigation water is a potentially valuable management principle under conditions of water scarcity. The aim is to reduce the risk of crop failure and to stabilize yields where rainfall is normally sufficient, but vulnerability to drought is considerable. In the dryland farming area of northern Syria the International Center for Agricultural Research in Dry Areas (ICARDA) found substantial increases in crop yield in response to the application of relatively small amounts of irrigation water. The impact of SI goes beyond yield increases to substantially improve water productivity. Both the productivity of irrigation water and that of rainwater are improved when both are used together. The technology is considered to have large potential across West Asia and North Africa if combined with efficient water harvesting and adequate training of farmers (Oweis and Hachum, 2006).

Characteristics of smallholder farms in developing countries

Smallholder farming systems in developing countries are very variable, but do have some general characteristics that distinguish them from industrial farming. Typically they are characterized by small farm sizes, fragmented holdings and multiple production objectives. Integrated production for food, fodder, cash crops, fuel and housing often lead to complex systems that involve interactions amongst trees, crops and livestock. Levels of mechanization are generally low, and production is, therefore, labour-intensive. Smallholders are exposed to a multitude of risks, such as high variance in rainfall amount and distribution, and pests and diseases of crops and animals. Furthermore, agricultural production is also affected by flooding, frost, illness of household members, war and crime, all of which can have major effects on rural livelihoods. Investment and production decisions by smallholders are made within very unpredictable environments (Table 3.1).

Smallholder farmers often live in areas with little infrastructure and face high transaction costs that significantly reduce their incentives and opportunities

Table 3.1 Comparison between typical smallholder or family farms and commercial agriculture

Characteristics	Family farms	Commercial agriculture
Role of household labour	Major	Little or none
Community linkages	Strong – based on solidarity and mutual help between households and broader groups	Weak – often based on social connections between entrepreneur and local community
Priority objectives	Consume Stock Sell	Sell Buy Consume
Diversification	High, to reduce exposure to risk	Low – specialization in very few crops and activities to benefit from economies of scale
Size of holdings	Small, average 2 ha[a]	Large – exceeding 100 ha
Land access	Inheritance and social arrangements	Purchase

Source: adapted from Toulmin and Guèye, 2005.

Note
a Von Braun (2005) reports that the average farm size both in Africa and Asia is about 1.6 ha, whereas farm averages are 27 ha in Western Europe, 67 ha in Latin America and the Caribbean and 121 ha in Canada and the United States.

for market participation. In addition, small farms with few assets often have limited access to services, including extension and rural credit, which are often important prerequisites for increasing productivity (Fischer and Qaim, 2011). Smallholders have often been slow adopters of what scientists and extension staff consider optimal use of fertilizer, improved seeds and other production inputs (Shiferaw et al., 2009). One contributing factor for this is that recommendations rarely take within- and between-season variability of rainfall into account. Farmers are aware that this contributes to risk associated with investment and are often unwilling to adopt interventions that have high expected outcomes (as estimated by economists) but also have inherently greater risk (Donovan and Casey, 1998; Akponikpè et al., 2011).

Smallholder farms are hugely variable in terms of soil fertility status, as well as other biophysical determinants of production. This heterogeneity is evident at a range of scales: amongst farms within a locality, amongst fields within a farm that are not necessarily contiguous and within fields (Dobermann et al., 2002; Vanlauwe et al., 2010; Tittonell et al., 2011). The between-field variability at the individual farm level may be as large as differences between different agro-ecological zones, with obvious consequences for crop productivity (Table 3.2).

Variability of smallholder farms is not confined only to biophysical properties, they also vary in their availability of labour, livestock ownership, income, production orientation, cultural norms and wealth (Ojiem et al., 2006). In general, household income increases with the size of the landholding, despite

Table 3.2 Topsoil fertility status for different agroecological zones and for various fields within a farm in Burkina Faso

Area	Organic C $g\,kg^{-1}$	Total N $g\,kg^{-1}$	Available P $mg\,kg^{-1}$	Exchangeable K $mmol\,kg^{-1}$
Agroecozones				
Equatorial forests	24.5	1.6	–	–
Guinea savanna	11.7	1.4	–	–
Sudan savanna	3.3	0.5	–	–
Fields within a village				
Home garden	11–22	0.9–1.8	20–220	4.0–24
Village field	5–10	0.5–0.9	13–16	4.1–11
Bush field	2–5	0.2–0.5	5–16	0.6–1

Source: adapted from Vanlauwe et al., 2010.
Note: Home gardens are near the homestead, bush fields furthest away from the homestead and village fields are at intermediate distances.

the fact that the proportion of non-farm income tends to be greater among land-poor households (Jayne *et al.*, 2005). Limited land resources result in household members engaging in non-farm work, with returns per hour often lower or less reliable than those from work within the smallholding. Off-farm activities with the greatest potential for income generation are also those with the highest barriers to entry so tend to be concentrated among wealthier rural households (Rigg, 2006). Cultural differences condition how households differ in resource endowment, production orientation and objectives, education, past experience, management skills and attitude towards risk. These culturally conditioned differences lead to different natural resource management strategies and fine-scale variation in farmer practice and productivity (Tittonell *et al.*, 2010). Increasing population density and pressure on land amplifies livelihood constraints so that they become ever more predominant in driving observed variability.

Using principles of precision agriculture to customize interventions for smallholder farmers

Smallholder agriculture is the basis for food security and rural livelihoods in most developing countries and it is generally underperforming. Given the fine-scale variability of smallholder farming systems, and the fact that PA is designed to optimize farming under heterogeneity, PA principles could be expected to contribute to their improvement. In this section we review what can be learnt from the adoption of PA in commercial farming and combine this with an understanding of smallholder systems to make suggestions for the effective use of PA principles to enhance productivity of smallholder farms.

There has been no comprehensive study of PA adoption rates in developing countries, but it appears that interventions based on PA technologies follow the same fate as other natural resource management strategies (Tey and Brindal,

2012). While economic and environmental benefits are demonstrated in intensively supported projects, wide-scale adoption outside these projects is rarely observed (Shiferaw *et al.*, 2009). In Europe and the USA, despite positive net returns from PA experiments on commercial farms (Lowenberg-DeBoer, 2001), strong scientific approval and massive outreach campaigns by agro-industry, adoption rates have fallen short of expectations (Tey and Brindal, 2012). Studies in Germany, Denmark and the USA agree that high adoption rates are strongly linked to large farm sizes, a high level of mechanization and overall dependency on large agrochemical inputs. This is especially true where products are heading for markets with strong market differentiation based on quality, such as viticulture and horticulture and where there is access to consultants and extension workers (Lowenberg-DeBoer, 2001; Tey and Brindal, 2012).

Low rates of adoption are associated with the lack of awareness of PA technology among farmers, lack of access to sources of information, insufficient quality of information, time requirements, lack of technical knowledge, problems with the incompatibility of different hardware devices and the high cost of the technology (Kutter *et al.*, 2011; Tey and Brindal, 2012). One of the fundamental problems of PA is that its benefits are greatest when analysed using a holistic systems approach that includes putting values on environmental protection, food safety and other external benefits, while at a farm level, without realizing these values, the costs often outweigh perceived benefit (Lowenberg-DeBoer, 2001; Srinivasan, 2006).

In summary, the European and American experience with PA teaches us that a set of conditions have to be met before farmers are able and willing to adopt PA. These are:

- yield-limiting factors that can be addressed with PA;
- access to agronomic data;
- perceived economic benefits;
- access to extension services and or consultants.

A summary of the challenges in meeting these requirements in smallholder systems, and potential solutions, is provided in Table 3.3. This leads us to propose the following four main elements of a strategy for using PA principles to enhance smallholder productivity:

1 Build on farmers' knowledge and expertise by facilitating local experimentation, observation and learning.
2 Use high-resolution spatial and temporal data to inform farmers and target interventions.
3 Match extension methods to local circumstances and demand.
4 Manage social and economic factors within the PA framework at a range of scales.

Together these elements constitute widening the classical PA focus from concentrating only on yield-limiting factors to embrace development inhibiting

Table 3.3 Precision agriculture's requirements, enabling factors, challenges and potential solutions in smallholder systems

Requirements	Enabling factors	Challenges in smallholder systems	Solutions
Perceived economic benefits	• access to information • demonstration plots and or working examples • payments for ecosystem services	• limited access to resources, such as fertilizers, labour and land leading to unsustainable land uses • lack of economic incentives for environmental stewardship • social capital required to manage ecosystem services manifest at landscape scale is often lacking • institutional barriers	• create opportunities for income diversification to allow higher inputs into land • provide better support for long-term sustainable strategies versus quick gains • support environmental stewardship with PES schemes • build institutions for collective management of landscape-scale processes
Access to agronomic data	• technology for data collection • functional dissemination systems	• relevant data not readily available to farmers and extension staff • most farmers rely on radio and telephone usage rather than Internet • weak social networks beyond friends and relatives	• investment in national agricultural databases • work at larger spatial scale in which mapping is feasible • make use of local knowledge • develop capacity for local informal experimentation by farmers, farmer groups and extension staff • improve communication flows within farming communities as well as between farmers, extension workers and research organizations
Extension services and or consultants	• demand-driven extension support or service providers • well-functioning social networks both vertical between research institutions, extension workers and farmer groups, and horizontal amongst farmers	• small-scale farmers lack capacity and mechanisms to articulate their demands • low levels of organization amongst farmers and weak negotiation power • technologies not available and low priority to supply them • low capacity to use these technologies at farm or research level	• develop tools that facilitate extension workers and farmer groups to employ principles in their specific contexts rather than pushing prescriptions • identify which extension methods are appropriate for different messages and contexts • facilitate farmer-to-farmer dissemination where appropriate • invest in private and public rural support centres where input and knowledge supply are combined • connect research and extension by embracing local experimentation, observation and learning • scale up local learning through embedding research within development and generalizing about what interventions work, where and for whom
Identifying yield-limiting factors that can be addressed with PA	• access to fertile land • understanding macro-level economic factors that influence farm gate prices • understanding the complex interaction of soil biochemical and plant physiological interactions	• insecure tenure • very small farm sizes • rainfed agriculture • market imperfections • no access to credit • strong link between culture, ethnicity and farming practices	• widen the classical PA focus from biophysical factors to also include social and economic factors • use advances in remotely sensed data availability and processing to characterize and map variability of smallholder farm systems

factors that are the ultimate causes of underperformance in smallholder agriculture. These four factors are each examined in detail below together with their integration to improve the precision of agricultural development.

Build on farmers' knowledge and expertise by facilitating local experimentation, observation and learning

There are numerous examples of variable management technologies that follow the fundamental principles of PA and that evolved as a result of farmers' local knowledge (Vanlauwe *et al.*, 2010; Heijting *et al.*, 2011). Tailoring soil and crop management to match local within-field variation has been a common strategy for Asian farmers. The growers traditionally noted yield variability both in space and time, and adjusted farm practices according to local site conditions (Srinavasan, 2006). In Malaysia, for example, this is reflected in small farm and field sizes of traditional agricultural communities (Mohd Noor *et al.*, 2005). In sub-Saharan Africa smallholder farms often consist of multiple plots managed differently in terms of allocation of crops, fertilizers and labour resources (Vanlauwe *et al.*, 2010). Three well-documented examples of such traditional PA practices are set out below.

Field dispersion in Niger

The subsistence mixed millet production system in Niger is characterized by intercropping with a range of secondary crops, either dual-purpose legumes (cowpea) or cash crops (sesame, sorrel). Smallholder farmers have adopted a variety of management strategies to secure at least a minimum yield each year by reducing agro-climatic risk. One such strategy is the dispersion of fields cultivated by a single household throughout the village territory, with farmers using varieties of differing time to maturity in order to distribute labour and other resource use across a longer period. Small and remote fields, in particular, are sown with early maturing varieties because these are the ones where seeding may be delayed as a result of labour shortages and problems of access. Using the APSIM crop simulation model Akponikpè *et al.* (2011) were able to provide evidence that field dispersion does indeed reduce inter-annual yield variability at household level and hence reduces the risk of severe household food deficits.

Management of microvariability in communal land in Zimbabwe

Ecologists and social scientists have documented the complex ways in which Zimbabwean farmers have coped with environmental variability since the early 1980s. As in other parts of Africa, input in the form of labour or nutrient amendments has to be adjusted carefully to both climatic as well as edaphic variability to maximize returns to labour. In a case study from the Mutoko area in northeastern Zimbabwe, Carter and Murwira (1995) point out that the different farm-management strategies used by households also depend strongly on their wealth status. Wealthier farmers, with access to cows and manure, tend to outbalance natural soil variation by applying nutrients to low fertility patches in the field, whereas farmers with few assets will concentrate their efforts on the more promising and fertile areas. Resources used are cattle, goat and chicken manure, composted crop residues and leaf litter, undecomposed leaf litter or termite-mound soil. At the beginning of the season farmers plant as much as they can as part of a deliberate strategy to give flexibility in the face of uncertain rainfall. Once the nature of the season is clear, they are able to concentrate on those crops, niches and patches within fields where successful yields are most likely. Farmers tend to grow cash crops (maize) on the best fields and patches and, if labour shortage becomes a problem during the season, concentrate management on these fields (Vanlauwe *et al.*, 2010).

Rice-production zones in Senegal and Gambia

Carney (1991) provides a detailed account of the traditional rice production in Senegal and Gambia. Over the past millennia farmers in Senegal and Gambia have fine-tuned rice cultivation to a range of agroecological zones with differing edaphic properties and moisture regimes. The system recognizes six micro-environments, each one combining hydrological regimes (rainfed, tidal and rainfall combined with marine tides) and soil properties in unique ways. Cultivation schedules, agronomic practices and seed selection are adjusted to the specific characteristics of each micro-environment. Diola women in southern Senegal plant as many as 15 rice varieties throughout the production zones, whereas Mandinka women in central Gambia name nearly 30 varieties, local as well as introduced, that are cultivated.

It is clear from these examples that smallholder farmers often recognize and address environmental variability at patch, field and landscape scales in their cropping practices. It is also well established that in many traditional systems, significant nutrient transfers are made via livestock at landscape scales to concentrate fertility on crop fields (Vanlauwe *et al.*, 2010) and that farmers have detailed understanding of tree–crop interactions and how different species affect yield and other ecosystem services, and their variability in space and time (Cerdan *et al.*, 2012). Increasingly, farmers' knowledge is being found to be dynamic and explanatory, consistent with the now well-established notion that farmers actively observe and experiment (Shiferaw *et al.*, 2009; Cerdan *et al.*, 2012). A key requirement to address heterogeneity in smallholder farming systems is for research and extension staff to acknowledge the limitations of their ability to match interventions to sites and farmer circumstances and the resulting risk that farmers face in adopting them (Tittonell *et al.*, 2011). This can be addressed by acknowledging that agricultural development at a fine scale involves local experimentation, observation, risk-taking and learning by farmers that build on their local knowledge and expertise.

Use of high-resolution spatial and temporal data to inform farmers and target interventions

Classical PA is measurement- and knowledge-intensive (Srinivasan, 2006; Tey and Brindal, 2012). This creates two challenges in a development context. First getting relevant information to famers, and second building their capacity to use this information. Mohd Noor *et al.* (2005) suggested that implementation of PA without major investment in building the capacity of the smallholder sector would widen social divisions and marginalize smallholders even more, as only commercial farms are presently capable of taking advantage of this knowledge-intensive technology. There are several examples that demonstrate that appropriate technological solutions to provide real-time data to smallholders do now exist.

In 2008, IRRI implemented a computer-based decision tool to address these issues, disseminated as a CD, web-based and mobile phone-based application. The tool consists of 10 to 15 questions regarding crop performance, easily answered within 15 minutes by an extension worker and farmer. Based on responses to the questions, a field-specific guideline with amounts of fertilizer by crop growth stage is provided (Global Rice Science Partnership, 2010; Timsina *et al.*, 2010). There are other examples of knowledge-based systems being developed and used to customize extension information across heterogeneous smallholder farm environments, for example FORMAT (Thorne *et al.*, 1997), LEGINC and LEXSYS (Moss *et al.*, 2003).

In 2005 Esoko, an agricultural market information platform, launched a new initiative to provide information on farming and agricultural produce. Farmers can request information such as produce price alerts, bids and offers, and news and advisories. MTN is sponsoring training of 500 farmers on the use and benefits of the Esoko Information Product. These farmers will also enjoy free SMS subscription to Esoko's market information for one year. Currently, the Esoko platform has registered over 14,000 contacts (users of Esoko), 847,000 prices, 517 trade groups and 480 markets (David-West, 2011).

The Grameen Foundation started the Community Knowledge Worker (CKW) initiative, which is building on a self-sustaining, scalable network of rural information providers. By disseminating and collecting relevant information via mobile phones the CKW provides access to up-to-date information on best farming practices, market conditions, pests and disease control, weather forecasts and market access. Upon request from a farmer, a CKW will use his or her mobile to access actionable information to meet farmer needs. In Uganda, CKWs have proved to be a vital link between farmers, government programmes, non-governmental organizations and other entities (e.g. Kiiza and Pederson, 2012).

The Africa Soil Information Service (AfSIS, http://africasoils.net) is a pioneering effort funded by the Bill and Melinda Gates Foundation (BMGF), and the Alliance for a Green Revolution in Africa (AGRA) to fill one of the major gaps in spatial information worldwide. The AfSIS produces timely, cost-effective, soil health surveillance maps at a scale useful to smallholders and rural development practitioners (Terhoeven-Urselmans *et al.*, 2010).

The Seeing Is Believing – West Africa (SIBWA) project provides farmers with real-time data about the spatial and temporal variation of the landscape they are in. Teams on the ground verify the data and update the database of information that they can use to develop an accurate map of each farm. The SIBWA partners translate the information into local languages and take the detailed maps back to the individual farmers, who can use them to plan and manage their crops for the coming growing season (Traore, 2009).

It is clear from these examples that progress is being made in using modern remote sensing and information technology to provide farmers and research and extension staff with fine-scale data on biophysical and socio-economic variables. More challenging is the need to develop tools and build capacity to use this information effectively in farmer decision-making and in targeting intervention options to sites and farmer circumstances. Examples exist of knowledge-based systems tools for customizing extension messages to local circumstances, but mainstreaming the development and use of such approaches remains in its infancy.

Match extension methods to message and context

The conventional wisdom in PA has been that smallholders lack process-based knowledge concerning agroecosystem function, creating uncertainty that obstructs sound decision-making under conditions of change. Therefore, providing farmers with spatial information about how best to use their resources can improve their practice (Cook *et al.*, 2003). But this view that smallholders have largely descriptive rather than explanatory agroecological knowledge does not stand up to scrutiny, with accumulating evidence that smallholder farmers in a range of contexts in Africa, Asia and Latin America display a well-developed understanding of agroecosystem function (Shiferaw *et al.*, 2009; Cerdan *et al.*, 2012). This farmer knowledge of ecosystem function, discussed in the previous sections, is bounded by their means of observation and comparison, and is often largely complementary to that of scientists and extension workers. Appreciating the sophistication of farmers' local knowledge, while recognizing both gaps in this knowledge and that farmers are often looking for innovations, opens the way towards the extension of principles that farmers can incorporate into their practice, rather than prescriptions that have to be customized to local circumstances, as discussed in the preceding section.

Many African and Asian countries are undergoing a progressive policy change towards more demand-driven and market-orientated agricultural services. This includes a policy shift from centralized extension systems, for example, Training and Visit (T&V), to decentralized, demand-driven agricultural advisory systems (Friis-Hansen and Duveskog, 2012). Traditional research and extension systems view farmers as end-users who must be persuaded into adopting research outputs, rather than as partners in the process. Advisory services for PA technologies can only be demand-driven if there is both a choice of advisers who are able to offer quality advisory services at an appropriate price as well as farmers that are capable of articulating their needs (Friis-Hansen and Duveskog, 2012). Farmers also need to be well informed about the different services and service providers, as well as being capable of recognizing quality services. Three examples of extension methods aiming to address a demand-led agenda are set out below.

Farmer field schools

The Farmer Field School Extension (FFS) approach originated in the context of integrated pest management in wet paddy fields in the Philippines and Indonesia. The success in these two countries has since been documented and used to promote and expand FFS and FFS-type activities to other countries and to other crops. The FFS is a group approach to agricultural technology development, focusing on adult, non-formal education through hands-on field-discovery learning (Friis-Hansen and Duveskog, 2012). These activities consist of simple field experiments, regular field visits and participatory analysis. A typical group of trainees includes 20–25 participants; the duration is about 8–12 weeks within a single crop-growing season. A facilitator leads the programme, conveying knowledge of and facilitating discussion of good crop-management decision procedures and practices. The knowledge gained from these activities enables participants to make their own locally specific decisions about crop-management practices. The success of FFS depends strongly on the dissemination of the knowledge and experience gained by participants to other farmers outside the FFS (Feder *et al.*, 2004). It therefore goes hand in hand with farmer-to-farmer dissemination.

Farmer-to-farmer dissemination

Using farmers as extension agents follows the theory of agents of change. The idea is that if one farmer adopts a technology successfully, other farmers may learn the innovation from him or her and share with others, thereby developing a multiplier effect. Rather than simply being agents for technologies imposed from outside, champion farmers are expected to become catalysts, mobilizing other farmers to experiment, recognizing local innovations and helping to assess and encourage innovation. Farmer-to-farmer dissemination builds on strong, socially based processes of learning, promulgating innovations through informal social networks such as friendships, kinships and farmer groups. This concept is being formalized and refined in the use of volunteer farmer trainers as a novel extension approach (Lukuyu *et al.*, 2012).

Farmer participatory research and innovation systems

Farmer participatory research describes a process that is based on a dialogue between farmers and researchers to develop improved technologies that are practical, effective, profitable and will address identified agricultural production constraints. Collaboration and communication between farmers and scientists ensures that research findings are relevant to farmers' needs and applicable within their biophysical and socio-economic environments. With assistance from moderators, farmers themselves discover solutions to their problems during informal discussions (Nain *et al.*, 2012). This sort of approach sits at the centre of the use of innovation platforms that bring stakeholders together to address problems at a range of scales from national to local (Hounkonnou *et al.*, 2012). Innovation system approaches are in use both for generally positioned integrated rural development and research-led projects that focus on particular aspects of smallholder productivity, such as the N2Africa project that focuses on the use of legumes (http://www.n2africa.org/).

Extension approaches themselves need to be evaluated on the basis of a continuously refined understanding of what works where and for whom. This requires systematic evaluation of different methods across a range of messages and contexts rather than reliance on the *post hoc* comparative analyses of case studies that has been more commonly used in this field. It requires research embedded within development praxis. Global networking exists that could facilitate this, but it remains to be seen whether systematic studies will be conducted and precision in the use of extension methods improved (Veldhuizen and Wettasinha, 2010).

Incorporate social and economic dimensions within the precision agriculture framework at a range of scales.

The concept of tailor-made crop-management interventions did influence the thinking of scientists working within the development domain in the late 1990s. The classical PA approach focuses mainly on biophysical factors, with anthropogenic factors recognized as drivers of farm-level heterogeneity but not targeted by interventions. The International Food Policy Research Institute (IFPRI) is trying to address this shortcoming by combining the idea of PA with those of economic geography (Chamberlin *et al.*, 2006). The concept of the agricultural domain was developed in the early 2000s and has been widely used by IFPRI to assist African governments in developing strategic priorities (Adeogun, 2009).

Looking from a broader development perspective rather than a narrow agronomic perspective, IFPRI is asking whether there is a set of indicators that explains the comparative advantage of one location over another location in terms of opportunities and constraints for sustainable agricultural development. Pender and colleagues proposed that the main factors that describe these localized comparative advantages are agricultural potential, access to markets and population density (Chamberlin *et al.*, 2006). Areas that are similar with respect to these three factors are called development domains and development interventions should be designed for each within a country. This approach thus uses PA ideas at a scale beyond the farm, producing options while having a much more local relevance than the traditional agricultural ecozone approach.

Tittonell *et al.* (2010) grouped smallholder farms in the highland and midland humid zones of East Africa based on resource endowment, dependence on off-farm income and production objectives into five farm types. They argue that efforts to enhance farmers' livelihoods can be successful only if these different farm typologies are taken into consideration when designing technological innovations and or development efforts. Households with a more agriculture-based livelihood strategy are more likely to implement and eventually adopt proposed technologies for agricultural intensification, whereas poorer farmers may be the major beneficiaries of social promotion (policy and or development) interventions. Compared to IFPRI's approach, Tittonell *et al.* (2011) bring the use of PA principles down a scale to account for between-farm variation within a development domain.

Another concept that incorporates a social dimension within an agronomic-centred PA frame is the concept of the socio-ecological niche (Ojiem *et al.*, 2006), which defines a multidimensional space of environmental, economic and social factors that affect the success of a farm-level intervention (Figure 3.1). This recognizes that farmers' decision-making depends largely on the farmer's evaluation frame of reference. This in turn is determined mainly by their belief in the technical and socio-economic consequences of decisions, their perception of the likelihood that these consequences will emerge and their evaluation of such consequences in relation to a set of aspirations (Donovan and Casey, 1998; Shiferaw *et al.*, 2009).

Applying precision agriculture principles to rural development

In classical agronomic theory, the limiting factor is that which prohibits a crop attaining its full yield potential when all other factors are optimal (Liebig, 1840). In a rural development context there are many interacting and overlapping factors constraining livelihoods, and there is controversy about how best to address rural poverty and the extent to which agricultural innovations can do so (Rigg, 2006; Harris and Orr, in press). Widening our focus on limiting factors from the crop to the livelihood involves traversing scales from the plant, field, farm and landscape to incorporate wider social networks and markets. This leads to fundamental questions in which agronomic interventions may or may not be appropriate, for example:

- What hinders a smallholder moving from subsistence to commercial production?
- Are the main constraints institutional or political, economic, agroecological or socio-cultural?
- Can these barriers be addressed by farm-level interventions or do they need much larger political interventions?
- What is the site-specific 'development' potential?
- Will farmers ever be able to produce sufficient products from their farm to ensure a life above the poverty line or should they be supported to abandon the farm and turn to more economically sustainable livelihoods?

Investigations of development interventions often reveal that the solution to rural poverty lies in the invigoration of farming and the redistribution of land. But patterns and associations of wealth and poverty have become more diffuse

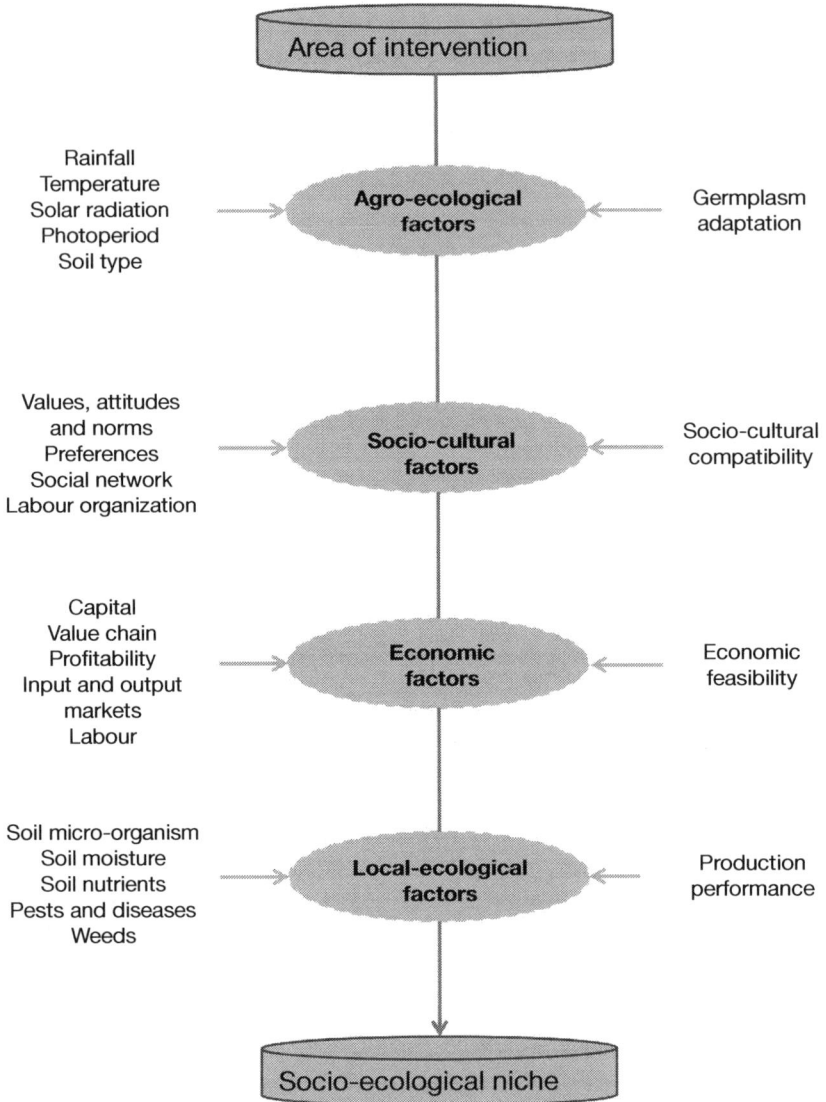

Figure 3.1   Schematic diagram depicting the concept of the socio-ecological niche, the hierarchical arrangement of factors that influence the delineation of the niche and the functions and outputs of the factors (Ojiem *et al.*, 2006).

and diverse as non-farm opportunities have expanded and heightened levels of mobility have led to livelihoods becoming less locally situated (Rigg, 2006). Investing in agriculture may preferentially support the wealthier households and thereby widen inequalities in the countryside (Tittonell *et al.*, 2011). Ersado (2006) cites research showing that in more remote areas of Zimbabwe, off-farm income sources increase income inequality because only the better-off and well-connected farmers can diversify, whereas in areas better connected to the major urban markets, it decreases income inequality because opportunities are more widely available. Understanding both the change in rural farming communities in terms of resource access as well as socio-economics and the variability between households or regions is essential to identify where and what the limiting factors are for farmers to reach prosperity. Rural communities operate within a fast-changing environment, both biophysically and socio-economically. Common drivers of change include the following:

1 declining soil health and increasing shortages of fertile land,
2 the erosion of profitability and returns to smallholder agricultural production,
3 the emergence of new opportunities in the non-farm sector, both local and distant,
4 high levels of mobility leading to livelihoods with increasing dependence on remittances from elsewhere,
5 increasing population and declining landholding size,
6 climate change.

Practitioners of PA have recognized the importance of addressing temporal variation in crop performance (Odgaard *et al.*, 2011). Plant and soil properties that depend on climate, such as nutrient availability and severity of pests and diseases, can have large inter-annual variations. Time-series of yield maps, either produced from modelling or based on annual measurements, are an integral tool of PA. Distinctive spatial and temporal trends in yield maps can often be identified by eye. Spatial trend maps are used to visualize consistently high- and low-yielding areas of fields or landscapes and temporal yield stability maps to identify distinctive management zones (McKinion *et al.*, 2010). The same approach could be applied to livelihood metrics to produce spatial trend and stability maps for livelihoods as opposed to crop yield.

Development practitioners have struggled to link information on social capital to that on natural capital. Often-stated reasons are the problem of integrating socio-economic and biophysical data because they have usually been collected at different scales (Thornton *et al.*, 2006; Carletto *et al.*, 2011) and the fundamentally different understanding of scale between social and environmental research (Gibson *et al.*, 2000). However, the spatial dimension of social processes and the context that defines them have come back into sharp focus among social and behavioural scientists. Social network and spatial analytic strategies in particular are placing social phenomena in relational and physical contexts. Social networks are described by the propinquity effect, the phenomenon that people who are located closer together in physical space have a higher probability of forming relationships (Adams *et al.*, 2012). This spatial autocorrelation of social ties between farmers provides bias-free criteria for domain selection. In addition it provides a social variable with the same spatial dimension as other biophysical variables measured, resolving the scale problem.

Conclusions

Although long practised intuitively by subsistence farmers, precision agriculture came of age in the 1990s as a technically driven means to improve industrialized agriculture. By analysing the multiple crop and soil physiological factors that lead to spatial and temporal yield variations and then designing site-specific management strategies to close yield gaps, PA promised benefits to both farmers and society through increasing production efficiency while improving stewardship of the environment. Although PA has not delivered the technological revolution in the agricultural sector that was predicted, it has succeeded in highlighting the importance of locally adapted interventions to both agricultural practitioners and scientists.

In the context of intensification of smallholder production in developing countries the ability to adapt interventions locally is critically important. With average farm sizes in both Africa and Asia well below 2 ha, there can be little doubt that meeting the rising global demand for food will require closing yield gaps on smallholder farms. These farms and the contexts in which they operate are highly heterogeneous at fine scales, so that interventions to improve productivity need to be tailored to sites, farmer circumstances and institutional settings. This involves a focus on applying the principles of PA at a range of spatial scales to improve the precision of agricultural development, rather than directly trying to support the field- and farm-level agronomic decisions of millions of smallholder farmers. To achieve this, the classical PA approach that focuses mainly on biophysical factors clearly has to be broadened to include variability in resource endowment, culture, market access and gender realities. This means using PA concepts at a range of scales, not just for variable within-field management. Such approaches show promise and are being implemented, but remain in their infancy.

Further application of PA concepts to bring benefits to smallholder farmers requires (a) increased understanding of the processes and principles determining farm performance, (b) increased capacity for local experimentation, monitoring and learning and its aggregation across scaling domains and (c) increased access to real-time information on both biophysical and socio-economic factors. Precision agriculture principles of using and adapting to spatial heterogeneity need to be made more important within research and extension thinking. It also requires adjustment of the research development continuum from research for development to research in development. The concept of eco-efficiency is emerging as a dominant paradigm for smallholder agricultural development. It stresses the need to tighten nutrient and water cycles to intensify production sustainably, without increasing the risk to which smallholder farmers are exposed. The principles of PA, if applied appropriately, readily lend themselves to improving resource use efficiency and reducing leakage at field, farm and landscape scales. They can also contribute to reducing the risk that farmers face in adopting and adapting innovations when applied across large scaling domains, by involving integrated research and extension teams, using interdisciplinary and participatory approaches to agricultural innovation.

References

Adams, J., Faust, K. and Lovasi, G. S. (2012) 'Capturing context: integrating spatial and social network analyses', *Social Networks*, 34, 1–5.

Adeogun, G. (2009) *Validation Workshop on Agriculture Development Domains in Nigeria*, NSSP Workshop Report 14, Abuja, Nigeria: International Food Policy Research Institute (IFPRI).

Akponikpè, P. B. I., Minet, J., Gérard, B., Defourny, P. and Bielders, C. L. (2011) 'Spatial fields' dispersion as a farmer strategy to reduce agro-climatic risk at the household level in pearl millet-based systems in the Sahel: a modeling perspective', *Agricultural and Forest Meteorology*, 151, 215–27.

Barrios, E., Sileshi, G. W., Shepherd, K. and Sinclair, F. (2012) 'Agroforestry and soil health: linking trees, soil biota, and ecosystem services', in D. H. Wall, R. D. Bardgett, V. Behan-Pelletier, J. E. Herrick, H. Jones, K. Ritz, J. Six, D. R. Strong and W. H. van der Putten (eds) *The Oxford Handbook of Soil Ecology and Ecosystem Services*, New York: Oxford University Press, pp. 315–30.

Bouma, J. (1997) 'Precision agriculture: introduction to the spatial and temporal variability of environmental quality', in J. V. Lake, G. R. Bock and J. A. Goode (eds) *Precision Agriculture: Spatial and Temporal Variability of Environmental Quality*, CIBA Foundation Symposium 210, Chichester, UK: John Wiley & Sons, pp. 5–13.

Carletto, C., Savastano, S. and Zessa, A. (2011) *The Impact of Measurement Errors on the Farm-Size–Productivity Relationship*, Policy Research Working Paper 5908, Washington, DC: World Bank.

Carney, J. (1991) 'Indigenous soil and water management in Senegambian rice farming systems', *Agriculture and Human Values*, 8, 37–58.

Carter, S. and Murwira, H. K. (1995) 'Spatial variability in soil fertility management and crop response in Mutoko Communal Area', *Zimbabwe Ambio*, 24, 77–84.

Cassman, K. G. (1999) 'Ecological intensification of cereal production systems: yield potential, soil quality, and precision agriculture', *Proceedings of the National Academy of Sciences*, 96, 5952–9.

Cerdan, C. R., Rebolledo, M. C., Soto, G., Rapidel, B. and Sinclair, F. L. (2012) 'Local knowledge of impacts of tree cover on ecosystem services in smallholder coffee production systems', *Agricultural Systems*, 110, 119–30.

Chamberlin, J., Pender, J. and Bingxin, Y. (2006) *Development Domains for Ethiopia: Capturing the Geographical Context of Smallholder Development Options*, EPTD Discussion Papers 159, Washington, DC: International Food Policy Research Institute (IFPRI).

Cook, S. E., O'Brien, R., Corner, R. J. and Oberthur, T. (2003) 'Is precision agriculture irrelevant to developing countries?', in J. V. Stafford and A. Werner (eds) *Precision Agriculture*, Wageningen: Wageningen Academic Publishers, pp. 115–20.

David-West, O. (2011) *Esoko Networks: Facilitating Agriculture through Technology*, GIM Case Study No. B061, New York: United Nations Development Programme.

Dobermann, A., Witt, C., Dawe, D., Gines, G. C., Nagarajan, R., Satawathananont, S., Son, T. T., Tan, P. S., Wang, G. H., Chien, N. V., Thoa, V. T. K., Phung, C. V., Stalin, P., Muthukrishnan, P., Ravi, V., Babu, M., Chatuporn, S., Kongchum, M., Sun, Q., Fu, R., Simbahan, G. C. and Adviento, M. A. A. (2002) 'Site-specific nutrient management for intensive rice cropping systems in Asia', *Field Crops Research*, 74, 37–66.

Donovan, G. W. and Casey, F. (1998) *Soil Fertility Management in Sub-Saharan Africa*, World Bank Technical Paper No 408, Washington, DC: World Bank.

Ersado, L. (2006) *Income Diversification in Zimbabwe: Welfare Implications from Urban and Rural Areas*, Policy Research Working Paper Series 3964, Washington, DC: World Bank.

Feder, G., Murgai, R. and Quizon, J. B. (2004) 'The acquisition and diffusion of knowledge: the case of pest management training in farmer field schools, Indonesia', *Journal of Agricultural Economics*, 55, 221–43.

Fischer, E. and Qaim, M. (2011) 'Linking smallholders to markets: determinants and impacts of farmer collective action in Kenya', *World Development*, 40, 1255–68.

Friis-Hansen, E. and Duveskog, D. (2012) 'The empowerment route to well-being: an analysis of farmer field schools in East Africa', *World Development*, 40, 414–27.

Gebbers, R. and Adamchuk, V. I. (2010) 'Precision agriculture and food security', *Science*, 327, 828–31.

Gibson, C., Ostrom, E. and Ahn, T. K. (2000) 'The concept of scale and the human dimensions of global change: a survey', *Ecological Economics*, 32, 217–39.

Giller, K. E., Corbeels, M., Nyamangara, J., Triomphe, B., Affholder, F., Scopel, E. and Tittonell, P. (2011) 'A research agenda to explore the role of conservation agriculture in African smallholder farming systems', *Field Crops Research*, 124, 468–72.

Godfray, H. C. J., Beddington, J. R., Crute, I. R., Haddad, L., Lawrence, D., Muir, J. F., Pretty, J., Robinson, S., Thomas, S. M. and Toulmin, C. (2010) 'Food security: the challenge of feeding 9 billion people', *Science*, 327, 812–17.

Global Rice Science Partnership (GRiSP) (2010) CGIAR Thematic Area 3: 'Sustainable Crop Productivity Increase for Global Food Security', Proposal for a CGIAR research program on rice-based production systems, September 2010. Available online at <http://www.grisp.net/uploads/files/x/000/068/9c1/GRiSP%20Full%20Proposal.pdf?1312766027> (accessed 8 June 2013).

Harris, D. and Orr, A. (in press) 'Is rainfed agriculture really a pathway from poverty?', *Agricultural Systems*.

Heijting, D., de Bruin, S. and Bregt, A. K. (2011) 'The arable farmer as the assessor of within-field soil variation', *Precision Agriculture*, 12, 488–507.

Hounkonnou, D., Kossou, D., Kuyper, T. W., Leeuwis, C., Nederlof, S., Roling, N., Sakyi-Dawson, O., Traoré, M. and van Huis, A. (2012) 'An innovation systems approach to institutional change: smallholder development in West Africa', *Agricultural Systems*, 108, 74–83.

Jayne, T. S., Mather, D. and Mghenyi, E. (2005) 'Smallholder farming in difficult circumstances: policy issues for Africa', in IFPRI (International Food Policy Research Institute), *The Future of Small Farms: Proceedings of a Research Workshop, Wye, UK, June 26–29, 2005*, Washington, DC: International Food Policy Research Institute, pp. 103–23.

Jerich, M. (2011) 'Potential of precision conservation agriculture as a means of increasing productivity and incomes for smallholder farmers', *Journal of Soil and Water Conservation*, 66, 171A–174A.

Keating, B. A., Carberry, P. S., Bindraban, P. S., Asseng, S., Meinke, H. and Dixon, J. (2010) 'Eco-efficient agriculture: concepts, challenges, and opportunities', *Crop Science*, 50, 109–19.

Kiiza, B. and Pederson, G. (2012) 'ICT-based market information and adoption of agricultural seed technologies: insights from Uganda', *Telecommunications Policy*, 36, 253–9.

Kutter, T., Tiemann, S., Siebert, R. and Fountas, S. (2011) 'The role of communication and co-operation in the adoption of precision farming', *Precision Agriculture*, 12, 2–17.

Liebig, J. (1840) *Die organische Chemie in ihrer Anwendung auf Agrikultur und Physiologie*, 1st edn, Braunschweig: Vieweg.

Lowenberg-DeBoer, J. (2001) 'Global adoption of precision agriculture technologies: who, when and why?', in G. Grenier and S. Blackmore (eds) *Third European Conference on Precision Agriculture*, Montpellier: Agro, pp. 557–62.

Lukuyu, B., Place, F., Franzel, S. and Kiptot, E. (2012) 'Disseminating improved practices: are volunteer farmer trainers effective?', *Journal of Agricultural Education and Extension*, 18, 525–40.

McBratney, A. B. and Whelan, B. M. (1999) 'The "null hypothesis" of precision agriculture', in J. V. Stafford (ed.) *Precision Agriculture '99*, Sheffield, UK: Sheffield Academic Press, pp. 947–57.

McKinion, J. M., Willers, J. L. and Jenkins, J. N. (2010) 'Comparing high density LIDAR and medium resolution GPS generated elevation data for predicting yield stability', *Computers and Electronics in Agriculture*, 74, 244–9.

Mohd Noor, M. F., Gassner, A. and Schnug, E. (2005) '15 years of precision farming: lessons to be learned for Malaysia', in S. Haneklaus, R. M. Rietz, J. Rogasik and S. Schroetter (eds) *Recent Advances in Agricultural Chemistry*, Landbauforschung Völkenrode, Sonderheft 286, Braunschweig: Federal Agricultural Research Centre (FAL), pp. 55–64.

Moss, C., Doores, J., McDonald, M. A., Nolte, C. and Sinclair, F. L. (2003) *Decision Support Tools (LEXSYS and LEGINC)*. Annex D of the Final Technical Report of Project R7446. Bangor: School of Agricultural and Forest Sciences, University of Wales. Available online at <http://www.dfid.gov.uk/r4d/Output/173111/Default.aspx> (accessed January 2013).

Mueller, N. D., Gerber, J. S., Johnston, M., Ray, D. K., Ramankutty, N. and Foley, J. A. (2012) 'Closing yield gaps through nutrient and water management', *Nature*, 490, 254–7.

Nain, M. S., Singh, R., Vijayraghavan, K. and Vyas, A. K. (2012) 'Participatory linkage of farmers, technology agricultural researchers for improved wheat production in national capital region of India', *African Journal of Agricultural Research*, 7, 5198–207.

Nord, A. K. and Luckscheiter, J. (2011) 'Editorial', *Perspectives: Political Analysis and Commentary from Africa*, 1.11, *Food Security in Africa*, 3. Available online at <http://www boell org za/downloads/Perspectives_1 11%281%29 pdf/> (accessed January 2013).

Odgaard, M. V., Bøcherb, P. K., Dalgaarda, T. and Svenning, J. C. (2011) 'Climatic and non-climatic drivers of spatiotemporal maize-area dynamics across the northern limit for maize production: a case study from Denmark', *Agriculture, Ecosystems and Environment*, 142, 291–302.

Ojiem, J. O., de Ridder, N., Vanlauwe, B. and Giller, K. E. (2006) 'Socio-ecological niche: a conceptual framework for integration of legumes in smallholder farming systems', *Agricultural Systems*, 4, 79–93.

Oweis, T. and Hachum, A. (2006) 'Water harvesting and supplemental irrigation for improved water productivity of dry farming systems in West Asia and North Africa', *Agricultural Water Management*, 80, 57–73.

Rigg, J. (2006) 'Land, farming, livelihoods, and poverty: rethinking the links in the rural South', *World Development*, 34, 180–202.

Shiferaw, B. A., Okello, J. and Reddy, R. V. (2009) 'Adoption and adaptation of natural resource management innovations in smallholder agriculture: reflections on key lessons and best practices', *Environment, Development and Sustainability*, 11, 601–19.

Srinivasan, A. (ed.) (2006) *Handbook of Precision Agriculture*, New York: Food Products Press.

Taddesse, G., Peden, D., Abiye, A. and Wagnew, A. (2003) 'Effect of manure on grazing lands in Ethiopia, East African Highlands Mountain', *Research and Development*, 23, 156–60.

Terhoeven-Urselmans, T., Vagen, T. G., Spaargaren, O. and Shepherd, K. D. (2010) 'Prediction of soil fertility properties from a globally distributed soil mid-infrared spectral library', *Soil Science Society of America Journal*, 74, 1792–9.

Tey, Y. S. and Brindal, M. (2012) 'Factors influencing the adoption of precision agricultural technologies: a review for policy implications', *Precision Agriculture*, 13, 713–30.

Thorne, P. J., Sinclair, F. L. and Walker, D. H. (1997) 'Using local knowledge of the feeding value of tree fodder to predict the outcomes of different supplementation strategies', *Agroforestry Forum*, 8, 45–9.

Thornton, P. K., Jones, P. G., Owiyo, T., Kruska, R. L., Herrero, M., Kristjanson, P., Notenbaert, A., Bekele, N. and Omolo, A., with contributions from Orindi, V., Otiende, B., Ochieng, A., Bhadwal, S., Anantram, K., Nair, S., Kumar, V. and Kulkar, U. (2006) *Mapping Climate Vulnerability and Poverty in Africa*, Report to the Department for International Development (UK), Nairobi: International Livestock Research Institute (ILRI).

Timsina, J., Jat, M. J., and Majumdar, K. (2010) 'Rice-maize systems of South Asia: current status, future prospects and research priorities for nutrient management', *Plant and Soil*, 335, 65–82.

Tittonell, P., Muriuki, A., Shepherd, K. D., Mugendi, D., Kaizzi, K. C., Okeyo, J., Verchot, L. and Coe, R. (2010) 'The diversity of rural livelihoods and their influence on soil fertility in agricultural systems of East Africa: a typology of smallholder farms', *Agricultural Systems*, 103, 83–97.

Tittonell, P., Vanlauwe, B., Misiko, M. and Giller, K. E. (2011) 'Targeting resources within diverse, heterogeneous and dynamic farming systems: towards a "uniquely African green revolution"', in A. Bationo, B. Waswa, J. M. M. Okeyo, M. Fredah and J. M. Kihara (eds) *Innovations as Key to the Green Revolution in Africa*, Dordrecht: Springer, pp. 747–58.

Toulmin, C. and Guèye, B. (2005) 'Is there a future for family farming in West Africa?', *IDS Bulletin*, 36, 23–9.

Traore, P. C. S. (2009) 'Agcommons Quickwin: Seeing is Believing'. Presented at the CGIAR–CSI Annual Meeting 2009: Mapping Our Future, 31 March–4 April 2009, ILRI Campus, Nairobi, Kenya. Available online at <http://www slideshare net/csi2009/day3-agcommons-quickwin-seeing-is-believing> (accessed January 2013).

Twomlow, S. J., Rohrbach, D. D., Dimes, J., Rusike, J., Mupangwa, W., Ncube, B., Hove, L., Moyo, M., Mashingaidze, N. and Mahposa, P. (2010) 'Micro-dosing as a pathway to Africa's Green Revolution: evidence from broad-scale on-farm trials', *Nutrient Cycling in Agroecosystems*, 88, 3–15.

Vanlauwe, B., Bationo, A., Chianu, J., Giller, K. E., Merckx, R., Mokwunye, U., Ohiokpehai, O., Pypers, P., Shepherd, K. D., Smaling, E. M. A., Woomer, P. L. and Sanginga, N. (2010) 'Integrated soil fertility management: operational definition and consequences for implementation and dissemination', *Outlook on Agriculture*, 39, 17–24.

Von Braun, J. (2005) 'Small-scale farmers in liberalised trade environment', in T. Huvio, J. Kola, and T. Lundström (eds) *Small-Scale Farmers in Liberalised Trade Environment: Proceedings of the Seminar, October 2004 in Haikko, Finland*, Department of Economics and Management Publications 38, Agricultural Policy, Helsinki, Finland: University of Helsinki. Available online at <http://www.fearp. usp.br/fava/pdf/pdf247.pdf> (accessed January 2013).

Veldhuizen, L. van and Wettasinha, C. (2010) *Towards a Network that Learns: Mapping Long-Term Options for GFRAS*. Leusden, The Netherlands: ETC EcoCulture. Available online at <www.g-fras.org/en/knowledge/gfras-publications> (accessed January 2013).

Part 2
Techniques

4 Rational management decisions from uncertain spatial data

Ben P. Marchant, A. Gordon Dailey and Richard Webster

Introduction

Precision agriculture (PA) attempts to vary efficiently the management of farming in a manner that accounts for the inherent spatial variation of the land. For example, when deciding on the quantity of fertilizer to apply across a farm, a farmer might consider the amount of nutrient that is available from other sources, the soil type, the efficiency of nutrient uptake by the crop and whether other factors such as weather or disease are likely to limit the yield. If these factors vary within the farm then the optimal nutrient additions are likely to vary. If farmers know accurately the spatial variation of such factors then they might use PA techniques to decide how to manage their land locally.

Farmers wishing to vary their management within individual fields to maximize their profit or minimize damage to the environment need to understand how their crops respond to applications of fertilizer and to know how the soil itself, and in particular its most relevant properties, varies within their fields. The yield responses of a crop to fertilizer might be investigated through field trials which lead to mathematical models. Knowledge of the spatial variation within fields might come from the farmers' experience or from instruments such as electrical conductivity meters that sense the soil at short intervals. There are as yet no sensors that can measure nutrient concentrations cheaply in this way, though Viscarra Rossel *et al.* (2005) have devised a machine for measuring the soil's pH on the run in the field. So the utility of data from these sensors depends on farmers' or their advisers' knowing how the sensor data relate to nutrient status and to measurements of sparse samples of soil in the laboratory. More often only data from sparse sampling of the soil are available, and so they must be interpolated to all sites where farmers wish to make management decisions. Usually farmers want maps, and inevitably the interpolated values from which they are made will be sources of uncertainty. Farmers should therefore be able to quantify and account for this uncertainty when they make their decisions.

Geostatistical methods and in particular kriging have become popular for mapping from sparse data. The reasons are several: the interpolated values are unbiased, they have minimum variance and that variance can also be estimated and so express the uncertainty of the interpolated predictions. The standard geostatistical methods are now widely available to precision farmers and their advisers. They are described in comprehensive textbooks such as those by Goovaerts (1997), Chilès and Delfiner (2012), Olea (1999) and Webster and Oliver (2007). They are available in software such as GenStat (Payne, 2011) and GSLIB (Deutsch and Journel, 1998), geoR (Ribeiro and Diggle, 2001) and ArcGIS (ESRI, 2011). Some packages process data fairly automatically at the press of a few buttons on menus, and practitioners are tempted to trust naïvely the output. Practitioners should therefore be aware that there are circumstances in which the default geostatistical models are unsuitable and give rise to misleading results. For example, they often assume implicitly that the statistical distributions of the variables of interest are Gaussian (normal), whereas the concentrations of plant nutrients are often more or less strongly positively skewed with distributions that are approximately lognormal.

The interpretation and use of spatial information can be further complicated by a disparity between the spatial scales over which (i) soil properties are observed, (ii) crucial processes are understood and (iii) management decisions are made. The soil's properties might be measured on single cores of, say, 7.5 cm diameter; alternatively they might be measured on bulked material from several such cores from a wider area, which we refer to as the sample's support. Our understanding of the relation between available nutrients in the soil and crop yield might be based on experiments with plots of 10 m^2 and larger on which responses are recorded. The scale over which applications of fertilizer can be varied might be limited by farming technology or the physical dimensions of machinery. If the variable of interest has an approximately Gaussian distribution one can change the support from the point of measurement to larger blocks and know that the errors are themselves Gaussian. The methods are well known. Changing the support for non-Gaussian properties is more problematic, and solutions are still being sought (Paul and Cressie, 2011).

In this chapter we examine methods for quantifying the uncertainty of spatial predictions of variables that are not Gaussian. We also discuss strategies to present this information over several spatial scales and how to decide on management in the presence of such uncertain information. We discuss these matters in relation to the management of soil potassium (K) in an arable system. We describe what we assume to be an accurate mathematical model of the yield response of the arable crop to K fertilizer. We use geostatistical methods to predict the available K prior to the application of fertilizer at scales that a farmer would want and the uncertainty of these predictions. Finally, we devise a framework within which the uncertain spatial information can be combined with our understanding to lead to rational and cost-effective applications of fertilizer. Although we focus on the mapping of soil properties, one can use the same approaches when analysing spatial observations of other types such as pest or weed populations.

A case study: selecting variable inputs of potassium, K, on Broom's Barn Farm

The general principles outlined above can all be illustrated by data from the survey of Broom's Barn Farm in eastern England made in 1960. In that survey

J. A. P. Marsh and S. N. Adams measured the exchangeable potassium, K^+, of the topsoil (0–23 cm) at 434 places at 40-m intervals on a square grid (Figure 4.1). Each measurement was made on a bulked sample of 25 cores collected from a 16 m × 16 m square. The support of the data was thus a quadrat of this size. The data were first analysed geostatistically by Webster and McBratney (1987), and they have been used since in several more advanced studies and in the text by Webster and Oliver (2007) to illustrate methods. Cressie (2006) also used the data to demonstrate an algorithm for block kriging lognormal variables.

In this chapter we examine some of the complexities that can arise when spatial data of this kind are analysed. We show how to predict from point or punctual data a variable in blocks of the size a farmer would want for precise application of fertilizer when the variable is not Gaussian. We therefore follow Cressie (2006). We treat each observation as if it were punctual, and we krige to predict over 24 m × 24 m management units, 24 m being the width of a standard modern fertilizer spreader. The concentration of K^+ was originally measured in mg litre^{-1}. We convert the measurements to mg kg^{-1} by assuming the soil has a bulk density of 1 kg litre^{-1}.

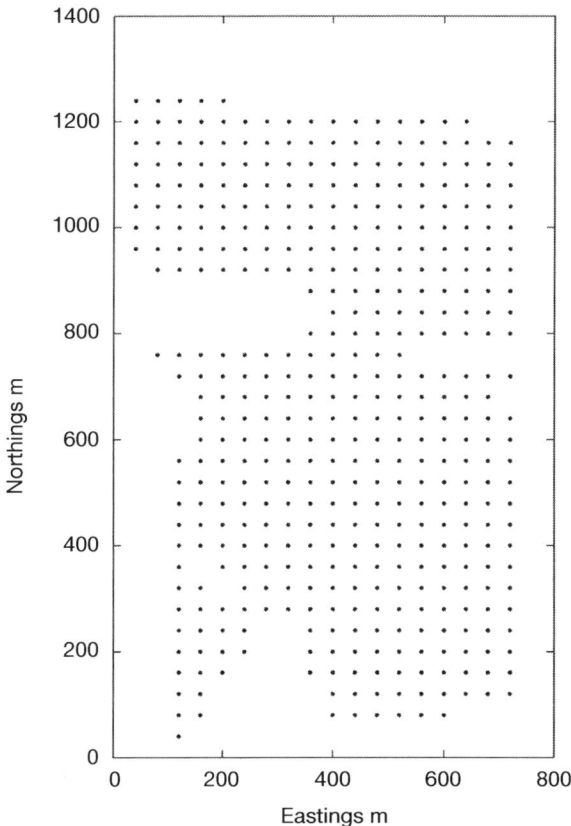

Figure 4.1 Sampling locations at Broom's Barn Farm.

Again for present purposes we assume that the survey was made to enable a farmer to decide how much K fertilizer to apply for a crop of winter wheat to maximize profit in the next season. We assume also that we have a thorough understanding of the relations between crop yield, the concentration of exchangeable K⁺ in the soil and additions of fertilizer at the 24-m scale. So if in each management unit the exchangeable K⁺ concentration were known exactly then the optimal amount of fertilizer to add could be calculated. However, the concentrations of K⁺ in the management units must be predicted from the data, and we want to account for the uncertainty of these predictions.

A model of the yield response of arable crops to fertilizer potassium

Our understanding of the relationships between the amounts of exchangeable K⁺ in the soil, the K added to the soil in fertilizer, the yield response of arable crops to this K and the profitability in British arable fields is derived from existing knowledge that was reviewed by Marchant et al. (2012). Johnston and Goulding (1990) observed the accumulation of K in the soil after fertilizer had been added. They found that for each 1 kg of K added in fertilizer, 0.34 kg was later detected in the soil as exchangeable K⁺. We assume that at Broom's Barn this additional K was contained in the top 30 cm of soil. Therefore addition of fertilizer at a rate of 1 kg K ha⁻¹ leads to an increase in the concentration in this layer of $0.34 \times 0.3 = 0.11$ mg kg⁻¹. Thus the total K available after addition of a quantity K_f is

$$K_a = K + 0.11 K_f, \tag{4.1}$$

where K is the initial concentration of exchangeable potassium in the soil.

Johnston and Goulding (1988) and Milford and Johnston (2007) did seven experiments on the yield response of crops to added K. Marchant et al. (2012) fitted a single yield response curve to the results of these experiments. The curve had the form

$$\text{Yield} = \text{Yield}_t \{1 + B(C^{K_a})\}, \tag{4.2}$$

where Yield_t denotes the target yield and B and C are model parameters. The yield response from each experiment was normalized such that the target yield was 8.8 t ha⁻¹. The fitted parameter values were $B = -2.01$ and $C = 0.96$. We assume a typical wheat price W_p of £150 t⁻¹ and that the price of K_2O fertilizer is £0.54 kg⁻¹ (Nix, 2011). This corresponds to a price of £0.65 kg⁻¹ of K which we denote K_p. Therefore the profit margin per ha from wheat can be approximated by

$$P = W_p \times \text{Yield} - K_p K_f. \tag{4.3}$$

Thus, if the concentration of exchangeable K⁺ is known for a particular management zone then the amount of fertilizer to add, K_f, can be optimized to maximize the profit. For example, Figure 4.2 plots the profit against added fertilizer when prior to fertilization the concentration of exchangeable K^+ is 20 mg kg^{-1}.

Geostatistical analysis

We need to predict the average concentration of K^+ within each 24 m × 24 m block and the uncertainty attached to it. This is standard procedure, for which you can find full details in Webster and McBratney (1987) and Webster and Oliver (2007).

Geostatistical methods acknowledge that variables are often spatially correlated, that is, measurements made a small distance apart are more likely to be similar than more distant measurements. We assume that the variable of interest, which we denote $z(\mathbf{x})$ at a place $\mathbf{x}$ is a realization of an intrinsically stationary random function $Z(\mathbf{x})$. This function has constant mean and the variogram,

$$\gamma(\mathbf{h}) = \tfrac{1}{2}E[\{Z(\mathbf{x}) - Z(\mathbf{x} + \mathbf{h})\}^2], \tag{4.4}$$

is independent of the absolute location $\mathbf{x}$. Here, $\gamma(\mathbf{h})$ is a function of $\mathbf{h}$, the lag separating two observations. The variogram describes the spatial correlation of the variable which we assume to follow the same model throughout the field.

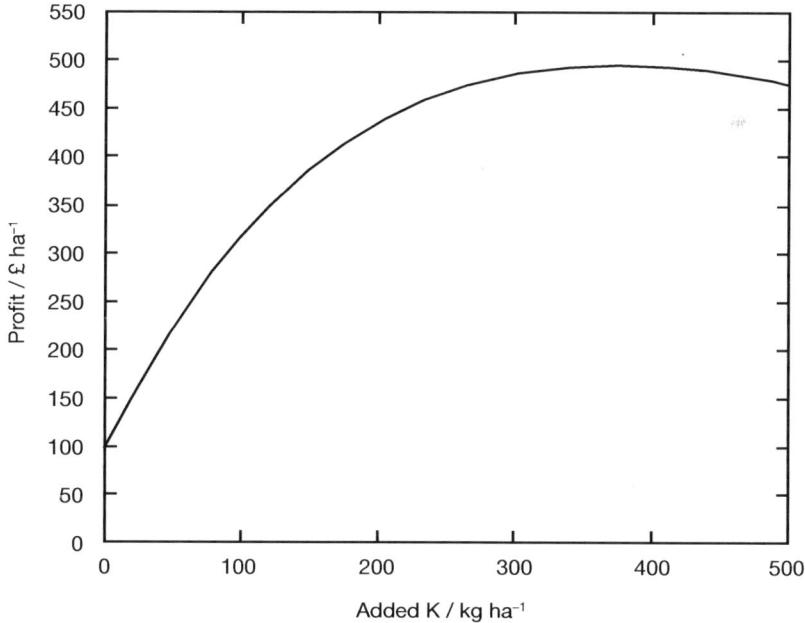

Figure 4.2   Graph of profit against K (kg ha^{-1}) additions for an initial K^+ concentration of 20 mg kg^{-1}.

In general the variogram can depend on both the distance and direction of **h**. Strictly, the form of the distribution of $Z(\mathbf{x})$ is immaterial, but the analysis is more efficient if one can assume that the distribution is Gaussian because then all higher-order moments of the distribution can be determined from the mean and variance, and, as we shall see, it is easier to interpret spatial predictions. Webster and McBratney (1987) found that the data from Broom's Barn were strongly positively skewed (Figure 4.3a) and took logarithms to stabilize the variances, that is, they computed $z(\mathbf{x}) = \log_{10}K(\mathbf{x})$ where K is the observed concentration of exchangeable K^+. This transformation approximately normalized the data (Figure 4.3b), and the skewness coefficient decreased from 2.04 to 0.39.

Webster and McBratney (1987) estimated the variogram from the observed data by the method of moments, for which the formula is

$$\hat{\gamma}(\mathbf{h}) = \frac{1}{2m(\mathbf{h})} \sum_{i=1}^{m(\mathbf{h})} \{z(\mathbf{x}_i) - z(\mathbf{x}_i + \mathbf{h})\}^2, \tag{4.5}$$

where $z(\mathbf{x}_i)$ and $z(\mathbf{x}_i + \mathbf{h})$ are the values of z at places $\mathbf{x}_i$ and $\mathbf{x}_i + \mathbf{h}$ and $m(\mathbf{h})$ is the number of such pairs separated by **h**. They considered whether the variogram varied according to the direction of the lag vector. They found only small differences in the different directions and therefore treated the lag as a scalar in distance only, so that $h = |\mathbf{h}|$, and we do the same here. Figure 4.4 shows estimates of $\gamma(\mathbf{h})$ for $h = 40, 80, \ldots , 520$ m, which constitute the experimental variogram.

For prediction and simulation we need to express the variogram as a mathematical function. Such a function must ensure that the predictions do not have negative variances. There are several popular admissible or authorized ones, and these can be found in the standard texts. Webster and McBratney (1987) fitted

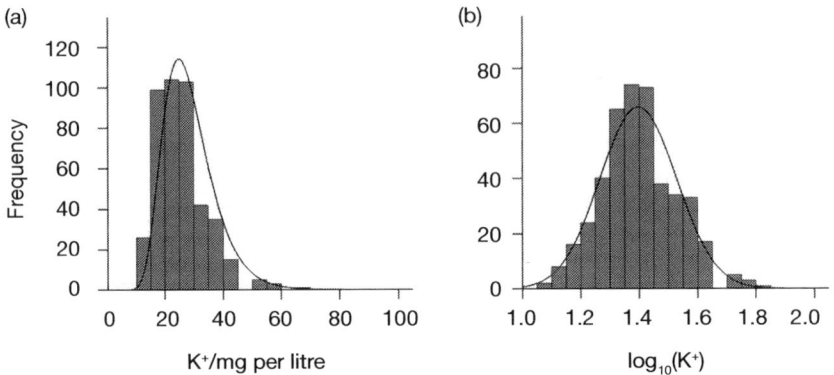

Figure 4.3　Histograms of (a) K^+ (mg l^{-1}) and (b) $\log_{10}(K^+)$ with fitted distribution function.

several of these to their experimental variogram by weighted least squares with weights proportional to $m(h)$. They found that the best fit (Figure 4.4) resulted from a nested spherical and nugget model:

$$\gamma\,(h) = \begin{cases} c_0 + c_1 & \text{if } h > a \\ c_0 + c_1 \left\{ \frac{3h}{2a} - \frac{1}{2}\left(\frac{h}{a}\right)^3 \right\} & \text{if } 0 < h \leq a\,, \\ 0 & \text{if } h = 0 \end{cases} \tag{4.6}$$

where c_0 is the nugget variance, that is, the intercept on the ordinate, c_1 is the variance of the autocorrelated component and a is the range of the variogram. The fitted parameter values were $c_0 = 0.0048$, $c_1 = 0.01519$ and $a = 439.2$ m.

Having estimated the variogram, Webster and McBratney (1987) predicted $z(\mathbf{x}) = \log_{10} K(\mathbf{x})$ across the farm by ordinary kriging as the weighted sum of n observations:

$$\hat{Z}(\mathbf{x}_0) = \Sigma_{i=1}^n \lambda_i\, z(\mathbf{x}_i), \tag{4.7}$$

where $\hat{Z}(\mathbf{x}_0)$ is the predicted value at unobserved site $\mathbf{x}_0$, and the λ_i are the kriging weights. The kriging weights sum to 1 to ensure that the predictor is unbiased, that is:

$$E[\hat{Z}(\mathbf{x}_0) - Z(\mathbf{x}_0)] = 0, \tag{4.8}$$

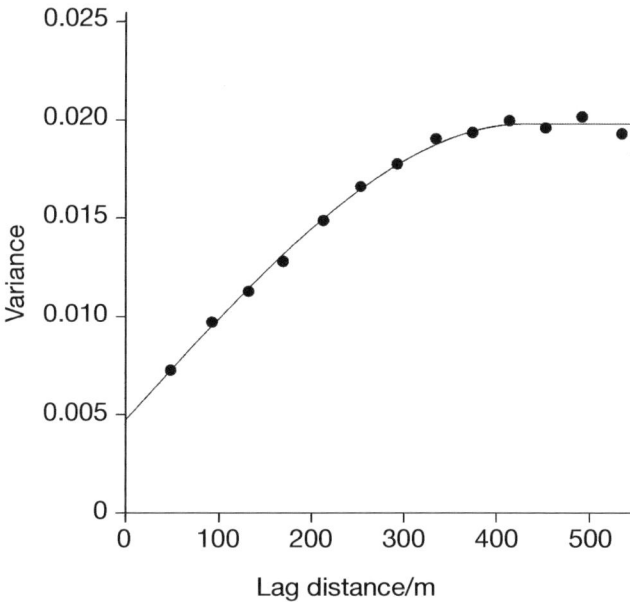

Figure 4.4 Spherical variogram model (continuous line) fitted to point estimates (dots) of the variogram of $\log_{10}(K^+)$ by the method of moments.

and they are selected to minimize the kriging variance

$$\hat{\sigma}^2(\mathbf{x}_0) + E[\{\hat{Z}(\mathbf{x}_0) - Z(\mathbf{x}_0)\}^2]. \tag{4.9}$$

The weights are largest for observations close to the sample points, and the variogram determines how quickly weights decrease with increasing distance between an observation and the target site. One can include all of the N observations in the weighted sum ($n = N$), but it is often advisable to reduce the number to around $n = 20$. This reduces the computational load and it usually avoids instability in the inversion of the variance matrix, below.

The kriging weights are determined from the matrix equation

$$\boldsymbol{\lambda} = \mathbf{A}^{-1}\mathbf{b}, \tag{4.10}$$

where

$$\mathbf{A} = \begin{bmatrix} \gamma(\mathbf{x}_1,\mathbf{x}_1) & \gamma(\mathbf{x}_1,\mathbf{x}_2) & \cdots & \gamma(\mathbf{x}_1,\mathbf{x}_n) \\ \gamma(\mathbf{x}_2,\mathbf{x}_1) & \gamma(\mathbf{x}_2,\mathbf{x}_2) & \cdots & \gamma(\mathbf{x}_2,\mathbf{x}_n) \\ \vdots & \vdots & \cdots & \vdots \\ \gamma(\mathbf{x}_n,\mathbf{x}_1) & \gamma(\mathbf{x}_n,\mathbf{x}_2) & \cdots & \gamma(\mathbf{x}_n,\mathbf{x}_n) \\ 1 & 1 & \cdots & 1 \end{bmatrix}, \tag{4.11}$$

$$\boldsymbol{\lambda} = \begin{bmatrix} \lambda_1 \\ \lambda_2 \\ \vdots \\ \lambda_n \\ \psi(\mathbf{x}_0) \end{bmatrix} \text{ and } \mathbf{b} = \begin{bmatrix} \gamma(\mathbf{x}_1,\mathbf{x}_0) \\ \gamma(\mathbf{x}_2,\mathbf{x}_0) \\ \vdots \\ \gamma(\mathbf{x}_n,\mathbf{x}_0) \\ 1 \end{bmatrix}$$

Here $\psi(\mathbf{x}_0)$ is a Lagrange multiplier which is introduced to minimize the kriging variance.

As above, one reason that kriging is favoured over other methods of interpolation is that it not only minimizes the variance of predictions but also provides a measure of the uncertainty of each estimate. This measure is the prediction variance or kriging variance:

$$\hat{\sigma}^2(\mathbf{x}_0) = \mathbf{b}^T\boldsymbol{\lambda}. \tag{4.12}$$

Since we have assumed that the distribution of Z is Gaussian, our prediction at site $\mathbf{x}_0$ has a Gaussian probability density function (pdf) with mean $\hat{Z}(\mathbf{x}_0)$ and variance $\sigma^2(\mathbf{x}_0)$. From this pdf one can calculate any specified confidence limit for $Z(\mathbf{x}_0)$ or calculate the probability that $Z(\mathbf{x}_0)$ exceeds a critical threshold.

The spatial variation of $\log_{10}(K^+)$ across the farm is evident from a map of kriged predictions (Figure 4.5a). For example, in the north of the farm concentrations are relatively small, whereas further south they are much larger. The map of kriging variances (Figure 4.5b) follows a more regular pattern. Notice how the kriging variances are controlled by the sampling configuration. The kriging variance is zero at sample points, and it increases away from these locations. It is largest on the boundaries of the farm because there are no data beyond them. The kriging variance at a site is independent of the kriged prediction at the same site. This is because the formula for the kriging variance does not include the values of the observations.

Back-transformation of kriged predictions

The map of predicted $\log_{10}(K^+)$ gives some indication of the variation of the amounts of K in the soil of the farm. The predictions could be easier for the farmer to understand if they were back-transformed and expressed in the original units. At the observation locations $z = \log_{10}(K^+)$ is known exactly and the kriging variance is zero. Hence the prediction at these sites can be back-transformed by the inverse of the $\log_{10}$ transform, which is 10^z. Such a simple back-transformation is biased if the kriging variance is greater than zero because the pdf of the predicted K concentration is not symmetric. Webster and Oliver (2007) noted that in this situation the back-transformation for a punctual prediction is

$$\widehat{K}(\mathbf{x}_0) = \exp\{\hat{Z}(\mathbf{x}_0) \times \ln 10 + 0.5\hat{\sigma}^2(\mathbf{x}_0) \times (\ln 10) - \psi(\mathbf{x}_0) \times (\ln 10)\}. \qquad (4.13)$$

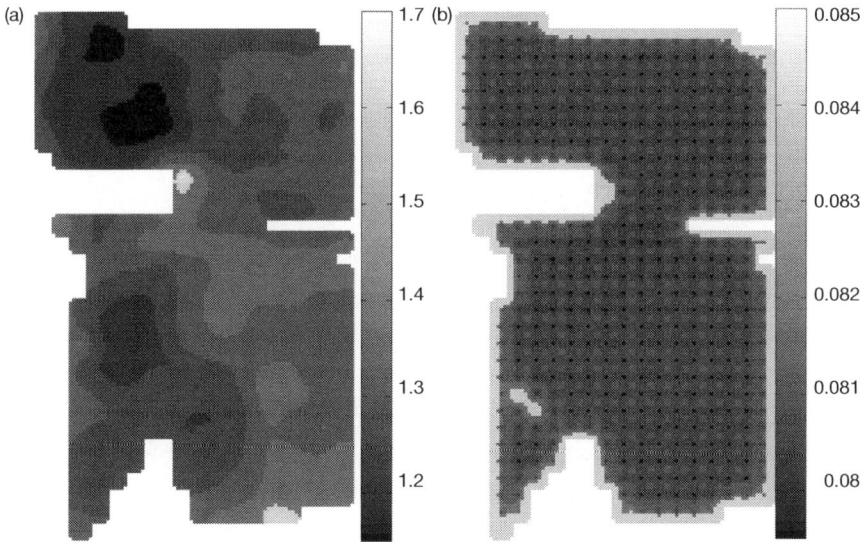

Figure 4.5 (a) Kriged prediction of $\log_{10}(K^+)$ across Broom's Barn Farm and (b) corresponding kriging standard deviation.

Transforming the ordinary prediction variance of $\widehat{K}(\mathbf{x}_0)$ is more complex, and is perhaps best estimated by an alternative approach.

Let us consider a specific location within the farm, say the site 952.5 m north and 192.5 m east of the origin of Figure 4.1. The kriged prediction of $z = \log_{10}(K^+)$ there is 1.1739. If we simply inverted the transformation without taking the uncertainty of the prediction into account then our predicted concentration would be 14.92 mg kg^{-1}. However, the kriging variance of the prediction is 0.0066 and the Lagrange multiplier is -1.2×10^{-6}. Hence, from Equation (4.13), $\widehat{K}(\mathbf{x}_0) = 15.19$ mg kg^{-1}. Our alternative approach simulates a large number of plausible realizations of $z(\mathbf{x}_0)$. Since we have assumed that $Z(\mathbf{x})$ is a Gaussian random function we know that the distribution of these realizations will be Gaussian with mean $\hat{Z}(\mathbf{x}_0)$ and variance $\hat{\sigma}^2(\mathbf{x}_0)$. Such a distribution can be generated easily by numerical software packages.

The familiar bell-shaped curve of such a set of simulated realizations of the Gaussian distribution is plotted for our specified site in Figure 4.6a. Each of these realizations of $z(\mathbf{x}_0)$ corresponds to a realization of $K(\mathbf{x}_0) = 10^{z(\mathbf{x}_0)}$. The histogram that results (Figure 4.6b) is no longer symmetric. The mean or variance of $\widehat{K}(\mathbf{x}_0)$ can be determined from the mean or variance of the set of realizations of $10^{z(\mathbf{x}_0)}$.

At the location considered above, the mean calculated by this method is 15.19 mg kg^{-1}, the same as that from Equation (4.13), and the variance is 8.24 (mg kg^{-1})^2. One can determine the probability that any specified threshold is exceeded or any specified confidence limit for the prediction. The method can be used to back-transform predictions after any transformation provided that the inverse of the transform is known.

The map of predicted exchangeable K$^+$ following this back transformation is shown in Figure 4.7a. Its basic pattern is similar to that of the map of $\log_{10}(K^+)$. However, the pattern of the prediction variance has changed greatly. The prediction variance now increases with the predicted K$^+$. This is because for skewed distributions, large predictions are more uncertain than small ones. One must recognize this when using such predictions to decide how much fertilizer to apply.

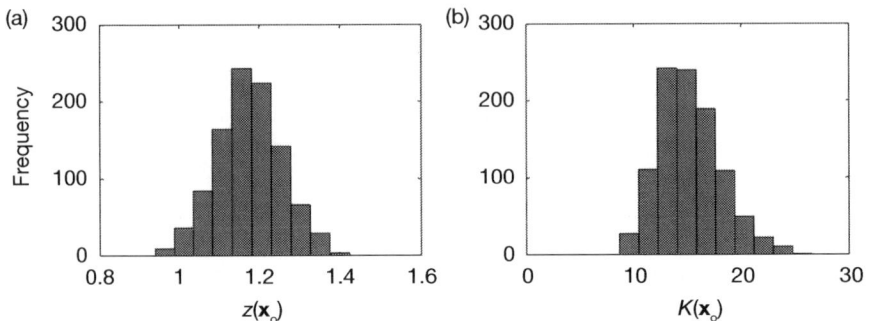

Figure 4.6   (a) Histogram of kriged prediction of $\log_{10}(K^+)$ at coordinates (952.5 m, 192.5 m) and (b) corresponding histogram for K$^+$.

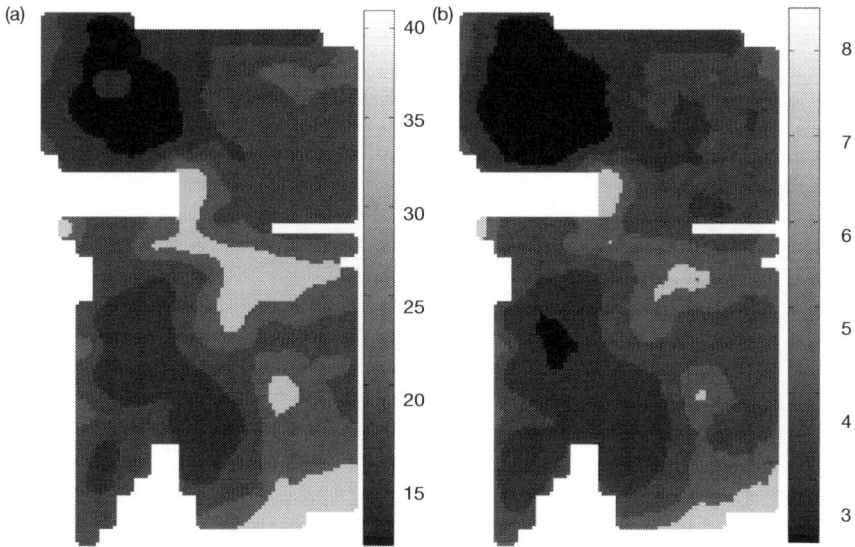

Figure 4.7 (a) Kriged prediction of K^+(mg kg^{-1}) across Broom's Barn Farm and (b) corresponding kriging standard deviation.

Validation

Geostatistical practitioners assume that the spatial variation of a variable can be described by a fairly simple model. In the model described above we assume that the observations of $\log_{10}(K^+)$ are a realization of a multivariate Gaussian distribution and that the mean and variogram of this distribution do not vary across the region. Such assumptions are not always plausible. For example, the variogram of Z might vary from one soil type to another, or the distribution of Z might resemble some other parametric model.

Before the results of a geostatistical analysis are accepted the assumed models should be validated. Validation requires either that some observations are held in reserve and not used in the estimation of the variogram and kriging or that additional observations are made for the validation. Predictions are made at the sites of these validation data, and the consistency of these predictions with the actual observations is tested. If data are plentiful or extra data readily obtained then validation is a good test of the model because it guards against over-fitting. This is where too detailed a model is fitted to data so that the model matches the variation of the data well but predicts poorly at independent validation sites.

Where the costs of individual measurements are large, surveyors might be unable to afford separate validation data. In these circumstances cross-validation may be used to confirm that the assumed model is appropriate. The model is fitted to all of the measurements, and then one datum, $z(\mathbf{x}_i)$ say, is removed and the remaining data are used to predict $\hat{Z}(\mathbf{x}_i)$. The process is repeated for all N observations and then quantities such as the mean error (ME),

$$\text{ME} = \frac{1}{N} \sum_{i=1}^{n} \{z(\mathbf{x}_i) - \hat{Z}(\mathbf{x}_i)\}, \tag{4.14}$$

the mean squared error (MSE),

$$\text{MSE} = \frac{1}{N} \sum_{i=1}^{n} \{z(\mathbf{x}_i) - \hat{Z}(\mathbf{x}_i)\}^2, \tag{4.15}$$

and the standardized squared prediction error at each and every site,

$$\theta_i = \frac{\{z(\mathbf{x}_i) - \hat{Z}(\mathbf{x}_i)\}^2}{\hat{\sigma}^2(\mathbf{x}_i)}, \tag{4.16}$$

are calculated. The mean of $\boldsymbol{\theta} = (\theta_1, \theta_2, \ldots, \theta_N)$ is called the mean square deviation ratio, abbreviated to MSDR.

Kriging is an unbiased predictor, so if the chosen model is indeed appropriate then the ME should be close to zero. As we also choose the kriging weights to minimize the variances the MSE should also be minimal. Further, if the variogram model is accurate then the expected square error should be equal to the kriging variance, and the mean of $\boldsymbol{\theta}$ should be 1. This is a fairly weak diagnostic since it is often possible to ensure that the mean of $\boldsymbol{\theta}$ is close to 1 through careful selection of the variogram parameters even when the assumed model is not appropriate. If the MSDR = 1 then the kriging variances are unbiased predictors of the squared errors across the region. It is still possible, however, that the kriging variances are poor measures of the uncertainty at individual sites. For example, if the data contain a few outliers then inappropriate models often under-estimate the uncertainty at the sites where they occur but compensate by over-estimating the uncertainty elsewhere. This situation can be identified through calculation of the median of $\boldsymbol{\theta}$. If the property is a realization of a Gaussian distribution then the median of $\boldsymbol{\theta}$ should be 0.45. A median less than 0.45 suggests that some observations are too far removed from the main body of the data to be realized by the fitted model.

Some cross-validation results for the Broom's Barn data are listed in Table 4.1. We initially compare a fitted model in which we assume that K has a Gaussian distribution with one in which K is a realization of a lognormal one. Both models have small MEs, and the MSE of the lognormal variable is approximately 3 per cent less than for the Gaussian model. This is a fairly small improvement and illustrates how insensitive a kriged prediction tends to be to the fitted model. The kriging variance is much more sensitive to the model and hence a much larger improvement in the median of $\boldsymbol{\theta}$ is observed. We discuss the results for the Box–Cox distribution later.

Table 4.1 Cross-validation results for spatial models of the Broom's Barn K data with three assumed distribution functions

Distribution	ME/mg kg⁻¹	MSE/(mg kg⁻¹)²	Mean of θ	Median of θ
Gaussian	−0.04	47.41	1.16	0.21
Lognormal	−0.16	45.93	1.03	0.31
Box–Cox	−0.05	43.95	1.01	0.36

Block predictions

Equations (4.10)–(4.12) lead to predictions of a variable on areas or supports the same size as the supports on which the original observations are made, that is, they are punctual. Farmers have to treat larger blocks, and with modern machinery these are at least 24 m across, and so we want to predict conditions of these larger blocks. If the variable is Gaussian then the punctual equations can be easily generalized to a block B. Equations (4.10)–(4.11) are unaltered except that the vector **b** becomes

$$\mathbf{b} = \begin{bmatrix} \bar{\gamma}(\mathbf{x}_1, B) \\ \bar{\gamma}(\mathbf{x}_2, B) \\ \vdots \\ \bar{\gamma}(\mathbf{x}_N, B) \\ 1 \end{bmatrix}, \tag{4.17}$$

where $\bar{\gamma}(\mathbf{x}_i, B)$ is the average semi-variance between the ith sampling point and block B. The kriging variance becomes

$$\hat{\sigma}^2(B) = \mathbf{b}^T\boldsymbol{\lambda} - \bar{\gamma}(B,B), \tag{4.18}$$

where the additional term $\bar{\gamma}(B,B)$ is the average semi-variance between points within block B, that is, the within-block variance.

Paul and Cressie (2011) note that block kriging introduces a complication for skewed properties. The kriged estimates are still unbiased, but because values from the upper tail of the distribution have a large effect on the expected block, the kriging variances cannot be readily back-transformed. Paul and Cressie use formulae for the kriged prediction and kriging variance of a lognormal random variable. This formula does not generalize to other non-Gaussian distributions, however, and even if the distribution of the punctual observations is lognormal it is not necessarily clear what the nature of the distribution will be across a block.

These difficulties can be overcome by simulation. We use our estimated variogram for $z(\mathbf{x}) = \log_{10}(K^+)$ to simulate values of z at 1,000 evenly dispersed positions within each 24 km × 24 km block of the farm. Each simulated value can then be back-transformed to a concentration as 10^z, and then these values can be averaged to give the mean for that block. If this process is repeated many times, say 5,000, then a histogram of the block mean can be formed.

We note that within each of the 5,000 realizations the 1,000 values of $z(\mathbf{x})$ in the block will be correlated. Therefore, a simulation algorithm that can account for this correlation must be used. We generated values by the standard LU simulation, details of which can be found in GSLIB (Deutsch and Journel, 1998) and in the texts by Goovaerts (1997) and Webster and Oliver (2007). These realizations can be conditioned on existing observations of the variable, and we do this for the Broom's Barn example.

The mean and variance of the block means for Broom's Barn Farm are mapped in Figure 4.8a and b. The pattern of variation reflects the punctual estimates, but the variances (Figure 4.8b) are smaller since these predictions are averaged over a larger support and, also, the nugget variance disappears from the kriging variances.

Rational decision-making

We have described how geostatistical methods can be used to create histograms of average exchangeable K^+ across each 24 m × 24 m block on Broom's Barn Farm. The histograms reflect the uncertainty of the predictions. This is the scale at which our understanding of yield and profit response to added K is valid. Therefore the final step in our decision-making is to determine the optimal addition of fertilizer for each block. Figure 4.2 shows the profit plotted against the amount of added K when the initial concentration of K^+ is known. Rather than assume that the initial concentration of exchangeable K^+ is known, however, we calculate the profit for each element of the histogram and assume that the expected profit is equal to the average of these profits.

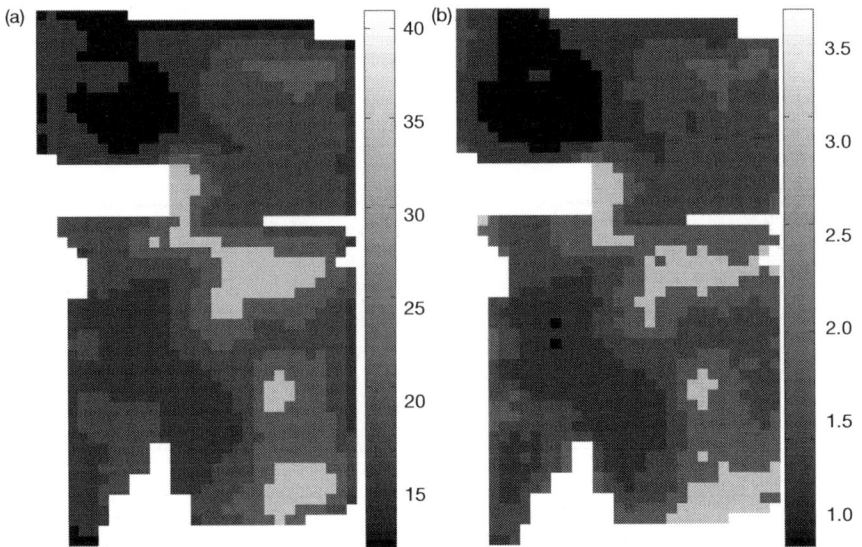

Figure 4.8   Mean (a) and standard deviation (b) of average K^+ within 24 m² blocks at Broom's Barn Farm.

Figure 4.9 shows two examples of this procedure. At both locations the distri-bution of K⁺ concentrations is slightly skewed. In the block corresponding to the left-hand graphs, the predicted concentrations of K^+ are fairly small, and an addition of more than 400 kg ha⁻¹ is required. In the other block the K^+ is more concentrated and only 200 kg ha⁻¹ is required. The optimal additions across the farm are mapped in Figure 4.10.

Potential extensions to the decision-making framework

The illustration above simplifies the farm management system, of course. For this reason the optimal additions of fertilizer might not accord with the indus-try's recommendations. The illustration is based on the assumption that the single objective of the system is to maximize the expected profit over the next season, and it uses a simple mathematical model of the relations between the nutrient stock in the soil, the added fertilizer and yield.

Rather than adding the amount of fertilizer to maximize the expected profit the manager might wish to include other factors such as the potential for leaching of surplus nutrient. The manager might also wish to account for the nutrient stock expected to remain in the soil after harvest. Even if profitability is the manager's primary objective, there might still be more appropriate descriptors

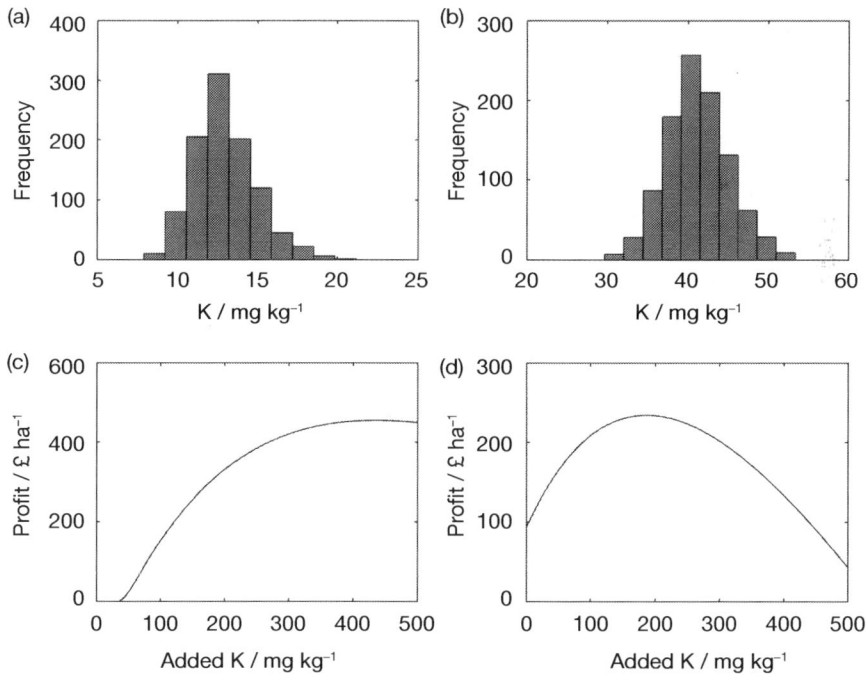

Figure 4.9 (a) and (b) Exemplar histograms of simulated predictions of K^+ (mg kg⁻¹) over 24 km² blocks and (c) and (d) corresponding graphs of profit against added K.

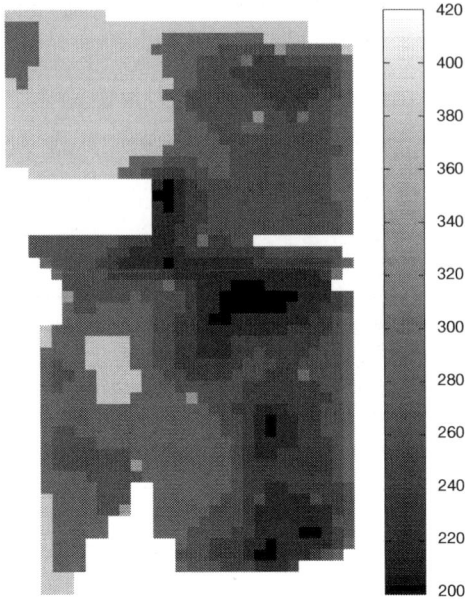

Figure 4.10 Optimal additions of K(kg ha^{-1}) to Broom's Barn Farm.

of profitability than the mean. For example, the continued viability of the enterprise might be achieved only if a certain amount of profit is achieved every year, and therefore the manager's objective function should be expressed in terms of the probability that this threshold on the profit is exceeded.

The mathematical models of the system could be expanded to include other uncertain factors such as weather or other spatially varying properties of the farm such as elevation. Leaf-area index measured remotely or electrical conductivity of the soil could also be incorporated.

If one understands how these additional factors affect crop performance one could incorporate them into one's decision-making. It would still be possible to simulate thousands of realizations of the potential outcomes of the management practices and to select the optimal strategy.

We have devoted much of our attention to the statistical analysis of spatially varying properties. We could improve the predictions resulting from such analyses by using some recent innovations in geostatistics. We briefly describe below two kinds of innovation, namely model-based geostatistics and optimized sample designs. The full details of these methods are beyond the scope of this chapter, but readers can consult the works we cite.

Model-based geostatistics

We estimated variograms by the method of moments, Equation (4.5). This approach requires the practitioner to make decisions that can influence the estimated model. For example, he or she must choose the lag distances over which the experimental variogram is estimated and the weights allocated to each estimated semi-variance when fitting the model. An alternative approach known as 'model-based geostatistics' (Lark *et al.*, 2006) does not require these decisions. Instead one selects the values of the variogram model to maximize an expression of the likelihood that the observed data would have occurred according to the model. Lark *et al.* (2006) assumed that the property of interest is a realization of a multivariate Gaussian distribution, but Marchant *et al.* (2011) have shown how the approach can be extended to fit models with other distributions.

We used the approach of Marchant *et al.* (2011) to fit a model with a Gaussian Box–Cox distribution to the data on K⁺ at Broom's Barn. If a variable z is a realization of this distribution then after the transformation,

$$z^* = \begin{cases} \dfrac{z^c - 1}{c} & \text{if } c \neq 0 \\ \log(z) & \text{if } c = 0 \end{cases}, \tag{4.19}$$

z^* is also a realization of a Gaussian distribution and c is a parameter. This distribution is often used for skewed variables because the inclusion of the additional parameter c means that one has more flexibility to select a distribution that matches the observed data. Table 4.1 lists the cross-validation statistics when this model is fitted to the Broom's Barn data. Both the MSE and median of θ improve, that is, the MSE is smaller and the median of θ is closer to 0.45, on those from the lognormal model fitted by the method of moments.

Sample design

The sample design for the survey at Broom's Barn is a regular square grid. It is easy to locate sampling points on a square grid in the field, and the grid ensures that the observations are evenly spread across the region. One might expect such a grid to be optimal, therefore. However, as Burgess *et al.* (1981) showed, a regular triangular grid is somewhat better. More importantly, sampling needs to be intensified near boundaries to achieve optimality.

Van Groenigen *et al.* (1999) adapted the method of spatial simulated annealing (SSA) to optimize the location of observations in a sample design. Spatial simulated annealing is a numerical algorithm that efficiently experiments with different sample designs to find the optimal one according to some criterion prescribed by the user. If the variogram of a property is specified then SSA can be used to select the sample design that has the smallest average kriging variance across a region. We use this method to optimize a 75-point survey of $z(\mathbf{x}) = \log_{10} K(\mathbf{x})$ at Broom's Barn Farm (Figure 4.11a). The design approximates a triangular grid, but more intense sampling occurs close to the boundaries of the farm. In practice the variogram would not be known at the beginning, but expert knowledge and previous studies of the property might permit one to use an approximate variogram.

We have mentioned two reasons why a regular square grid is not quite optimal. There is, however, a more important reason why such a grid is not optimal if one has to estimate the variogram before kriging. It is that the optimal grid for kriging does not provide information about the variation over distances less than the grid interval, and that information is needed to obtain accurate estimates of the variogram at those distances for kriging. This is recognized in practice. Some surveyors add points between grid nodes on their grids (see Webster, 2008; Webster and Lark, 2013, for examples); others add nested random samples at some grid nodes, for example Atteia *et al.* (1994). Marchant and Lark (2007) extended the SSA algorithm to select sample designs that minimized the total error in a geostatistical survey. This total error includes uncertainty because of both variogram estimation and kriging. A 75-point sample design for Broom's Barn optimized by this method is shown in Figure 4.11b. The design still has good spatial coverage, but it includes seven close pairs of observations to allow the short-range variation to be quantified.

The requirement to know the variogram prior to sampling and the computational requirements of the SSA algorithm when N is large mean that it is not always feasible to optimize the sample design. Nevertheless, a design in which most of the sampling points lie at the nodes of a regular grid will be efficient. The practitioner could choose either a square or triangular grid depending on whether to maximize efficiency or make location in the field easy. Adding further observations close to grid nodes should enable the practitioner to estimate the variogram effectively.

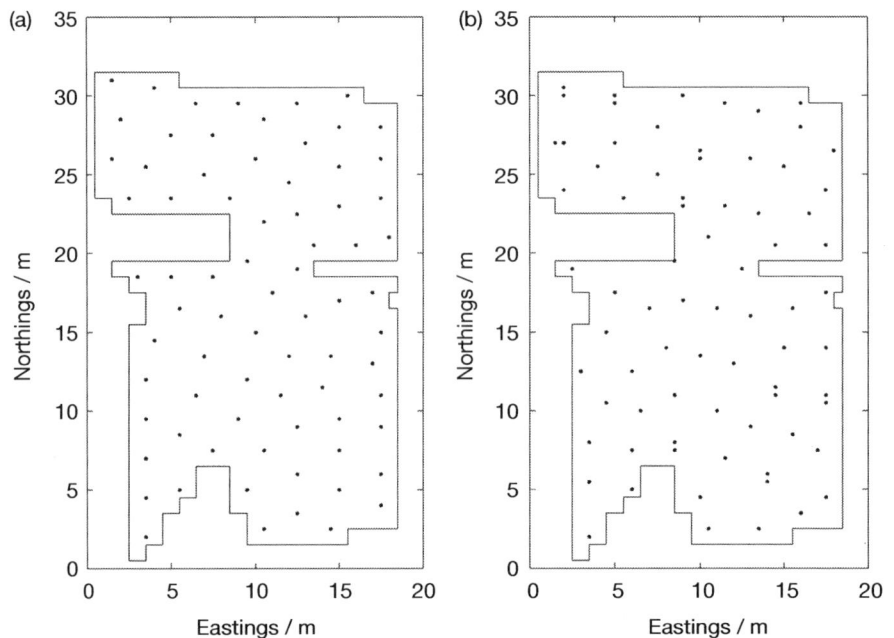

Figure 4.11 Optimized 75 observation sample designs for (a) kriging and (b) variogram estimation and kriging of $\log_{10}(K^+)$ across Broom's Barn Farm.

There remains the question of how many observations are required for a survey. Some rules of thumb have been suggested. Webster and Oliver (1992) recommended at least 100 observations on a square grid to estimate the variogram. Fewer observations could suffice if some close pairs are included in the design. Kerry *et al.* (2010) considered the most efficient grid-spacing for kriging if the variogram were known. Their suggested rule of thumb was that the grid-spacing should be less than half the distance over which the variable is spatially correlated. If sampling is intensified then the rate of decrease in the average kriging variance tails off. Of course, the variogram is not known when the sample scheme is designed. Therefore an a priori estimate of the variogram range is required to follow this suggestion. This could be based on previous experience of mapping the same soil property in similar environments. Some close pairs of observations additional to the grid would be required to ensure that the actual variogram was estimated adequately. In particular, the design should be such that semi-variances can be estimated accurately at 5 to 10 lag distances between zero and the expected range of the variogram, that is, over the increasing part of the curve. These values should enable the practitioner to model the variogram reasonably well.

The required sampling intensity will depend on the variability itself, the purpose of the survey and the required accuracy. Marchant *et al.* (2012) developed a framework to determine the sampling required to estimate the mean nutrient status of arable fields cost-effectively. Where this is not feasible, the sample size might be determined by a combination of financial constraints and previous experience.

Discussion

We have described a framework within which models and the results of geostatistical analyses can be combined to make decisions that take into account uncertain information. We have described fairly simple modifications to standard geostatistical methods so that punctual predictions of soil properties with non-Gaussian distributions can be made and somewhat more elaborate procedures for predictions on larger blocks. The methods are based on the assumption that the type of non-Gaussian distribution is known. It is important that the fitted models are validated to confirm that the assumed distribution is plausible. Our case study emphasizes the benefits of geostatistics to quantify the uncertainty of the predictions.

The decision-making framework is based on the maximization of a profit function, and we considered a simple system. A major challenge will be to develop cost functions for more complex systems. In particular, the farming system might need to achieve more than one goal. The goal might be to maximize profitability subject to constraints on the amount of nutrient that is expected to be leached from the soil. This will require more detailed process modelling and a method of weighing these competing goals against one another.

Both the process and geostatistical models can be expanded to include other sources of information such as measurements at close intervals from remote sensors or detailed elevation maps. Further experimentation might be required for this information to be included in process models. If the information is correlated with nutrient concentrations then the maps of these can be improved by quantification of these spatial correlations with a linear mixed model (Lark *et al.*, 2006).

New innovations will almost certainly improve the predictions that result from geostatistical analyses. Model-based methods can lead to models that fit the observations better than traditional methods based on the method of moments, particularly if a wider class of non-Gaussian models is used. As we have shown, one can use spatial simulated annealing to optimize designs. However, the method requires prior knowledge of the variogram and can take a long time to compute. An adequate design can be achieved through the addition of five or six close pairs to a regular grid. We note that despite these innovations, the cross-validation statistics suggest that our non-Gaussian models do not fit the K$^+$ data perfectly at Broom's Barn. Expert judgement is required to determine if the models are adequate. If not, further generalizations of the non-Gaussian spatial model will be required.

Acknowledgements

The research described in this chapter was funded by the Biotechnology and Biological Sciences Research Council (BBSRC) and the Home-Grown Cereals Authority (HGCA). The authors are grateful for the advice received from Dr Murray Lark.

References

Atteia, O., Webster, R. and Dubois, J.-P. (1994) 'Geostatistical analysis of soil contamination in the Swiss Jura', *Environmental Pollution*, 86, 315–27.

Burgess, T. M., Webster, R. and McBratney, A. B. (1981) 'Optimal interpolation and isarithmic mapping of soil properties', *Journal of Soil Science*, 32, 643–54.

Chilès, J.-P. and Delfiner, P. (2012) *Geostatistics: Modeling Spatial Uncertainty*, 2nd edn, Hoboken, NJ: John Wiley & Sons.

Cressie, N. (2006) 'Block kriging for lognormal spatial processes', *Mathematical Geology*, 38, 413–52.

Deutsch, C. V. and Journel, A. G. (1998) *GSLIB Geostatistical Software Library and User's Guide*, 2nd edn, New York: Oxford University Press.

ESRI (2011) *ArcGIS Desktop*, Release 10, Redlands, CA: Environmental Systems Research Institute.

Goovaerts, P. (1997) *Geostatistics for Natural Resources Evaluation*, New York: Oxford University Press.

Johnston, A. E. and Goulding, K. W. T. (1988) 'Rational potassium manuring for arable cropping systems', *Journal of the Science of Food and Agriculture*, 46, 1–11.

Johnston, A. E. and Goulding, K. W. T. (1990) 'The use of plant and soil analyses to predict the potassium supplying capacity of soil', in *Proceedings of 22nd Colloquium of the International Potash Institute*, Bern: International Potash Institute, pp. 177–204. Available online at <http://www.ipipotash.org/udocs/development_of_k_fertilizer_recommendations.pdf> (accessed May 2013).

Kerry, R., Oliver, M. A. and Frogbrook, Z. L. (2010) 'Sampling in precision agriculture', in M. A. Oliver (ed.) *Geostatistical Applications for Precision Agriculture*, Dordrecht: Springer, pp. 35–63.

Lark, R. M., Cullis, B. R. and Welham, S. J. (2006) 'On optimal prediction of soil properties in the presence of spatial trend: the empirical best linear unbiased predictor (E-BLUP) with REML', *European Journal of Soil Science*, 57, 787–99.

Marchant, B. P. and Lark, R. M. (2007) 'Optimized sample schemes for geostatistical surveys', *Mathematical Geology*, 39, 113–34.

Marchant, B. P., Saby, N. P. A., Jolivet, C. C., Arrouays, D. and Lark, R. M. (2011) 'Spatial prediction of soil properties with copulas', *Geoderma*, 162, 327–34.

Marchant, B. P., Dailey, A. G. and Lark, R. M. (2012) *Cost-Effective Sampling Strategies for Soil Management*, Project Report 485, Stoneleigh, Kenilworth, UK: Home-Grown Cereals Authority.

Milford, G. F. J. and Johnston A. E. (2007) 'Potassium and nitrogen interactions in crop', in *Proceedings 61*, Leek, UK: International Fertiliser Society, pp. 1–22.

Nix, J. (2011) *Farm Management Pocketbook 2012*, Melton Mowbray, UK: Agro Business Consultants.

Olea, R. A. (1999) *Geostatistics for Engineers and Earth Scientists*, Boston: Kluwer Academic Publishers.

Paul, R. and Cressie, N. (2011) 'Lognormal block kriging for contaminated soil', *European Journal of Soil Science*, 62, 337–45.

Payne, R. W. (ed.) (2011) *The Guide to GenStat Release 13. Part 2: Statistics*, Hemel Hempstead, UK: VSN International.

Ribeiro Jr., P. J. and Diggle, P. J. (2001) 'geoR: a package for geostatistical analysis', *R-News*, 1.2, 15–18.

Van Groenigen, J. W., Siderius, W. and Stein, A. (1999) 'Constrained optimisation of soil sampling for minimisation of the kriging variance', *Geoderma*, 87, 239–59.

Viscarra Rossel, R. A., Gilbertson, M., Thylen, L., Hansen, O., McVey, S. and McBratney, A. B. (2005) 'Field measurements of soil pH and lime requirement using an on-the-go measurement system', in J. V. Stafford (ed.) *Precision Agriculture 05*, Wageningen: Wageningen Academic Publishers, pp. 511–20.

Webster, R. (2008) *El muestreo en los estudios del suelo*, Coyoacán, Mexico City: Universidad Nacional Autónoma de México.

Webster, R. and Lark, R. M. (2013) *Field Sampling for Environmental Science and Management*, Abingdon, UK: Routledge.

Webster, R. and McBratney, A. B. (1987) 'Mapping soil fertility at Broom's Barn by simple kriging', *Journal of the Science of Food and Agriculture*, 38, 97–115.

Webster, R. and Oliver, M. A. (1992) 'Sample adequately to estimate variograms of soil properties', *Journal of Soil Science*, 43, 177–92.

Webster, R. and Oliver, M. A. (2007) *Geostatistics for Environmental Scientists*, 2nd edn, Chichester, UK: John Wiley & Sons.

5 Applications of remote sensing in precision agriculture for sustainable production

Paul G. Carter and Stephen L. Young

Introduction

Precision farming incorporates the key elements of geo-technology (one of the mega-technologies of the twenty-first century) in an integrated farm-management programme. Key components of geo-technology include global positioning systems (GPS), geographical information systems (GIS), remote sensing data, data-collection devices and variable-rate controllers or robotics. Remote sensing (data and data-collection devices) provides elements critical to the success of the technology-rich management system. The information derived is analysed in a systems approach and used to provide economic sustainability for the production of food and fibre while preserving the environment.

Remote sensing is an extensive science that has experienced explosive growth in all areas of society and has advanced from basic science to include many commercial hardware and software applications. Remote sensing has been defined as the art and science of obtaining information about an object (target) without being in physical contact with that object and separated from it by some distance (Figure 5.1) (Lillesand and Kiefer, 1994). This distance might be many miles, as in spacecraft platforms, or a few microns, as with some handheld devices and equipment-mounted sensors. This chapter will provide a discussion of remote sensing technology tools currently in use in agriculture and provide an overview of their application in precision agriculture (PA). It is not possible in these few pages to present an extensive and in-depth review of the science and all of the possible applications. The focus of this chapter will therefore be on basic concepts and the applications of satellite, aircraft and other imagery in PA. The presentation and discussion of proximal sensors are addressed in Chapters 6 and 8 of this book.

Historical information

An extensive history of remote sensing is documented in the *Manual of Remote Sensing*, vol. I, published in 1975 by the American Society of Photogrammetry. Jensen (2005) also lists a detailed historical account showing continuous periods of remote sensing development.

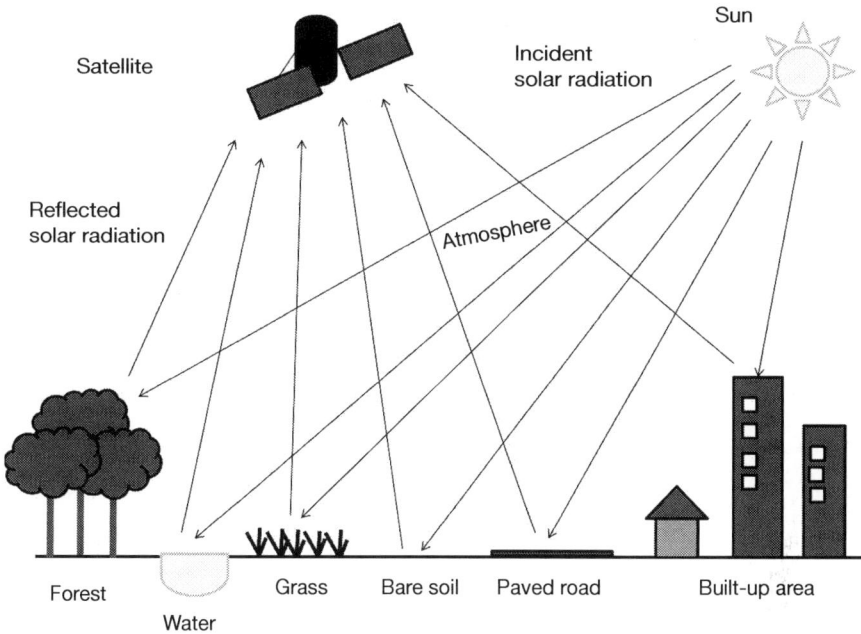

Figure 5.1   Sun's radiation energy reflected and captured with a passive remote sensing system and recorded for later processing and analysis. © NASA, USA.

Remote sensing, as understood today, began in the form of photographic observations in the 1800s with the development of photography. Early photographs of the landscape were taken from platforms including balloons, kites and even pigeons. The first aerial photograph was taken in 1858 by Gaspard-Félix Tournachon (later known as 'Nadar') from a tethered balloon at a great risk to life. Exactly 30 years later, Arthur Batut took the first aerial photographs using a kite as the aerial platform. In 1903, Julius Neubranner designed and patented a breast-mounted aerial camera, weighing 70 g, for use on carrier pigeons. The camera took automatic exposures at 30-second intervals along the flight line flown by a pigeon. Although faster than balloons they were not always reliable in following their flight paths. About this same time, the successful flight of heavier-than-air aircraft added another available aerial platform. From then until the early 1960s, the aerial photograph remained the single standard tool for depicting the Earth's surface from a vertical or oblique perspective.

By the mid 1900s, space-flight Earth observations became the next platform for photographing the Earth's surface. 'Remote sensing' as a term was first applied to the science in 1960 by Evelyn Pruitt of the US Office of Naval Research (Cracknell and Hayes, 2007, p. 12). It is now commonly used to describe the science and the art of observing and measuring objects on the Earth's surface.

In the USA, Earth observation from space began with the National Aeronautics and Space Administration (NASA) programme with the launch of V-2 rockets acquired from Germany in 1946 after World War II. At the same time, Russia was also experimenting with rockets and Earth observations. These rockets contained cameras that recorded still and moving images while the rockets ascended. The first US Earth observation satellite to reach orbit, TIROS 1, was launched in 1960 and was followed by many other TIROS meteorological satellites. The first of eight Earth land-observing satellites (ERTS-1) in that programme was launched in 1972, later renamed Landsat 1. The Landsat programme was born and continues to provide seamless data for the surface of the Earth.

Since the development of space vehicles, Earth observing platforms have been used by the governments of Russia, the United States of America, Canada, Europe, India, Germany, Brazil and Argentina, to name a few. More recently, private commercial enterprises such as SPOT Image, DigitalGlobe, GeoEye and RapidEye AG have developed specialized satellites for high-resolution imaging of the Earth's surface.

Remote sensing in agriculture began in the 1930s following the camera and film technology advances developed during World War I. Agricultural applications quickly became of interest as a possible method to monitor and quantify crop production and crop health with the application of colour infrared (CIR) and later the near-infrared (NIR) technology. At this time, a focus of much research was on the ability to discriminate one crop from another based on reflectance.

Examples of this research effort are the Large Area Crop Inventory Experiment (LACIE), Agriculture and Resources Inventory Surveys through Aerospace Remote Sensing (AgRISTARS) and Crop Identification Technology Assessment for Remote Sensing (CITARS) programme of field experiments (Bauer, 1985). A particular project of interest in the USA included the Corn Blight Watch (CBW) experiment in 1971 (Bauer, 1985), which used an aircraft-mounted multispectral sensor using 12 individual spectral wavelength band sensors between 400 and 1,000 nm to track a crop disease. This project was the first evidence that the presence and progression of an agricultural crop disease could be monitored using aerial sensor data. Following this pioneering work, other projects were initiated that focused on specific characteristics, such as soil reflectance and geologic components.

From a traditional perspective, remote sensing is viewed by many as simply a tool for displaying satellite and aircraft data as imagery. In fact, on the contrary, remote sensing and related technology have continued to advance beyond just satellite and aerial imagery to additional and more user-friendly tools, such as handheld and vehicle-mounted chlorophyll greenness meters and plant thermal scanning sensors. The use of the range of specific wavelengths in the visible, near-infrared and thermal spectra is a particularly striking development.

The first groups to use data generated by remote sensing extensively were those associated with land, atmosphere, ocean, communication, navigation and support systems. Today, there are very few groups and individuals who do not use remote sensing data in some way. Farmers are one of the many groups working in the earth and environmental sciences that use the data from remote sensing. Crop consultants and agronomists that have expertise in remote sensing technology usually assist them. The objectives are to increase efficiency in the use of labour and other inputs such as fertilizers and water to optimize production costs, scout fields for disease and pest outbreaks, and to manage overall production operations better.

Remote sensing is commonly discussed in terms of four types of resolution. These include spatial, spectral, temporal and radiometric resolution. Spatial resolution refers to the image pixel size in terms of the ground surface it represents and is usually referred to in terms of metres. Radiometric resolution is the measure of the emitted energy that the sensor can detect or the sensitivity of the radiation measurements. Radiometric resolution is affected by the size of the spatial resolution and has an inverse relationship with the spatial resolution. Spectral resolution is the width of the wavebands that the sensor is designed to detect or the ability to discriminate fine spectral differences and is measured in nanometres (nm). Trade-offs must be made between spatial, spectral and radiometric resolutions as each one affects the others in its ability to discriminate fine measurements. The fourth is temporal resolution, which is the frequency of measurements or overpasses by the sensor or scanner.

The use of multispectral or hyperspectral satellite and aerial imagery has proved to be of great benefit to PA because it records many spectral wavelengths ranging from the visible (350 to 700 nm), to the near infrared (NIR) (700 to 1,000 nm) and to the short-wave infrared (SWIR) (1,000 to 2,500 nm) reflectance (Figure 5.2). With remote sensing, conditions in an entire field can be observed and recorded quickly using imagery and the data are available for later analysis at relatively little cost. The spatial and temporal variation in landscape, soil and crops can be assessed instantly and over large scales to allow for analysis and corrective measures to be applied in seemingly real time.

Introduction to vegetation and soil indices

The use of remote sensing, beyond just visual observation, requires processes that record and analyse data in a systematic framework. This is typically done with the assistance of computers and software, as the computations can be complicated and lengthy without such tools. The processed data can then be applied to relatively simple tasks, such as applications and monitoring activities of precision agriculture, and to more complicated decision-making processes such as controlling variable-rate fertilizer applications or seeding rates. Early applications of remote sensing to PA involved the use of imagery to monitor vegetation development, plant stress levels and irrigation water management. The recording of these data is based on variation in colour intensity, texture and patterns.

Since the advent of digital image data, scientists have worked to extract vegetation and other Earth surface information from remotely sensed data by analysing the reflectance digital numbers. Much of this effort has been applied

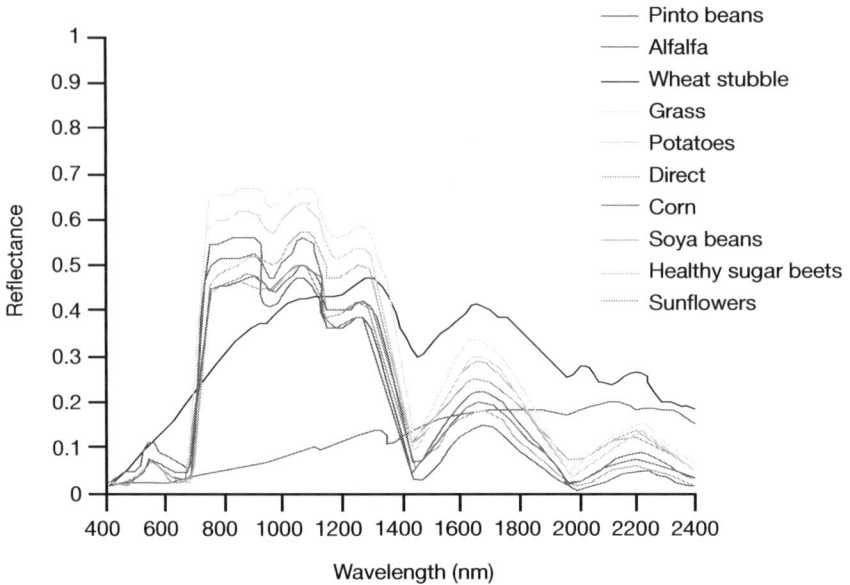

Figure 5.2   Remote sensing data showing the range of the visible and near-infrared electromagnetic energy curves for different agricultural crops. © NASA, USA.

to developing vegetation indices that help to detect changes in vegetation over time. Jensen (2005) defined these as dimensionless, radiometric measures that function as indicators of relative abundance and activity of green vegetation, often including leaf-area index, percentage green cover, chlorophyll content, green biomass and absorbed photosynthetically active radiation. There are many (more than 20) indices in use today and a select few are listed in Table 5.1.

Running *et al.* (1994) and Huete and Justice (1999) stated that a vegetation index should maximize sensitivity to plant biophysical parameters, normalize or model external effects (such as sun angle), normalize internal effects (canopy background) and be coupled to some specific measurable biophysical parameter such as biomass as part of the validation effort and quality control. Indices are categorized as simple ratio and orthogonal transformations and were developed to reduce 'noise' from the data, introduce corrections for atmosphere and normalize results.

Through the development of indices and ratios such as the vegetation index (VI), normalized difference vegetation index (NDVI), enhanced vegetation index (EVI), greenness vegetation index (GVI) and the soil-adjusted vegetation index (SAVI), it is now possible to compare and treat separate areas within fields quantitatively.

With the aid of stereo-plotters and more recently GPS, photographs and digital imagery of an area can be measured and boundaries defined for a wide range of landscapes. For agriculture, many applications have been developed

Table 5.1 List of basic vegetation indices for the study of vegetation ground cover

Index	Equation	Applications
Vegetation index (VI)	NIR / RED	Healthy plants, amount of green biomass and chlorophyll content
Greenness vegetation index (GVI)	− (0.2848*BLUE) − (0.2435*GREEN) − (0.5436*RED) + (0.7243*NIR) + (0.0840*Mid IR) − (0.1800*Mid IR)	For making greater distinctions between vegetation
Normalized difference vegetation index (NDVI)	(NIR − RED) / (NIR + RED)	Most commonly used ratio, used for detection of plant chlorophyll content, amount of green biomass
Green normalized difference vegetation index (GNDVI)	(NIR − GREEN) / (NIR + GREEN)	Predictor of forage biomass and or forage N uptake
Soil-adjusted vegetation index (SAVI)	(NIR − RED) (1 + L) / (NIR + RED + L), where L is correction factor from 0 to 1, usually 0.5	Reduce the effect of soil background reflectance

that include variable-rate seeding and fertilizer applications, crop yield mapping, weed mapping and management, historical mapping (e.g. former fence lines, building sites and roadways), soil-management zones and insurance assessments of crop damage.

Types of remote sensing systems

Remote sensing systems can be grouped into two types, either passive or active, based on the energy source. Passive remote sensing systems do not have their own light energy source; they detect naturally occurring energy radiation that is reflected and or emitted from the target surface. Examples of passive sensor types (Table 5.2) range from cameras that take simple pictures to the more complicated hyperspectral sensors or thermal image scanner that can detect heat.

Passive systems can sense low-level microwave radiation given off by all objects in the natural environment. For all reflected energy, this can take place only during the time when the sun is illuminating the Earth. Only thermal energy available from the sun and naturally emitting surfaces (thermal infrared) can be detected by day or night, as long as there are sufficient amounts of energy to be detected. Some well-known examples of satellite systems that have operated for a considerable time are Landsat and SPOT.

An active remote sensing system provides its own source of energy. It is a system that measures the energy emanating from the target from which a signal was directed and then reflected back to the sensor. Active sensors then transmit

Table 5.2 NASA Earth Observatory (2011) list of passive sensors

Sensor type	Description
Radiometer	Measures the intensity of electromagnetic radiation in a band of wavelengths in the spectrum; identified by the portion of the spectrum covered (e.g. visible, infrared or microwave).
Imaging radiometer	Radiometer with mechanical or electronic scanning capabilities to provide a two-dimensional array of pixels from which an image may be produced.
Spectrometer	Detects, measures and analyses the spectral content of the incident electromagnetic radiation and uses gratings or prisms to disperse the radiation for spectral discrimination.
Spectroradiometer	Measures the intensity of radiation in multiple wavelength bands (i.e. multispectral and hyperspectral) and is designed for the remote sensing of specific properties (e.g. sea surface temperature, cloud characteristics, ocean colour, vegetation, trace chemical species in the atmosphere, and so on).

short bursts or 'pulses' of electromagnetic energy in the direction of the target and record the origin and strength of the backscatter received from objects within the system's field of view. For example, the Radarsat-1 synthetic aperture radar (SAR) system has an antenna that beams pulses of electromagnetic energy toward its target. A major advantage of active sensors is their ability to obtain measurements at any time, regardless of the time of day, season or amount of natural illumination (e.g. cloud cover).

Active sensors can be used for examining wavelengths that are not sufficiently provided by the sun, such as microwave, or to control better the way that a target is illuminated. A disadvantage of these systems is that they require a significant amount of energy to reflect adequate levels for measurement. Examples of other active sensors (Table 5.3) are the laser fluorosensor and SAR. Finally, the use of radar or LiDAR is becoming increasingly popular for a range of applications, including measurements of topography, plant canopy architecture and similar geographical categories.

In a brief review of the current literature, active systems used for plant canopy measurements range from maize and wheat to higher-value crops, such as sugar beet and broccoli. Several active systems are available and can be purchased as complete packages or individual original equipment manufacturer (OEM) components that require assembly by the user. Some systems are robust and can make very precise measurements, whereas others need further refining. The cost of a commercial system varies depending on features selected, intended application and complexity of design.

Table 5.3 NASA Earth Observatory (2011) list of active sensors

Sensor type	Description
Radar (radio detection and ranging)	Transmitter operating at either radio or microwave frequencies with emitter and receiver that measures time of arrival of reflected radiation from distant objects. Distance to object determined based on speed of light.
Scatterometer	High frequency microwave radar designed specifically to measure backscattered radiation (e.g. maps of surface wind speed and direction over ocean surfaces).
LiDAR (light detection and ranging)	Transmission of a light pulse or laser (light amplification by stimulated emission of radiation) to a receiver with sensitive detectors to measure the reflected light. Speed of light used to calculate the distance travelled and common for atmospheric profiles of aerosols, clouds and other constituents of the atmosphere.
Laser altimeter	Used by LiDAR to measure height of instrument platform above a surface; common for determining topography of many natural and man-made surfaces

Precision agriculture: applications of remote sensing

Since the early development of remote sensing technology, there have been many applications for agriculture. Some have proved to be effective, whereas others have failed to succeed in assisting farmers to solve problems and make management decisions. Historically, remote sensing images generated for cropping systems were often used as a visual tool and were not related to real-time field observations or locations. The field location of the area in question often involved a certain amount of guesswork at best. In these early applications, a more accurate use of the images was achieved by processing or 'digitizing' the data, which allowed close measurement at a discrete location for an item of interest. Digitizing was not an easy task and required considerable equipment, expertise and time. Later, the development of GPS made the use of image data easier and reduced the need for specialized equipment and highly trained personnel. The application of GPS to remote sensing significantly improved the usability of data and the technology.

Johannsen and Carter (2000) identified areas of potential applications of remote sensing for commercial farming. Some of the topics include crop inventory and yield prediction, crop stress, crop injury detection, vegetation change and mapping soil properties. Soil investigations, surveys and mapping can use remotely sensed images. They include three different approaches: the effects of soil properties on reflectance or image response, the influence of soil surface conditions on the response and the use of imagery for mapping soil patterns. Management zones can be derived as they can define logical boundaries for farming activities (see Chapter 8 of this book). The USDA's National

Agricultural Statistical Service (NASS) and the Foreign Agricultural Service (FAS) explored the use of remote sensing images for crop identification, inventory of planted crops and inventory of potential harvest quantities (Wade *et al.*, 1994). Current crop inventories are based on image vegetation indices such as NDVI. Crop stress or injury, a condition within the crop field that is different from what was planned, might include water stress, hail, chemical damage, weed patches, diseases and nutrient deficiencies. Imagery has been used to determine the extent of the damage for insurance assessment purposes.

For most agriculturally based groups and even individual growers, certain remote sensing applications have become well used, whereas others have yet to be widely adopted. This is probably due to the acceptance or understanding of the technology, how difficult it is to incorporate into current equipment and farming activities and the availability of commercial vendor-trained personnel. To be widely adopted, the technology needs to be financially feasible, simple to incorporate, easy to operate and supported by trained service technicians who can resolve issues arising during operation.

Recently, many remote sensing applications have been used to map certain aspects of a field (e.g. drainage-tile lines, historical fence lines and building locations, rock outcrops and weed patches) and in some cases to make assessments for regulatory management of government programmes (Johannsen *et al.*, 1999). As an example, farmers report the number of planted or seeded hectares in a field to the farm programme agency. The area reported can be verified with the aid of the aerial or satellite image using computer software such as ArcGIS. Specific applications of remote sensing to agricultural cropping systems include detailed soil mapping and crop scouting, variable-rate applications (seed and fertilizer), precision application of crop protection products, weed mapping, crop disease tracking and treatment, and crop insurance injury assessments.

In Columbia County, Washington, USA, a dryland wheat farmer who is also a pilot (Carter, pers. comm.) uses remote sensing (digital camera) to photograph many of the farm's fields each year to help develop farm management plans for the coming year. He sometimes assists his neighbours by recording aerial images and providing some interpretation for them. He uses data to track many aspects of his and his neighbours' cropping system, such as expanding or contracting weed patches, outbreaks or 'hot spots' of crop diseases such as wheat stripe rust, changing soil conditions for variable-rate nitrogen applications, directed soil sampling and planning for other miscellaneous activities (e.g. on-farm research plot locations). His efforts have had a direct impact on the improved management of over 10,000 ha in the Palouse region, which is known for its high risk of soil erosion and intensive wheat production.

Soil sampling by management zones

Following the adoption of GPS, soil sampling on a grid became popular, but the ability to aggregate samples based on common soil properties was still lacking. A more thorough assessment of field variability through the application of remote sensing imagery has led to targeted or zone sampling based on soil and vegetation spectral and spatial properties within the field (Mallarino and Wittry, 2001). The process of developing zones may incorporate other technologies such as the use of field equipment (electromagnetic sensors and other instrumentation – see Chapter 8 of this book) for real-time identification of fertility, wet and dry areas and soil types such as sand and clay. The integration of data through the seamless, on-the-go assessment and recording process of the newest farm implements has vastly improved management capabilities of growers. Further information related to these sensors and PA applications can be found in Chapter 6 of this book. The combination of these data with remote sensing can provide a better directed system of sampling the soil based on production capability zones in a field. This has led to more efficient use of time and resources (Mallarino and Wittry, 2001) and is likely to make crop production more sustainable. The soil sampler (agronomist or grower) can reduce the number of samples and the time to collect them.

Variable-rate applications

The collection of site-specific data in real time has created a distinct advantage that has allowed growers to respond to crop needs and design dynamic management strategies. Previously, the broad (and sometimes excessive) application of inputs (e.g. pesticides and fertilizers) limited management opportunities. The extra inputs added to crop production not only reduce profitability, but the sustainability of an agroecosystem. In particular, variable-rate application technology (Figure 5.3) provides growers with an alternative to make real-time decisions in their management operations. Variable-rate applications can be used for seeding, fertilizer applications and weed control. The strategy is to maximize the effectiveness and efficiency of inputs based on characteristics of specific and

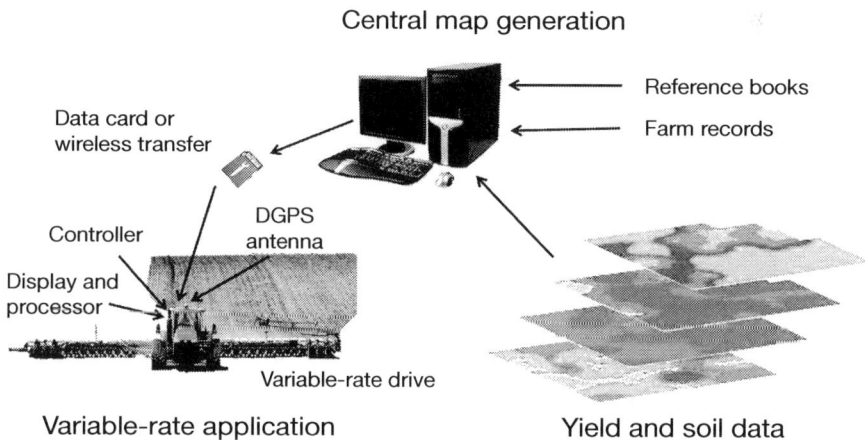

Figure 5.3 Example of how image data are used in a system for variable-rate application. © Paul G. Carter.

discrete locations in the field. For example, in areas of low capability, fewer seeds and fertilizer are applied, whereas in areas of greater potential farmers apply more seed and fertilizer to maximize production.

Variable-rate application is the practice of responding to crop needs based on field conditions. Fields that have been remotely sensed and the yield mapped with NIR will have fewer uncertainties and consequently a grower can make informed decisions allowing more planning and a better response to production capability. In addition, growers can implement new approaches, such as variable-rate technology, without disrupting their entire system. When variable-rate application is combined with existing technology and remotely sensed field information, the efficiency and sustainability of today's farm is significantly improved and the load on the environment is reduced.

Scouting agricultural fields

Specific applications of remote sensing, either passive or active, to agriculture have been increasingly common over the past decade. The ability to scout and map fields with handheld or tractor- (or other equipment) mounted units or using satellite and aircraft imagery has significantly advanced the utility of the technology and the management of time spent in the field. Previously, farmers or consultants walked fields recording data on paper notepads. Although this is still done, more frequently growers are doing several activities simultaneously from either the cab of a combine or tractor or while driving an all-terrain vehicle (ATV) to a field edge and recording or downloading field data from automated devices that can track conditions such as soil moisture, plant nutrient stresses and pest thresholds.

Current field crop scouting includes the use of remote sensing image data, made possible by the combination of GPS technology and small handheld electronic devices, such as personal digital assistants (PDAs). For the field scout, the ability to have a geo-referenced digital image uploaded to a PDA with GPS capabilities and the necessary software is greatly advancing how they work within various spatial and temporal scales. This technology allows the agronomist or field scout to identify anomalies within an image and then be able to walk directly to that anomaly within the field. Prior to this modern advancement, field scouting was conducted by walking in a designated pattern to identify abnormal crop conditions. The inadequacy of this earlier approach is the inability to view the entire field and hence miss certain anomalies or crop conditions.

The current remote sensing technologies that are allowing growers to direct scouting efforts in their fields are also making it possible to develop application maps. By annotating field observations on a remote sensing image, an application map can be generated for the precise use of fertilizers, herbicides, fungicides or other management tools. The geo-referenced map can be uploaded to application equipment that treats only marked areas and leaves the other areas untreated.

Other available technology can classify images (using computer software) into zones based on the levels of NDVI reflectance (see Chapter 8). These maps can then be used to apply treatments to protect and increase production efficiently in cropping systems. Yang and Anderson (1996) used remote sensing to determine vegetation responses and to help define management zones in an agricultural field. The management zones were also used as an aid to soil sampling (directed soil sampling) as well as to define field boundaries.

Disease management

In cropping systems, disease from insects, pathogens and other infectious organisms can become a serious problem. In some cases, disease development on crop plants occurs rapidly and results in entire fields incurring injury to various degrees. For example, navy beans often develop white mould during critical periods of the season and, therefore, are typically sprayed with fungicides as a routine practice. When remote sensing is used, areas of high NDVI reflectance from the bean canopy correlate with greater biomass accumulation, which is a preferred location for moulds and other fungi to become established. Case study (1) below describes how a grower can evaluate the crop condition at a specific growth stage with an NDVI map of the field and apply fungicides on only those areas that have a high risk of infection.

It has been reported that fungicide applications in many cropping systems have been reduced by more than 20 per cent using remote sensing imagery. This is often the case for fields of winter wheat that are susceptible to stripe rust fungus in the Pacific Northwest region of the USA (Carter, pers. obs.). By inspecting anomalies in remotely sensed imagery, locations of rust can be identified within certain fields and then targeted for application. This process of analysis, identification and targeted treatment can result in a significant reduction in the amount of overall area that receives treatment, reducing costs and total volume of fungicide applied. Reduced chemical applications mean less material that the environment (plants and soils) must degrade to neutralize their impact on natural systems. The environment is a constantly changing state of everything living and dead, and these chemicals may have some adverse long-term effects. The benefits of targeted applications are both economic and environmental.

In high-value crops (e.g. broccoli, strawberries and parsley), growers use imagery data to determine when to apply materials to protect their crops. The practice of making applications at discrete temporal and spatial scales allows for real-time response, reduction in inputs and maximizing an 'as needed' basis for operations.

Estimating crop yield

A significant application of remote sensing in agricultural systems is for the prediction and measurement of crop production. Yield measurements are essential in the process of sound management decisions and it is beneficial to analyse multiple years of data to reduce the effects of non-management influences. In fields where airborne imagery is available, crop assessments are made with NIR reflectance, which can then be correlated to biomass for estimating yields

(Doraiswamy *et al.*, 2003). Although some ground-truth scouting is required to verify yield estimates, this has been developed into a real-time application, which has been used for many years in the most progressive farming operations. It is rising in popularity as a result of the ease and accuracy of estimating crop yields over a broad area that it brings.

The USDA's National Agricultural Statistics Service (NASS) and the Foreign Agricultural Service (FAS) have explored the use of remotely sensed images for crop identification and inventory of areas planted and estimation of potential harvest yields (Wade *et al.*, 1994). To be able to ascertain the potential production for food grains for given areas provides the opportunity to meet the needs of the population more readily through transportation from areas of excess supply to areas of great need. For grain crops, images near the time of flowering (e.g. flag leaf emergence in wheat) are optimal for yield forecasting, whereas later stages of vegetative biomass development are difficult to decipher from the inflorescence. For other plants, such as soya beans, more accurate yield predictions may be obtained later in the growing season because of easily recognizable vegetative development that contributes to biomass for estimating yield.

Current methods of yield prediction are based on vegetation indices such as the NDVI. These indices combine image information from near-infrared and visible red bands into ratios that normalize or create a uniform measure of field variability. At the scale of farm and field, remotely sensed images subjected to NDVI analysis have been used to estimate yield variation prior to harvest. When compared to actual measurements, estimated yield maps are quite accurate.

Case studies: farming for sustainability and environmental benefit

Case study 1. Reduced fungicide application to crops

The use of Landsat satellite imagery for the application of fungicides to control white mould in navy beans reduced the area of fungicide application by 30 per cent. Navy bean plants are at risk of a condition called white mould when the biomass is greatest. Areas of less biomass or smaller plant populations are much less likely to develop this condition. By using the remote sensing image data (Figure 5.4a) and developing an NDVI of the field, the areas lacking sufficient biomass to sustain the crop infection were verified using ground investigations. The NDVI of the field provided the tool to develop a variable-rate application map (Figure 5.4b). This map provided a 'spray/no spray' indication that could be loaded into the spray computer to control when to turn nozzles on and off. This provided a saving of 30 per cent of chemical cost (US$972) and reduced the chemical load to the environment.

Case study 2. Assessing water damage of a sunflower crop

Weather is one of the greatest concerns in relation to crop damage for farmers, for example hail, flood, drought and wind. Water damage is one of the leading

(a)
(b)

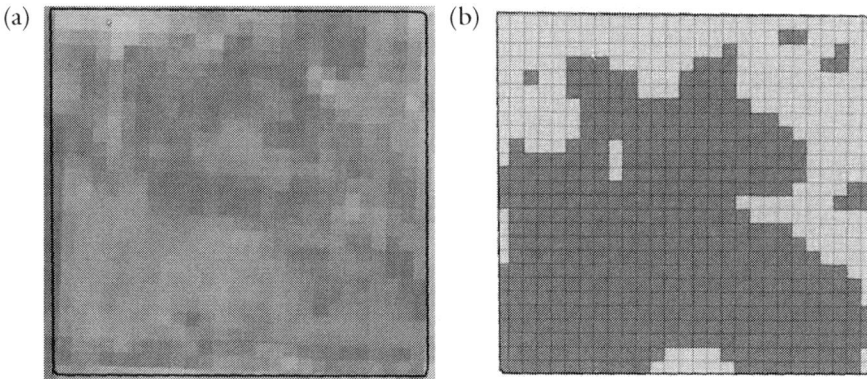

Figure 5.4 Landsat image data (NIR) providing vegetation reflectance (a) and a classified image of the reflectance with thresholds for spray/no spray decision-making (b). © Paul G. Carter.

causes of crop production losses. Insurance companies in the past often assessed the loss to the entire field even though the damage was variable. Remote sensing imagery was used on fields of sunflowers to determine the extent of water damage resulting from heavy rainfall in a Minnesota (USA) field. Based on the NDVI of the biomass for five fields, it was determined that 8 per cent of the fields were drowned out (total loss), 9 per cent heavily damaged and 17 per cent moderately damaged. Figure 5.5 provides an example of one of the observed fields. The NDVI image was classified into four zones (levels of vegetation biomass) and labelled as drowned out, low vegetation, medium vegetation and high vegetation or no damage. The remainder of the fields in the area were considered undamaged and not reported in the totals. By using the imagery, it was possible to assess the damage completely. Payments could be assigned more accurately based on zones of damage.

Case study 3. Chemical off-target application damage to a sugar beet crop

Crop desiccants are often used for certain crops to reduce the time interval until harvest and the impact of weeds at harvest time. Applications are normally applied by aircraft as ground-based equipment cannot be used in the standing crop. A crop desiccant was used on a potato field and applied by a fixed-wing aeroplane. The weather conditions seemed favourable, but drift caused the chemical to damage a neighbouring sugar beet crop. Although the field appeared to sustain considerable damage, it was determined, on the basis of visual inspection guided by image NDVI analysis, that 31 per cent of the field received more than 50 per cent damage but that the remainder of the crop was unaffected. The insurance company settled the payment claim to the satisfaction of all involved based upon the image analysis. This could have been an expensive settlement had it gone to litigation. Neighbours were happy and no additional costs were incurred.

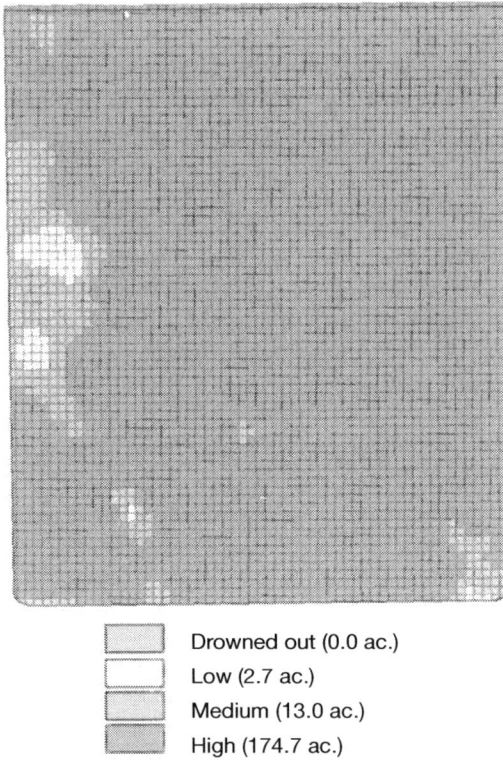

Drowned out (0.0 ac.)
Low (2.7 ac.)
Medium (13.0 ac.)
High (174.7 ac.)

Figure 5.5   Landsat NIR image data of one field classified into four zones of vegetation for water damage assessment. The categories of drowned out, low,
medium and high refer to the vegetation biomass of the classified NDVI
image. © Paul G. Carter.

Case study 4. Management of weeds in crop or fallow fields

Prior to chemical weed-control methods, entire crop and fallow fields were tilled
or cultivated several times a year to control or manage weeds. With the advent
of chemical weed controls, the use of tillage equipment was reduced but applications of chemicals replaced them. Both methods are detrimental to the soil,
crop and environment, but the newer tools helped farmers to be more efficient
and therefore sustainable at least for the short term. By including remote sensing
technologies weed patches can be mapped and targeted, so reducing the chemical applications. Scientists have documented that weed patches are fairly stable
over time and space (Izquierdo *et al.*, 2009), allowing for targeted management.

Remote sensing imagery was used to map the weed patches and target chemical applications of a 67-ha wheat field. Since it has been determined that weed
patches are stable over time, an application map was developed for spraying and
managing the weeds. This application map reduced the total amount of the
field where chemicals were applied. As a result, 35 per cent of the field was

not sprayed, thereby reducing the amount of chemical applied by 35 per cent with a saving of US$685 in chemical costs as well as reducing effects on the environment. Although this method did not control all of the weeds present, it was determined that what remained did not contribute to any yield losses. The 'escapes' were noted and added to the maps for the following year's application plan. This is not only good for the environment but also the economics of the farm by reducing costs. By using remote sensing imagery, farming and the environment can be more sustainable.

Summary and conclusions

Precision farming technology is a tremendous tool for increasing production efficiencies, increasing input return on investment (ROI), providing overall economic sustainability for continued food production and contributing to environmental stability. Many farmers and landowners have long tried to be good stewards of the land, concerned about conservation and sustainability, but also understanding the need to be economically sound for continued production. They have adopted soil-stewardship principles and devices as new tools and concepts have been developed and advanced, respectively. Unfortunately, the success of producing an abundant supply of food using some of these methods has come at a cost of increased risk to the environment. Nevertheless, producers continue to learn and adopt new methods in an effort to increase production to sustain a rapidly growing population, while reducing risk to and protecting our precious environment.

Technology has made sustainable crop production more of a reality, but we are not there yet. Being able to make targeted applications in response to crop needs reduces waste and allows cropping systems to move closer to individual plant management. Managing at the plant scale could be the way to higher yields, greater returns and less error in application timing and rates. Nevertheless, new technology has a price, both in the purchase cost of technology and in time investment for technology education. The critical challenge for the industry is the adoption of new, untested technology while continuing to increase production.

With new technology there is a requirement for some type of support to make sure its operation and performance are up to the standard. The bottleneck in advancing technology in cropping systems is user familiarity of capabilities, understanding operational principles, integration of dissimilar hardware and software systems and multiple data-layer management. Growers have a responsibility to learn how to use advanced technology and understand the application and value of the data. A whole new field of opportunity exists for the technology-aware crop consultant or equipment provider who can operate (or program) computers, provide information technology support and integrate technology systems, while understanding crop growth and development (agronomic principles). This is the next generation of cropping systems and managers of these cropping systems.

In an effort to feed and clothe a projected 9–12 billion people by 2050, adoption of precision technology in farming systems and crop management methods will need to continue. Newer and better production systems that are reliable, easily integrated and incorporate multiple data layers will be a critical component in achieving an abundant and safe supply of food for future generations.

References

Bauer, M. E. (1985) 'Spectral inputs to crop identification and condition assessment', *Proceedings of IEEE*, 73, 1071–85.

Cracknell, A. P. and Hayes, L. W. B. (2007) *Introduction to Remote Sensing*, New York: Taylor & Francis.

Doraiswamy, P. C., Moulin, S., Cook, P. W. and Stern, A. (2003) 'Crop yield assessment from remote sensing', *Photogrammetric Engineering and Remote Sensing*, 69, 665–74.

Huete, A. and Justice, C. (1999) *MODIS Vegetation Index (MOD 13) Algorithm Theoretical Basis Document*, Greenbelt, MD: NASA Godard Space Flight Center. Available online at <http://modis.gsfc.nasa.gov/data/dataprod/dataproducts. php?MOD_NUMBER=13> (accessed January 2013).

Izquierdo, J., Blanco-Moreno, J. M. Chamorro, L., Recasens, J. and Sans, F. X. (2009) 'Spatial distribution and temporal stability of Prostrate Knotweed (*Polygonum aviculare*) and Corn Poppy (*Papaver rhoeas*) seed bank in a cereal field', *Weed Science*, 57, 505–11.

Jensen, J. R. (2005) *Introductory Digital Image Processing*, 3rd edn, Upper Saddle River, NJ: Prentice-Hall.

Johannsen, C. J. and Carter, P. G. (2000) 'Remote sensing in agriculture', in *Precision Farming Profitability*, West Lafayette, IN: Purdue University, pp. 82–7. Available online at <http://www.agriculture.purdue.edu/ssmc/ PPBookunprotected/precision_farming2.htm> (accessed January 2013).

Johannsen, C. J., Carter, P. G., Morris, D. K., Erickson, B. J. and Ross, K. (1999) *Potential Applications of Remote Sensing*, Site Specific Management Guidelines SSMG-22, Norcross, GA: International Plant Nutrition Institute.

Lillesand, T. M. and Keifer, R. W. (eds) (1994) *Remote Sensing and Image Interpretation*, 3rd edn, New York: John Wiley & Sons, Inc.

Mallarino, A. P. and Wittry, D. (2001) *Management Zones Soil Sampling: A Better Alternative to Grid Soil Type Sampling?*, Extension Bulletin, Ames, IA: Iowa State University. Available online at <http://www.agronext.iastate.edu/soilfertility/ info/ICM_2001_ZoneSampling_Publ.pdf> (accessed May 2013).

NASA Earth Observatory (2011) 'Remote sensing'. Available online at <http:// earthobservatory.nasa.gov/Features/RemoteSensing/remote_08.php> (accessed September, 2011).

Running, S. W., Justice, C. O., Solomonson, V., Hall, D., Barker, J., Kaufman, Y. J., Strahler, A. H., Huete, A. R., Muller, J. P., Vanderbilt, V., Wan, Z. M., Teillet P. and Carneggie, D. (1994) 'Terrestrial remote sensing science and algorithms planned for EOS/MODIS', *International Journal of Remote Sensing*, 15, 3587–620.

Wade, G., Mueller, R., Cook, P. and Doraiswamy, P. (1994) 'AVHRR map products for crop condition assessment: a geographic information system approach', *Photogrammetric Engineering and Remote Sensing*, 60, 1145–50.

Yang, C. and Anderson, G. L. (1996) 'Determining within-field management zones for grain sorghum using aerial videography', in *Proceedings of the 26th Symposium on Remote Sensing of Environment, 25–29 March 1996, Vancouver, BC, Canada*, Kanata, ON, Canada: Canadian Aeronautic and Space Institute, pp. 606–11.

6 Proximal soil sensing

*Raphael A. Viscarra Rossel and
Viacheslav I. Adamchuk*

Introduction

Precision agriculture (PA) needs detailed information on the spatial variation of soil properties to achieve its goals of economic and environmental sustainability. To characterize soil spatial variation, many soil samples at fine spatial resolutions need to be collected and analysed. The spatial resolution might need to be closer to that of data collected with yield monitors. Conventional soil sampling is usually sparse; farmers often take only one sample per hectare. The sampling is generally based on a grid or some type of random sampling, but it often does not adequately represent the variation present in the field. Furthermore, samples have to be analysed at a later stage in a laboratory using conventional methods that are often slow and expensive. Proximal soil sensing (PSS) provides a practical solution to overcome these problems.

Proximal soil sensing refers to the use of field-based sensors to obtain signals from the soil when the sensor's detector is in contact with or close to (within 2 m) the soil (Viscarra Rossel and McBratney, 1998; Viscarra Rossel *et al.*, 2011). The sensors provide information on physical measures that can be related to the soil and its properties. This definition precludes remote sensing (see Chapter 5) and also laboratory measurements of soil properties on prepared samples with benchtop instruments. It is generally acknowledged, however, that the development of many proximal soil sensors starts in the laboratory, and that some (e.g. visible–near-infrared sensors) use calibrations derived from laboratory measurements. The rationale for using proximal soil sensors is that, although their results may not be as accurate as for conventional laboratory analysis per individual measurement, they facilitate the collection of soil data using cheaper, simpler and less labour-intensive techniques, which as an ensemble are very informative (Viscarra Rossel *et al.*, 2011). Moreover, the measurements are made at field conditions, they are taken from the surface or within the soil profile and information is produced almost instantly. Therefore, PSS offers advantages that cannot be achieved by remote sensing or destructive sampling and laboratory analyses.

Proximal soil sensors are described by how they measure and operate, the source of their energy and the inference used in the measurement of the target soil property. Figure 6.1 summarizes the possibilities.

MEASUREMENT	ENERGY	OPERATION	INFERENCE
non-invasive	passive	mobile	indirect
PROXIMAL SOIL SENSING			
invasive	active	stationary	direct
in situ ex situ			

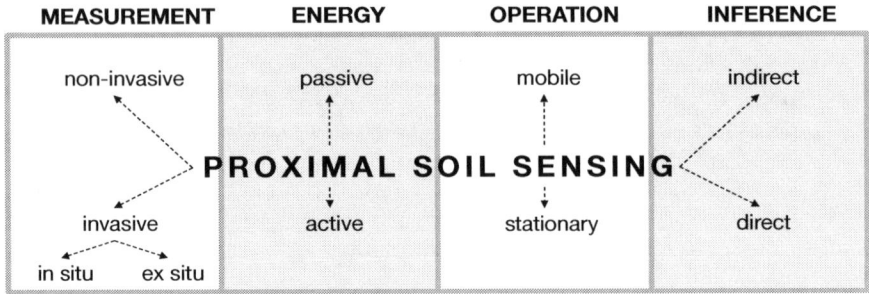

Figure 6.1   Categorization of proximal soil sensors. Proximal soil sensors may be described by their measurement as: invasive and *in situ* or *ex situ*, or non-invasive, their energy source as active or passive, their operation as stationary or mobile and their method for inferring soil properties as direct or indirect.

Source: from Viscarra Rossel *et al.*, 2011.

For instance, a proximal soil sensor is said to be invasive if during measurement there is sensor-to-soil contact, otherwise it is non-invasive. If measurements are invasive then the sensors may be further described as *in situ* (i.e. the measurements are made within the soil) or ex situ (i.e. the measurements are made on excavated soil, e.g. measurements on soil cores). Proximal soil sensors may be described as being mobile, in which case they measure soil properties while moving or 'on-the-go' (Adamchuk *et al.*, 2004a), or they may be stationary, whereby measurements are made in a fixed position and possibly at different depths. A proximal soil sensor that produces its own energy from an artificial source for its measurements is said to be active. It is passive if it uses naturally occurring radiation from the sun or earth. If the measurement of the target soil property is based on a physical process, then the proximal soil sensor is said to be direct, but when the measurement is of a proxy and inference is with a pedotransfer function, then the proximal soil sensor is indirect. For example, measurements with a resistivity proximal soil sensor that uses coulters inserted into the ground (e.g. VERIS, MUCEP, AARP) are invasive and *in situ*, the sensor uses an active source of energy, it has mobile operation and, depending on the soil property, inference might be either direct (e.g. electrical conductivity) or indirect (e.g. clay content). Measurements of an extracted soil core with a portable X-ray fluorescence proximal soil sensor might be, depending on the measurement setup, invasive and *ex situ* or non-invasive. The sensor uses an active source of energy, it has stationary operation and inference might be either direct or indirect. Measurements with a γ-radiometer are non-invasive, they use a passive source (naturally occurring radioisotopes of Cs, K, U, Th), operation is often mobile, although stationary measurements are also possible, and inference is mostly indirect.

The decision of where to measure using proximal soil sensors will depend on whether the proximal sensor is described as direct or indirect and stationary or mobile. If the sensor measures the target soil property directly and at fixed locations, the sampling problem will be the same as conventional spatial soil sampling because it requires optimization of the geographical extent of the measurements. If the sensor measurements are direct and are made with a mobile, on-the-go system, then the sampling problem might relate to the frequency (or resolution) of the measurements so as to optimize the amount of information collected. If the measurements are made indirectly, a calibration will need to be developed (using the sensor's measurements and soil samples collected and analysed in the laboratory) to predict the target property from the sensor measurements. In this case, a calibration sampling design that optimizes coverage of property (or feature) space will be required. Ideally, the sampling should also cover geographic space so that landscape position and other location-induced phenomena are included in the calibrations. De Gruijter *et al.* (2010) describe geographical and property space sampling with proximal soil sensors for fine-resolution soil mapping, and Adamchuk *et al.* (2011a) compare designs for mobile PSS that consider geographic and property space, field boundaries and other transition zones.

In this chapter, we provide a description of technologies that might be used for PSS to aid precision agriculture and illustrate the use of several in two case studies.

Proximal soil sensors

Here we describe some of the currently available technologies for PSS and follow the framework that uses the electromagnetic (EM) spectrum (Figure 6.2) for doing so (Viscarra Rossel *et al.*, 2011). The rationale for using the EM spectrum as the framework is that proximal soil sensors can measure the soil's ability to accumulate and conduct electrical charge; to absorb, reflect, and or emit EM energy; to release ions and to resist mechanical distortion.

Gamma rays

Gamma rays contain a very large amount of energy and are the most penetrating radiation from natural or man-made sources. Gamma-ray spectrometers measure the distribution of the intensity of γ radiation versus the energy of each photon. The sensors may be either active or passive. Active γ-ray sensors use a radioactive source (e.g. ^{137}Cs) to emit photons of energy that can then be detected using a γ-ray spectrometer. Passive γ-ray sensors measure the energy of photons emitted from naturally occurring radioactive isotopes of the element from which they originate. Passive γ-ray sensors have been used for mapping soil properties in PA (e.g. Viscarra Rossel *et al.*, 2007).

Inelastic neutron scattering relies on the detection of γ-rays that are emitted following the capture and re-emission of fast neutrons as the sample is bombarded with neutrons from a pulsed neutron generator. The emitted γ-rays are characteristic of the excited nuclide, and the γ-rays' intensity is directly related to the elemental content of the sample. The detectors are the same as those used in

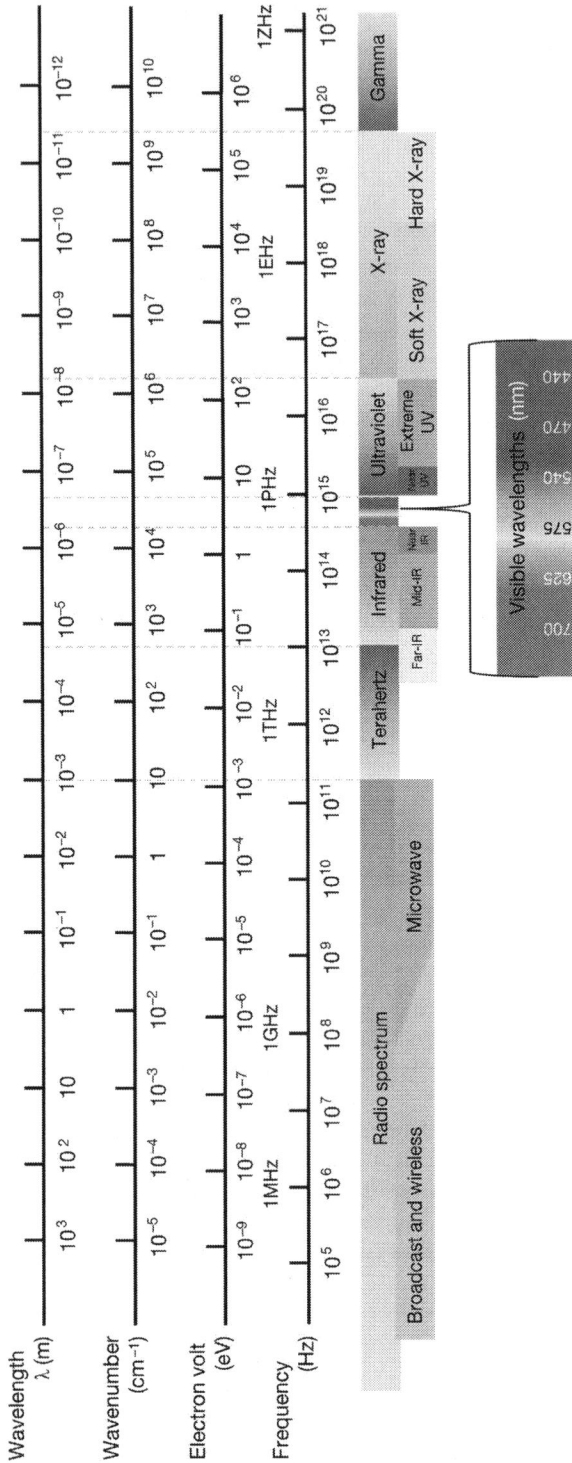

Figure 6.2 The electromagnetic (EM) spectrum.

γ-ray spectroscopy. Wielopolski *et al.* (2008) proposed it for the measurement of soil carbon and other elements.

X-rays

The two primary X-ray techniques for PSS are: X-ray fluorescence (XRF) and diffractometry (XRD). XRF relies on the fluorescence at specific energies of atoms that are excited when irradiated with X-rays. Detection of the specific fluorescent photons enables the qualitative and quantitative analysis of the elements in a sample. XRF spectroscopy has been used in the laboratory for many years and particularly for environmental analysis. XRD is a non-destructive technique used to acquire detailed information on the mineral composition of the soil. The use of portable XRD systems and research prototypes is starting to appear in the literature (Gianoncelli *et al.*, 2008). Sarrazin *et al.* (2005) describe the development and testing of a portable combined XRD and XRF system that can be used in the field. There is potential for the use of these sensors in PA, for example to measure total elemental concentrations and mineralogy, which might then be used to derive pedotransfer functions to estimate the nutrient status of soil.

Ultraviolet, visible and infrared spectra

Interest in the use of reflectance spectroscopy with ultraviolet, visible, near- and mid-infrared ranges to measure soil properties is widespread because the techniques are rapid, relatively inexpensive, need only a small amount of sample preparation, are non-destructive, require no hazardous chemicals and several soil properties can be measured from a single scan. The use of ultraviolet radiation for PSS has not been reported, but it has been used in combination with visible or infrared spectra to estimate soil attributes. Ultraviolet and visible spectra have been used to characterize inorganic species, such as iron oxides (Schwertmann and Taylor, 1989). There is a vast amount of literature on the use of visible–near-infrared (vis–NIR) and mid-infrared (mid-IR) spectra for soil analysis (e.g. Stenberg *et al.*, 2010) and, increasingly, on the use of these techniques for PSS (e.g. Ben-Dor *et al.*, 2008; Christy, 2008; Reeves *et al.*, 2010; Viscarra Rossel *et al.*, 2009; Waiser *et al.*, 2007). The mid-IR contains more information on soil mineral and organic composition than the vis–NIR, and its multivariate calibrations are generally more robust. The reason for this is that the fundamental molecular vibrations of soil components occur in the mid-IR, whereas only their overtones and combinations are detected in the NIR. Hence soil NIR spectra display fewer and much broader absorption features compared to mid-IR spectra.

Laser-induced breakdown spectroscopy (LIBS) uses an optically focused short-pulsed laser to heat the surface of the soil sample to the point of volatilization and ablation. This results in the generation of a high-temperature plasma on the surface of the sample. The plasma forms over a limited area so that only a very small amount of sample is measured during each event. As it cools, the excited atomic, ionic and molecular fragments produced in the plasma emit radiation characteristic of the elemental composition within the volatilized material. A spectrometer capable of resolving spectra in the range 200–900 nm is used to detect the emitted radiation. LIBS has been used for elemental analysis in geochemical exploration (e.g. Mosier-Boss *et al.*, 2002), for the analysis of soil carbon (e.g. Cremers *et al.*, 2001) and of other elements (Hilbk-Kortenbruck *et al.*, 2001).

Microwaves

Microwave sensors are more commonly used in remote sensing, but proximal sensors have been constructed for measuring soil water (e.g. Whalley, 1991). These sensors measure either changes in the emissivity of the soil or changes in microwave attenuation caused by changes in water content. This dependence of the soil's emissivity on its water content is due to the large contrast between the dielectric properties of free water (k' = 80), dry soil (k' = 2–5 depending on its bulk density) and air (k' = 1). As the water content of a soil increases, its dielectric constant (k') and attenuation increase, and changes in soil emissivity are produced. Therefore, microwave sensors measure the thermal radiation emitted by the soil, which is generated within the volume of the soil and is dependent on the water content (i.e. dielectric properties) and temperature of the soil.

Radiowaves

There are a number of proximal soil sensors that use radiowaves. For instance, well-established methods such as time domain reflectometry (TDR) and frequency domain reflectometry (FDR) (capacitance or electrical impedance sensors) rely on the dielectric properties of soils, that is, their permittivity, to measure water content. Noborio (2001) provides a good overview of the use of TDR for the measurement of soil water content and electrical conductivity, while Dean *et al.* (1987) describe FDR methods. Liu *et al.* (1996), Andrade-Sanchez *et al.* (2007) and Adamchuk *et al.* (2009) evaluated a dielectric-based moisture sensor for PA by incorporating the sensor into a nylon block attached to an instrumented tine.

Nuclear magnetic resonance (NMR) was used by Paetzold *et al.* (1985) to measure soil water content with a tractor-mounted NMR instrument, which detected and measured the NMR signal from the hydrogen in water. The researchers were able to measure soil water content at depths of 38, 51 and 63 mm. Their findings suggest that the NMR signal is a linear function of volumetric water content and is not affected by clay mineralogy, soil organic matter or texture.

Ground-penetrating radar (GPR) uses the transmission and reflection of high-frequency (106–109 Hz) EM waves in the soil using transmitter and receiver antennae that can be moved across the soil surface. The primary control on the transmission and reflection of the EM energy is the dielectric constant. Because of the large contrast between the dielectric constants of water, air and minerals,

GPR can be used to measure variations in soil water content (e.g. Lambot *et al.*, 2004). The GPR measurements are non-invasive and the sensors can measure soil water content of relatively large volumes of soil. The resolution of GPR images can be varied through the use of different antennae frequencies. Typically, higher frequencies increase the resolution at the expense of depth of penetration. Daniels *et al.* (1988) describe the fundamental principles of GPR. Knight (2001) provides an overview of GPR in environmental applications and Huisman *et al.* (2003) review its use for soil water determinations.

Electromagnetic induction (EMI) is a highly adaptable non-invasive technique that measures the apparent electrical conductivity of soil (EC_a). The instruments commonly have a transmitter and a receiver. Using a varying magnetic field of relatively low frequency (kHz), the technique induces currents in the ground in a way that ensures their amplitude is linearly related to the conductivity of the soil. The magnitude of these currents is determined by measuring the magnetic field that they generate; McNeill (1980) provides a good account. EMI has been used extensively in mapping soil since De Jong *et al.* (1979) first reported it. It has been particularly useful for mapping saline soil (Rhoades, 1993) and for PA (Corwin and Lesch, 2003). Because most soil and rock minerals are very good insulators, the electrical conductivity sensed by an EMI unit is electrolytic and it takes place through the pore water system. Therefore, the shape, size and connectivity of the pore system; water content; concentration of dissolved electrolytes in the soil water; temperature and phase of the pore water; and amount and composition of colloids all contribute to the signal. While clay content, electrical conductivity of the soil solution and water content are often recognized as the controlling factors that must be accounted for when calibrating EMI measurements, it is not that simple. It is the pore system and its contents rather than the clay content *per se* that should be considered. The bulk density of the soil should also be considered because it determines total porosity. Clay soil in most cropping areas usually has a substantial cation exchange capacity, CEC, and cations in solution are in equilibrium with the charged clay surface – these cations also contribute to the electrolyte concentration. Finally, colloids – particularly those associated with organic matter – may also contribute to the measured conductivity.

Magnetic, gravity and seismic sensors

Magnetic sensors, or magnetometers, measure variations in the strength of the Earth's magnetic field and the data reflect the spatial distribution of the gradient in the magnetic field near the surface. Magnetization of naturally occurring materials and rocks is determined by the quantity of magnetic minerals and by the strength and direction of the permanent magnetization carried by those minerals. Typically, magnetics has been used for the detection of geological bodies. However, there is increasing use of the technique for near-surface applications, for example, for mapping field drainage for hydrologic modelling (Rogers *et al.*, 2005), to provide a better understanding of soil genesis and formation (Mathe and Leveque, 2003), to detect anthropogenic pollution of the topsoil and for rapid identification and mapping of soil heavy metal contamination (Jordanovaa *et al.*, 2008). Similarly, there is good potential to use the technique in PA.

Gravity data can be collected using gravimeters (or gravitometers) and provide information on the local gravitational field. There are two types of gravimeter: relative and absolute. A relative gravimeter measures relative differences in the vertical component of the Earth's gravitational field based on variations in the extension of an internal spring in the gravimeter. The technique has typically been used to determine the subsurface configuration of structural basins, aquifer thickness and geological composition. An absolute gravimeter measures the acceleration of free fall of a control mass. Absolute gravimetry can be used to measure mass water balances at regional or local scales (Nabighian *et al.*, 2005).

Seismic reflection methods are sensitive to the speed of propagation of various kinds of elastic waves. The elastic properties and mass density of the medium in which the waves travel control the velocity of the waves and can be used to infer properties of the Earth's subsurface. Reflection seismology is used in exploration for hydrocarbons, coal, ores, minerals and geothermal energy. It is also used for basic research into the nature and origin of rocks that make up the Earth's crust, and can be used in near-surface applications for engineering, groundwater and environmental survey (Harry *et al.*, 2005).

Electrical resistivity and induced polarization

Electrical resistivity (ER) can be used to determine the resistivity distribution of the measured soil volume. This can be done through either galvanic contact or capacitive coupling techniques. Galvanic contact-based measurements of ER usually require four electrodes: two to inject the current (current electrodes) and two to measure the resulting potential difference (potential electrodes). The ER of the soil is determined from this and measurements of the apparent electrical conductivity (EC_a) are possible because resistivity is the reciprocal of conductivity. The soil properties that affect measurements of soil with EMI instruments also affect resistivity measurements. Samouëlian *et al.* (2005) review the use of ER in soil science.

Induced polarization (IP) is essentially an extension of the four-electrode resistivity technique. It operates by first injecting an electric current between a current electrode pair and the resulting voltage induced in the soil is measured between a potential electrode pair. However, IP captures both the charge loss (conduction) and the charge storage (polarization) characteristics of the soil.

Ion-sensitive electrodes (ISEs) and ion-sensitive field effect transistors (ISFETs)

Ion-sensitive electrodes (ISEs) are potentiometric sensors that use ion-selective membranes to measure the concentration of the target species. Ion-sensitive electrodes for measuring soil pH and many useful soil nutrients (nitrate, sodium, potassium, calcium) are commercially available and phosphate-selective electrodes for soil phosphorus are also being developed (Kim *et al.*, 2007).

Ion-sensitive field effect transistors (ISFETs) combine ISE technology with that of the field effect transistor (FET). Key advantages of pH ISFETs over standard glass pH electrodes are their small size, increased durability, fast response and the ability to mass produce them using microelectronic manufacturing techniques. They have been used for proximal sensing of soil pH (Viscarra Rossel and Walter, 2004) and lime requirement (Viscarra Rossel *et al.*, 2005). The ISFETs can be chemically modified by depositing membrane layers on the oxide surface of the FET to produce CHEMFETs selective for other ionic species. CHEMFETs selective for nitrate, calcium and potassium have been developed and evaluated for use in soil nutrient sensing (Artigas *et al.*, 2001; Birrell and Hummel, 2001).

Metal electrodes are also being explored for PSS applications to address a need for increased physical durability. Antimony electrodes are being researched as a durable alternative to glass electrodes in direct contact soil pH measurement (Adamchuk and Lund, 2008; Viscarra Rossel and McBratney, 1997). Kim *et al.* (2007) explored the use of cobalt rod-based ISEs for measuring soil phosphates.

Mechanical sensors

Another family of proximal soil sensors quantifies soil properties by measuring the mechanical interaction between the sensor and the soil (Hemmat and Adamchuk, 2008). Although there are no widely used commercial systems, a number of prototypes are being developed and include mechanical, acoustic and fluid permeability sensors.

Soil strength, or mechanical resistance to failure, has been widely used to estimate the degree of soil compaction. Soil compaction and soil strength can be measured using tine-based sensors (e.g. Hayhoe *et al.*, 2002). A method to determine soil physical properties using specific draft measurements was proposed by van Bergeijk *et al.* (2001). Penetration resistance of soil is relatively easy to measure and is governed by several soil properties, including shear strength, compressibility and friction between the soil and the metal. Numerous tip-based penetrometers have been developed including the standardized vertically operating cone penetrometer and other single- and multiple-tip horizontal and vertical soil impedance sensors. While a vertically operated sensor provides the conventional means for measuring soil strength, horizontally operated tip-based sensors have been used for mobile, on-the-go sensing.

Liu *et al.* (1993) tested an acoustic method for determining soil texture. A shank with a rough surface and hollow cavity was equipped with a microphone that recorded the sound produced through the interaction of soil and shank. The frequency of the resulting sound was used to distinguish different types of soil. In a system developed by Tekeste *et al.* (2002), sound waves were used to detect compaction layers. A small microphone installed inside a horizontal cone attached to a tine was pulled through the soil. The amplitude of sound in a selected frequency range was compared to the cone index obtained at different depths in the soil profile. The instrument could successfully detect a prepared hard pan at a particular depth; however, the authors needed to account for background noise in both studies.

Multisensor systems

Each of the sensing technologies described above has strengths and weaknesses and there is no one single sensor that can measure all soil properties. Therefore, before designing an experiment with proximal soil sensors, the selection of a complementary set of sensors to measure the required suite of soil properties is important. Integrating multiple proximal soil sensors into a single multi-sensor platform can provide a number of operational benefits over single-sensor systems, such as more robust operational performance, improved confidence as independent measurements are made on the same soil, extended attribute coverage and increased dimensionality of the measurement space (e.g. different sensors measure various portions of the EM spectrum).

Case study: sensing soil pH

Among different practices for applying agricultural inputs according to local needs, variable-rate liming is one of the most promising technologies (Bongiovanni and Lowenberg-DeBoer, 2000). However, the quality of variable-rate lime prescription maps remains one of the main considerations when it comes to non-uniform treatments of fields with spatially variable soil acidity (Bianchini and Mallarino, 2002). The most widespread practice used to prescribe variable-rate liming in the USA, for instance, has been grid sampling. Typically, one sample analysed by a laboratory represents at least 1 ha of land. As one sample per hectare can be insensitive to short-range soil variability, researchers have explored opportunities to map soil pH and lime requirement economically at a finer resolution (Viscarra Rossel *et al.*, 2005). As an alternative, on-the-go soil sensors offer the potential to increase the number of samples taken at a relatively low cost.

Based on the experience gained during the early development of a field prototype system for mapping soil nitrate content and pH (Loreto and Morgan, 1996), a follow-up research study was undertaken to investigate the applicability of flat-surface combination ion-selective electrodes (ISEs) to measure soil pH on moist soil samples directly (Adamchuk *et al.*, 2005). The initial results showed good correlation with conventional laboratory measurements ($r^2 > 0.92$), and a prototype automated system for mapping soil pH on-the-go was developed and tested by Adamchuk *et al.* (1999). A commercial implement that uses this approach was developed (Veris® Mobile Sensing Platform, Veris Technologies, Inc., Salina, Kansas, USA). While being pulled across a field, the system collects a core of soil at a predefined depth and brings it into direct contact with two moist ISEs used to conduct the measurements. As soon as electrode measurements become stable, the next sample is being collected and the electrodes are rinsed with water. This sensor (Figure 6.3) can be operated at 8 km h^{-1} using 20-m transects (distance between passes) while conducting measurements every 10 s, which results in more than 20 measurements per hectare. The complete system is typically equipped with additional sensors, such as apparent soil electrical conductivity (EC_a) and optical soil reflectance (not shown in Figure 6.3) sensors as well as an RTK-level (Real Time Kinematic) global navigation satellite system (GNSS) receiver for mapping field elevation.

Figure 6.3 Veris® Mobile Sensor Platform.

Adamchuk *et al.* (2007) showed that soil pH maps produced from data collected by on-the-go sensing are more accurate than those obtained using traditional grid-sampling methods. An agroeconomic analysis of automated soil pH mapping has shown that higher resolution maps (even with relatively low accuracy of individual measurements) can reduce overall errors considerably and result in potentially greater profitability with variable-rate liming (Adamchuk *et al.*, 2004b). Adamchuk *et al.* (2011b) have further evaluated the accuracy of lime prescription maps obtained using different mapping techniques. Several maps obtained for one production site used in this study are shown in Figure 6.4.

The 57.7-ha field (Figure 6.4) has relatively large variability. From the southeast corner to the northwest corner of the field the land is high and the soil comprises Belfore, Belfore-Moody and Nora-Moody silty clay loams. In the northeast and southwest corners are low-lying Judson silt loams, which are less well drained than the soil of the higher ground. The sloping interfaces in the northeast comprise Moody silty clay loams and Nora silt loams and in the southwest, eroded Crofton silt loams. The poorly drained lower reaches of the field contrast sharply with the higher centre band of the field. All the soil in the field is classified as prime farmland or farmland of statewide importance, with the exception of the Crofton silt loams. The typical soil pH of the upper 20 cm is 6.5 for Belfore, Belfore-Moody, Moody, Judson and Nora soil types. Nora-Moody silty clay loams on the extreme eastern edge of the field have a typical pH of 6.7 in the top 20 cm, whereas in the Crofton soil it is 7.9.

Acidity in the soil can be thought of as having two forms: active and reserve. Active acidity consists of free-moving hydrogen ions, which bond quickly with

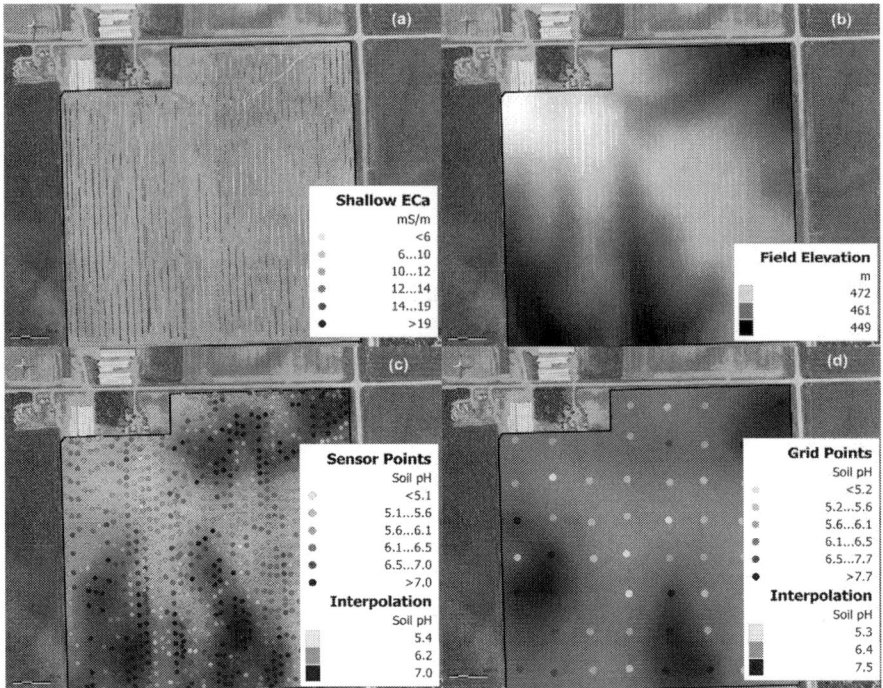

Figure 6.4   Maps of an agricultural field obtained using the sensing system shown in Figure 6.3: (a) EC$_a$, (b) elevation, (c) pH and (d) soil pH map produced from 1-ha square grid field sampling.

any available bases. Reserve acidity consists of hydrogen ions, which are bound only loosely. When the active acidity is neutralized with a base, the loosely bound hydrogen ions break their bonds and become active, preserving an acid equilibrium. Since a simple (water) pH measurement relates only to active acidity, more information is needed to determine a soil's lime requirement. When using on-the-go sensing, this is accomplished through multiple linear regressions that rely on combining sensor-based soil pH measurement and topsoil apparent electrical conductivity to predict buffer pH, which essentially relates to the lime requirement. Guided soil sampling and data analysis is an important part of the overall sensor calibration and validation practice (Adamchuk *et al.*, 2011a). Viscarra Rossel *et al.* (2005) in their prototype on-the-go sensing system made measurements of a lime requirement buffer to derive estimates of lime requirement directly to obtain a target pH.

Although this site represents a case with a reasonable correspondence between sensor-based and grid-based soil pH maps as well as between water and buffer soil pH, field locations with the greatest discrepancy between the maps revealed over 5 Mg ha^{-1} lime prescription error as a result of the lack of sampling density. There is marked loss of detail with the sparse sampling. When using validation samples, the mean absolute error was over 0.5 Mg ha^{-1} for grid-based mapping and 1.7 Mg ha^{-1} for uniform field treatment. Sensor-based mapping, on the other hand, produced a mean absolute error of 0.4 Mg ha^{-1}. In addition to variable-rate liming, both elevation and apparent soil electrical conductivity maps have been used for additional interpretation of field heterogeneity related to water and nutrient storage capacity, which typically (without irrigation) result in variable-rate seed and fertilizer management.

Case study: mapping soil organic carbon with a multisensor system

In this case study we demonstrate the use of a mobile multisensor system (MSS) equipped with an EM-38 (Geonics Limited, Mississauga, Ontario, Canada), a γ-radiometer RS700 (Radiation Solutions, Ontario, Canada) and a real-time kinematic global positioning system together with measurements of visible–near-infrared spectra with a portable spectrometer (Analytical Spectral Devices, Boulder, Colorado, USA) at selected sampling locations to map soil organic carbon of the 0–5 cm layer. The study site is located at CSRIO's Ginninderra experimental farm in the Australian Capital Territory. It is around 400 ha and is used commercially and for experimental research. The MSS is shown in Figure 6.5a. It was driven at around 3 m s^{-1} and data from the EM-38 and γ-ray sensors were recorded together with geographic position and elevation with the RTK-GPS at a frequency of 1 Hz. The EM-38 measured inphase and conductivity at two depths, 0–50 cm and 0–100 cm, whereas the γ-radiometer measures 512 channels of information from which the four regions of interest, total dose, potassium, uranium and thorium, are calculated. The locations of the fine-resolution sensor measurements are shown in Figure 6.5b.

The sensor data were analysed by deriving histograms and calculating the correlations between them. We removed outliers and other spurious data before further analysis. The uranium and thorium channels from the γ-ray sensor were filtered with a moving average to improve the signal to noise ratio. Variograms for the sensor data were derived and used for interpolation by using ordinary kriging (Webster and Oliver, 2001) on to a 5-m grid. The digital elevation map was used to derive terrain derivatives such as slope, aspect and plan curvature. The resulting maps are shown in Figure 6.6. They were used to design the soil sampling and as covariates for the mapping of soil organic carbon.

(a) γ-radiometer   Logger   RTK GPS   EM-38   (b)

Figure 6.5   (a) Mobile multisensor system (MMS) equipped with a γ-radiometer, a real-time kinematic global positioning system and an electromagnetic induction sensor and (b) sample of the sensor data collected by the system.

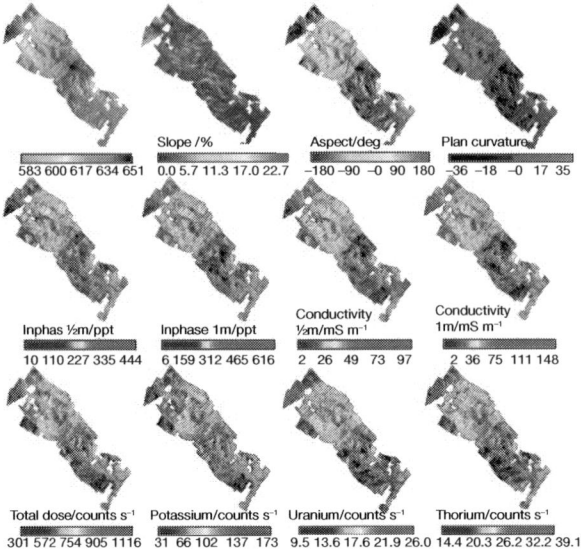

Figure 6.6   Interpolated maps of the sensor data collected by the mobile multisensor system (MMS) and derived terrain attributes.

A sampling design that helps with the calibration of soil sensors and also provides good spatial coverage was used to select 20 sampling sites (Figure 6.7a). At each location, soil was collected from the 0–5 cm layer, air-dried, crushed and measured with a portable vis–NIR spectrometer (Analytical Spectral Devices, Boulder, Colorado). The Australian vis–NIR spectroscopic database (Viscarra Rossel and Webster, 2012) was used to estimate soil organic carbon content from the 20 spectra (Figure 6.7b). The average soil organic carbon content of the 20 estimates was 1.33 per cent, their standard deviation was 0.92 per cent and the data ranged from 0.21 to 2.98 per cent.

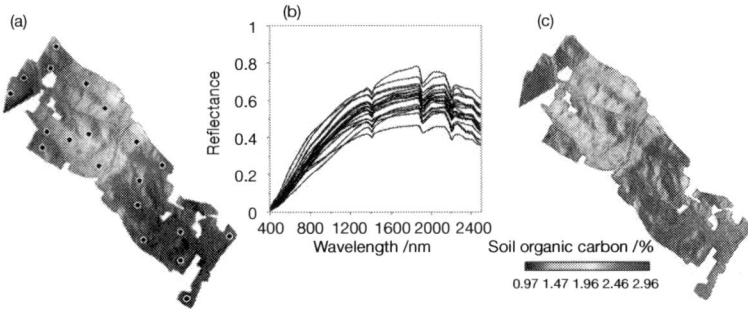

Figure 6.7   (a) Sampling locations, (b) the visible–near-infrared spectra of the 20 samples and (c) map of soil organic carbon content for the 0–5 cm layer.

Here we use a stepwise multiple linear regression (Equation 6.1) to model the 20 organic carbon estimates as a function of the covariates derived from the multisensor platform and the digital elevation map (Figure 6.6):

$$\hat{y} = b_0 + b_1\mathbf{x}_1 + b_2\mathbf{x}_2 + b_3\mathbf{x}_3 + \ldots + b_k\mathbf{x}_k, \tag{6.1}$$

where $\hat{y}$ is the soil organic carbon content being predicted at a pixel, the $\mathbf{x}$s are the independent variables representing the proximally sensed covariates 1,2,3, …, k and the bs are the coefficients of the multiple linear regression with b_0 being the constant. We used a combination of forward selection and backward elimination of the variables and selected a model that was parsimonious.

The selected variables for the multiple linear regression were elevation, total count, slope, potassium, plan curvature, inphase 0–50 cm and conductivity 0–50 cm, which produced an overall R^2 value of 0.46. The map of soil organic carbon content is shown in Figure 6.7c. It shows that there is more organic carbon present in the 0–5 cm layer at higher elevations where the soil is shallow and weathered, and where the land is used mainly for animal grazing on unimproved pastures. The average carbon content of the farm in the 0–5 cm layer was 1.46 per cent.

Discussion and conclusions

This chapter provides a summary of the most relevant technologies that might be used for proximal soil sensing in precision agriculture. The fundamental scientific principles of these sensors are not necessarily new, but recent developments have increased the range of potential applications. Some of the technologies have improved considerably, for example the development of vis–NIR array based detectors that increase instrument portability and ruggedness. Research has also refined our understanding of how we can best apply these sensors to measure soil properties. Our ability to extract useful information from the sensed data and to analyse them spatially and temporally has improved because of advances in mathematical, statistical and computational methods and also because soil scientists are now more quantitative.

The two case studies above illustrated how proximal soil sensors might be used to derive information on soil pH and organic carbon. The first case study uses a mobile platform equipped with ISEs to measure soil pH. The measurements with this system are invasive and *ex situ*, the energy used is passive, the operation is mobile and the inference is direct (Figure 6.1). The estimates of soil organic carbon were derived using a combination of sensors. The sensors on the MMS are non-invasive, passive sensors that are used both to derive the sampling design and as covariates in the mapping. The vis–NIR spectroscopy was conducted in the laboratory and estimates of organic carbon were derived using a large Australian spectroscopic database. The vis–NIR sensor might be classified as invasive and *ex situ*, having an active energy source, operation was stationary and inference indirect.

Proximal soil sensing provides soil scientists with an effective approach that can be used to learn more about the soil and so improve management in terms of economic benefits to the farmer and reduced environmental impacts from farming activities. Proximal soil sensors allow rapid and inexpensive collection of precise, quantitative data at fine (spatial and temporal) resolutions that can be used in more meaningful analyses to provide a better understanding of the soil and the spatio-temporal variation of its properties. The efficiency with which PSS can obtain soil data makes it naturally suited to many situations that require large amounts of quantitative data at fine spatial and or temporal resolutions, for example digital soil mapping, soil monitoring, precision agriculture, the assessment of contaminated sites, measurements of subsurface hydrology, archaeology and so on. We need proximal soil sensing to devise sustainable solutions to some of the issues that we face today: food, water and energy security, climate change, land degradation and so on. We hope that this chapter and the literature that we cite raise awareness about proximal soil sensing to further research and development as well as to encourage the use of proximal soil sensors in different applications.

References

Adamchuk, V. I., Morgan, M. T. and Ess, D. R. (1999) 'An automated sampling system for measuring soil pH', *Transactions of ASAE*, 42, 885–91.

Adamchuk, V. I., Hummel, J. W., Morgan, M. T. and Upadhyaya, S. K. (2004a) 'On-the-go soil sensors for precision agriculture', *Computers and Electronics in Agriculture*, 44, 71–91.

Adamchuk, V. I., Morgan, M. T. and Lowenberg-DeBoer, J. M. (2004b) 'A model for agro-economic analysis of soil pH mapping', *Precision Agriculture*, 5, 109–27.

Adamchuk, V. I. and Lund, E. D. (2008) 'On-the-go mapping soil pH using antimony electrodes', ASABE Paper 96–1087, St Joseph, MI: American Society of Agricultural and Biological Engineers.

Adamchuk, V. I., Lund, E., Sethuramasamyraja, B., Morgan, M. T. and Dobermann A. (2005) 'Direct measurement of soil chemical properties on-the-go using ion selective electrodes', *Computers and Electronics in Agriculture*, 48, 272–94.

Adamchuk, V. I., Lund, E. D., Reed, T. M. and Ferguson, R. B. (2007) 'Evaluation of on-the-go technology for soil pH mapping', *Precision Agriculture*, 8, 139–49.

Adamchuk, V. I., Hempleman, C. R. and Jahraus, D. G. (2009) 'On-the-go capacitance sensing of soil water content', ASABE Paper MC09–201, St. Joseph, MI: American Society of Agricultural and Biological Engineers.

Adamchuk, V. I., Viscarra Rossel, R. A., Marx, D. B. and Samal, A. K. (2011a) 'Using targeted sampling to process multivariate soil sensing data', *Geoderma*, 163, 63–73.

Adamchuk, V. I., Jonjak, A. K., Wortmann, C. S., Ferguson, R. B. and Shapiro, C. A. (2011b) 'Case studies on the accuracy of soil pH and lime requirement maps', in J. V. Stafford (ed.) *Precision Agriculture 2011, Proceedings of the 8th European Conference on Precision Agriculture*, Czech Centre for Science and Society, Prague, Czech Republic, pp. 289–301.

Andrade-Sanchez, P., Upadhyaya, S. K. and Jenkins, B. M. (2007) 'Development, construction, and field evaluation of a soil compaction profile sensor', *Transactions ASABE*, 50, 719–25.

Artigas, J., Beltran, A., Jimenez, C., Baldi, A., Mas, R., Dominguez, C. and Alonso, J. (2001) 'Application of ion sensitive field-effect transistor based sensors to soil analysis', *Computers and Electronics in Agriculture*, 31, 281–93.

Ben-Dor, E., Heller, D. and Chudnovsky, A. (2008) 'A novel method of classifying soil profiles in the field using optical means', *Soil Science Society of America Journal*, 72, 1113–22.

Bianchini, A. A. and Mallarino, A. P. (2002) 'Soil-sampling alternatives and variable-rate liming for a soybean–corn rotation', *Agronomy Journal*, 94, 1355–66.

Birrell, S. J. and Hummel, J. W. (2001) 'Real-time multi ISFET/FIA soil analysis system with automatic sample extraction', *Computers and Electronics in Agriculture*, 32, 45–67.

Bongiovanni, R. and Lowenberg-DeBoer, J. (2000) 'Economics of variable rate lime in Indiana', *Precision Agriculture*, 2, 55–70.

Christy, C. D. (2008) 'Real-time measurement of soil attributes using on-the-go near infrared reflectance spectroscopy', *Computers and Electronics in Agriculture*, 61, 10–19.

Corwin, D. L. and Lesch, S. M. (2003) 'Application of soil electrical conductivity to precision agriculture: theory, principle and guidelines', *Agronomy Journal*, 95, 455–71.

Cremers, D. A., Ebinger, M. H., Breshears, D. D., Unkefer, P. J., Kammerdiener, S. A., Ferris, M. J., Catlett, K. M. and Brown, J. R. (2001) 'Measuring total soil carbon with laser-induced breakdown spectroscopy (LIBS)', *Journal of Environmental Quality*, 30, 2202–6.

Daniels, D. J., Gunton, D. J. and Scott, H. F. (1988) 'Introduction to subsurface radar', *Radar Signal Processing: IEEE Proceedings F*, 135, 278–320.

Dean, T. J., Bell, J. P. and Baty, A. T. P. (1987) 'Soil moisture measurement by an improved capacitance technique: Part 1. Sensor design and performance', *Journal of Hydrology*, 93, 67–78.

de Gruijter, J. J., McBratney, A. B. and Taylor, J. (2010) 'Sampling for high resolution soil mapping', in R. A. Viscarra Rossel, A. B. McBratney and B. Minasny (eds) *Proximal Soil Sensing*, New York: Springer-Verlag, pp. 3–14.

De Jong, E., Ballantyne, A. K., Cameron, D. R. and Read, D. W. L. (1979) 'Measurement of apparent electrical conductivity of soils by an electromagnetic induction probe to aid salinity surveys', *Soil Science Society of America Journal*, 43, 810–12.

Gianoncelli, A., Castaing, J., Ortega, L., Dooryhee, E., Salomon, J., Walter, P., Hodeau, J. L. and Bordet, P. (2008) 'A portable instrument for in situ determination of the chemical and phase compositions of cultural heritage objects', *X-Ray Spectrometry*, 37, 418–23.

Harry, D. L., Koster, J. W., Bowling, J. C. and Rodriguez, A. B. (2005) 'Multichannel analysis of surface waves generated during high-resolution seismic reflection profiling of a fluvial aquifer', *Journal of Environmental and Engineering Geophysics*, 10, 123–33.

Hayhoe, H. N., Lapen, D. R. and McLaughlin, N. B. (2002) 'Measurements of mouldboard plow draft: I. Spectrum analysis and filtering', *Precision Agriculture*, 3, 225–36.

Hemmat, A. and Adamchuk, V. I. (2008) 'Sensor systems for measuring spatial variation in soil compaction', *Computers and Electronics in Agriculture*, 63, 89–103.

Hilbk-Kortenbruck, A., Noll, R., Wintjens, P., Falk, H. and Becker, C. (2001) 'Analysis of heavy metals in soils using laser-induced breakdown spectrometry combined with laser induced fluorescence', *Spectrochimica Acta Part B: Atomic Spectroscopy*, 56, 933–45.

Huisman, J. A., Hubbard, S. S., Redman, J. D. and Annan, A. P. (2003) 'Measuring soil water content with ground penetrating radar: a review', *Vadose Zone Journal*, 2, 476–91.

Jordanovaa, N., Jordanovaa, D. and Tsachevab, T. (2008) 'Application of magnetometry for delineation of anthropogenic pollution in areas covered by various soil types', *Geoderma*, 144, 557–71.

Kim, H. J., Hummel, J. W., Sudduth, K. A. and Birrell, S. J. (2007) 'Evaluation of phosphate ion-selective membranes and cobalt-based electrodes for soil nutrient sensing', *Transactions ASABE*, 50, 415–25.

Knight, R. (2001) 'Ground penetrating radar for environmental applications', *Annual Review of Earth Planetary Science*, 29, 229–55.

Lambot, S., Rhebergen, J., van den Bosch, I., Slob, E. C. and Vanclooster, M. (2004) 'Measuring the soil water content profile of a sandy soil with an off-ground monostatic ground penetrating radar', *Vadose Zone Journal*, 3, 1063–71.

Liu, W., Gaultney, L. D. and Morgan, M. T. (1993) 'Soil texture detection using acoustic methods', ASAE Paper 93-1015, St. Joseph., MI: American Society of Agricultural Engineers.

Liu, W., Upadhyaya, S. K., Kataoka, T. and Shibusawa, S. (1996) 'Development of a texture/soil compaction sensor', in P. C. Robert, R. H. Rust and W. E. Larson (eds) *Proceedings of the Third International Conference on Precision Agriculture*, Minneapolis: University of Minnesota, Precision Agriculture Center, pp. 617–30.

Loreto, A. B. and Morgan, M. T. (1996) 'Development of an automated system for field measurement of soil nitrate', ASAE Paper 961087, St. Joseph, MI: American Society of Agricultural Engineers.

McNeill, J. D. (1980) *Electromagnetic Terrain Conductivity Measurement at Low Induction Numbers*, Technical Note TN-6, Mississauga, Ontario, Canada: Geonics Ltd.

Mathe, V. and Leveque, F. (2003) 'High-resolution magnetic survey for soil monitoring: Detection of drainage and soil tillage effects', *Earth and Planetary Science Letters*, 212, 241–51.

Mosier-Boss, P. A., Lieberman, S. H. and Theriault, G. A. (2002) 'Field demonstrations of a direct push FO-LIBS metal sensor', *Environmental Science Technology*, 36, 3968–76.

Nabighian, M. N., Ander, M. E., Grauch, V. J. S., Hansen, R. O., LaFehr, T. R., Li, Y., Pearson, W. C., Peirce, J. W., Phillips, J. D. and Ruder, M. E. (2005) 'The historical development of the gravity method in exploration', *Geophysics*, 70, 63ND–89ND.

Noborio, K. (2001) 'Measurement of soil water content and electrical conductivity by time domain reflectometry: a review', *Computers and Electronics in Agriculture*, 31, 213–37.

Paetzold, R. F., Matzkanin, G. A. and De Los Santos, A. (1985) 'Surface soil-water content measurements using pulsed nuclear magnetic resonance techniques', *Soil Science Society of America Journal*, 49, 537–40.

Reeves, J. B., McCarty, G. W. and Hively, W. D. (2010) 'Mid- versus near-infrared spectroscopy for on-site analysis of soil', in R. A. Viscarra Rossel, A. B. McBratney and B. Minasny (eds) *Proximal Soil Sensing*, New York: Springer-Verlag, pp. 133–42.

Rhoades, J. D. (1993) 'Electrical conductivity methods for measuring and mapping soil salinity', *Advances in Agronomy*, 49, 201–51.

Rogers, M. B., Cassidy, J. R. and Dragila, M. I. (2005) 'Ground-based magnetic surveys as a new technique to locate subsurface drainage pipes: a case study', *Applied Engineering in Agriculture*, 21, 421–6.

Samouëlian, A., Cousin, I., Tabbagh, A., Bruand, A. and Richard, G. (2005) 'Electrical resistivity survey in soil science: a review', *Soil Tillage Research*, 83, 173–93.

Sarrazin, P., Blake, D., Feldman, S., Chipera, S., Vaniman, D. and Bish, D. (2005) 'Field deployment of a portable X-ray diffraction/X-ray fluorescence instrument on Mars analog terrain', *Powder Diffraction*, 20, 128–33.

Schwertmann, U. and Taylor, R. M. (1989) 'Iron oxides', in J. B. Dixon and S. B. Weed (eds) *Minerals in Soil Environments*, 2nd edn, Madison, WI: Soil Science Society of America, pp. 379–438.

Stenberg, B., Viscarra Rossel, R. A., Mouazen, A. M. and Wetterlind, J. (2010) 'Visible and near infrared spectroscopy in soil science', *Advances in Agronomy*, 107, 163–215.

Tekeste, M. Z., Grift, T. E. and Raper, R. L. (2002) 'Acoustic compaction layer detection', ASAE Paper 02-1089, St. Joseph, MI: American Society of Agricultural Engineers.

Van Bergeijk, J., Goense, D. and Speelman, L. (2001) 'Soil tillage resistance as tool to map soil type differences', *Journal of Agricultural Engineering Research*, 79, 371–87.

Viscarra Rossel, R. A. and McBratney, A. B. (1997) 'Preliminary experiments towards the evaluation of a suitable soil sensor for continuous "on-the-go" field pH measurements', in J. V. Stafford (ed.) *Precision Agriculture '97*, Oxford: Bios Scientific Publishers, pp. 493–502.

Viscarra Rossel, R. A. and McBratney, A. B. (1998) 'Laboratory evaluation of a proximal sensing technique for simultaneous measurement of clay and water content', *Geoderma*, 85, 19–39.

Viscarra Rossel, R. A. and Walter, C. (2004) 'Rapid, quantitative and spatial field measurements of soil pH using an ion sensitive field effect transistor', *Geoderma*, 119, 9–20.

Viscarra Rossel R. A. and Webster, R. (2012) 'Predicting soil properties from the Australian soil visible–near infrared spectroscopic database', *European Journal of Soil Science*, 63, 848–60.

Viscarra Rossel, R. A., Gilbertsson, M., Thylén, L., Hansen, O., McVey, S. and McBratney, A. B. (2005) 'Field measurements of soil pH and lime requirement using an on-the-go soil pH and lime requirement measurement system', in J. V. Stafford (ed.) *Precision Agriculture '05*, Wageningen, The Netherlands: Wageningen Academic Publishers, pp. 511–20.

Viscarra Rossel, R. A., Taylor, H. J. and McBratney, A. B. (2007) 'Multivariate calibration of hyperspectral g-ray energy spectra for proximal soil sensing', *European Journal of Soil Science*, 58, 343–52.

Viscarra Rossel, R. A., Cattle, S., Ortega, A. and Fouad, Y. (2009) 'In situ measurements of soil colour, mineral composition and clay content by vis–NIR spectroscopy', *Geoderma*, 150, 253–66.

Viscarra Rossel R. A., Adamchuk, V. I., Sudduth, K. A., McKenzie, N. J., Lobsey, C. (2011) 'Proximal soil sensing: updating the pedologist's toolkit', *Advances in Agronomy*, 113, 237–82.

Waiser, T. H., Morgan, C. L. S., Brown, D. J. and Hallmark, C. T. (2007) 'In situ characterization of soil clay content with visible near-infrared diffuse reflectance spectroscopy', *Soil Science Society of America Journal*, 71, 389–96.

Webster, R. and Oliver, M. A. (2001) *Geostatistics for Environmental Scientists*, Chichester, UK: John Wiley & Sons.

Whalley, W. R. (1991) 'The development and evaluation of a microwave soil moisture sensor for incorporation in a narrow tine', *Journal of Agricultural Engineering Research*, 50, 25–32.

Wielopolski, L., Hendrey, G., Johnsen, K. H., Mitra, S., Prior, S. A., Rogers, H. H. and Torbert, H. A. (2008) 'Nondestructive system for analyzing carbon in the soil', *Soil Science Society of America Journal*, 72, 1269–77.

7 Spatio-temporal analysis to improve agricultural management

Ana Horta and Thomas F. A. Bishop

Introduction

Site-specific crop management is concerned with adapting management to the variation of the crop-growing environment. For most applications and research (see Chapters 4 and 12, for example) the focus has been on the spatial variation of the crop-growing environment, for example different soil types may be present across a field. Each of these soil types may have different nutrient- and water-holding capacities, different yield expectations and therefore require different agronomic inputs. However, variation in time can be equally important for properties such as soil moisture, crop nutrient status and soil available nutrients. For example, the different soil types will, as a result of their different water-holding capacities, respond differently to rainfall or irrigation in terms of soil moisture and nutrient cycling, and ultimately yield (Lark *et al.*, 1998). Therefore, not only is it important to understand the spatial variation within fields, it is equally important to understand the temporal variation to achieve more precise management of the crop-growing environment.

Chapter 10 presents an example of variable-rate irrigation (VRI) where amounts of irrigation were varied according to the soil water-holding capacity of different management zones. The timing of the irrigation was not varied, that is, irrigation occurred at the same time for all locations. However, with good space-time predictions of soil moisture, irrigation amounts could be applied in the amount and at the time needed to maintain soil water at a particular level suitable for plant growth. This would improve the efficiency in the use of irrigation water.

Despite the potential use of space-time prediction, a problem with current remote and proximal sensors (Chapters 5 and 6) is that although they can record dense spatial data, recordings are not repeated often enough through time to provide intensive temporal data. This of course assumes that they actually measure a property of interest directly, for example crop nutrient status. In contrast, sometimes a few sensors (e.g. moisture sensors) are deployed in a field that measure at an intensive temporal resolution, but there is little spatial coverage (see Chapter 10). In the future it is expected that dense datasets in both space and time will be available, but complete spatial and temporal coverage of the crop-growing environment is unlikely, and some form of modelling and prediction in space and time will be required to provide that cover. Space-time geostatistics, as an extension to spatial geostatistics (Chapter 4) through the incorporation of a temporal component, can be applied to fill in the gaps.

In this chapter, a soil moisture dataset is used to illustrate how to apply space-time geostatistics to explain the joint variation in space and time of a property of interest. Using the space-time patterns of soil moisture as a basis for precision management is still a recent application in agriculture. This is particularly important for farmers who wish to adopt variable-rate irrigation.

Space-time modelling: a brief review

Here we define a space-time dataset as $z(s,t)$, which is a set of n observations of the continuous attribute z that is characterized by a spatial coordinate $s = (x,y)$ and a time coordinate t. The attribute z varies spatially for a time t when the measurements are made. These measurements are then repeated for other time points. The collection of all measurements is our space-time database. The attribute $z(s,t)$ can be any property of interest such as soil water content (Heuvelink *et al.*, 1996; Snepvangers *et al.*, 2003), nutrient concentration (Marchant *et al.*, 2012) or crop biomass (Heuvelink and van Egmond, 2010).

As in spatial geostatistics, we need a model to obtain a prediction of $z(s,t)$ at a location (s_0,t_0) at which z was not observed. To define our model, z is assumed to be a realization of a spatio-temporal random function (RF) model, $Z(s,t)$. In space-time geostatistics, $Z(s,t)$ can be decomposed into a trend component $m(s,t)$ and a residual component $R(s,t)$ (Kyriakidis and Journel, 1999):

$$Z(s,t) = m(s,t) + R(s,t). \tag{7.1}$$

The trend component models the underlying variation of the spatio-temporal process $Z(s,t)$ or the part of $Z(s,t)$ that can be explained using a physically based process model (one that can be explicitly stated as a known mathematical function) or described using available covariates, such as remotely (Chapter 5) or proximally (Chapter 6) sensed data. For most case studies, the easiest approach is to develop a regression-type model which relates the variable of interest (z) empirically to relevant covariates.

The residual component, $R(s,t)$, represents the difference between the observations and the predictions from the trend model. If the residuals are correlated in space and or time, then neighbouring observations in space and or time can be used to improve the final predictions (Heuvelink and van Egmond, 2010). To assess space-time correlation, the residuals are usually considered to be a realization of an intrinsically stationary random function, as explained in Chapter 4. This assumption implies that the space-time variogram depends solely on the separating lags between observations h in space and u in time. In Chapter 4 we saw that a spatial variogram model can be estimated by first making point estimates of the variogram for different lag distances and then fitting an authorized variogram model to these estimates. Space-time point estimates of the variogram can be estimated by:

$$\gamma\,(h,u) = \frac{1}{2 \cdot N(h,u)}\ \sum_{i=1}^{N(h,u)} [R(s,t) - R(s + h,t + u)]^2, \tag{7.2}$$

where $N(h,u)$ is the number of pairs of observations of z separated by lag (h,u) with h the separating distance between pairs of points in space and u the separation in time. One example of an authorized space-time variogram model is the sum-metric variogram model (Dimitrakopoulos and Luo, 1994), where $\gamma(s,t)$ is the sum of spatial (γ_s), temporal (γ_t) and spatio-temporal variogram (γ_{st}) structures (Equation 7.3).

$$\gamma\,(s,t) = \gamma_s(h) + \gamma_t(u) + \gamma_{st}\left(\sqrt{h^2 + (a \cdot u)^2}\right). \tag{7.3}$$

Each structure is itself an authorized variogram model expressed in terms of parameters such as the nugget variance, structural variance and range parameter as described in Chapter 4 and case study 1. The term γ_{st} represents the joint space-time structure for which a geometric anisotropy ratio a is computed to equate separation in space to separation in time. There are other space-time variogram models available in the literature as referred to in Kyriakidis and Journel (1999) and Cressie and Wikle (2011) (and references therein). Once the space-time variogram is modelled, we can predict the residuals by ordinary kriging (see Chapter 4). Finally, we can use Equation (7.1) to obtain the final space-time predictions by adding our trend predictions to the kriged residuals.

Case study: space-time modelling of soil moisture

The case study chosen is the Tarrawarra Soil Moisture Dataset (Western and Grayson, 1998). We first provide a description of the study site and then describe exploratory data analysis of the soil moisture data. A trend model is presented based on available covariates that are considered appropriate to explain soil moisture variation in space and in time. The space-time correlation of the residuals is modelled in order to improve the accuracy of predictions of soil moisture. Finally, the main patterns of soil moisture distribution are discussed based on the predicted maps for each time period.

Description of study site

The Tarrawarra catchment (southern Victoria, Australia; Figure 7.1) is an experimental site covering 10.8 ha that was monitored extensively for soil moisture variation between September 1995 and March 1997 (http://people. eng.unimelb.edu.au/aww/tarrawarra/tarrawarra.html). A detailed description of the dataset is published in Western and Grayson (1998). The Tarrawarra catchment is representative of landscapes with relatively shallow soil, in which topography plays a significant role in routing water through the landscape, and of climates ranging from temperate to sub-humid (Western *et al.*, 1999). Land use consists of perennial improved pastures for dryland dairy-cattle grazing.

The Tarrawarra dataset consists of 13 time-domain reflectometry (TDR) soil moisture surveys. All surveying was carried out using GPS total stations (Western and Grayson, 1998). Soil moisture surveys were performed using TDR probes inserted to a depth of 30 cm. The majority of the TDR surveys (11 out of 13 dates) were performed on the same 10 × 20 m grid with the 20 m axis oriented perpendicular to the main drainage direction. In this study we do not consider the two remaining surveys, as the measurements were not made at the same locations, that is, they are not collocated in space. The spatial distribution of moisture values per survey date (*t*) is shown in Figure 7.2. A graphical summary of soil moisture values per sampling period is presented in Figure 7.3 in the form of boxplots. The central box of the boxplot represents the middle 50 per cent of observations and the horizontal line within each box is the median soil moisture for each sampling period. Variation in soil moisture between sampling campaigns is clearly evident with larger moisture contents occurring from July to September and smaller values from November to March. Moisture was most variable in terms of the range (difference between maximum and minimum) in September of 1995 and 1996. The smallest individual moisture values were measured in February 1996 and in November 1996, whereas the largest were in September 1996.

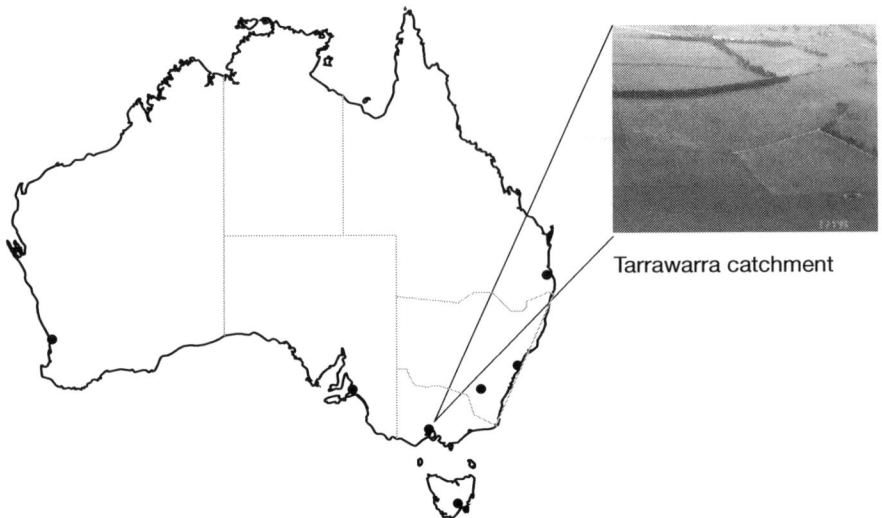

Tarrawarra catchment

Figure 7.1   Location of the Tarrawarra catchment (Western and Grayson, 1998).

Soil moisture

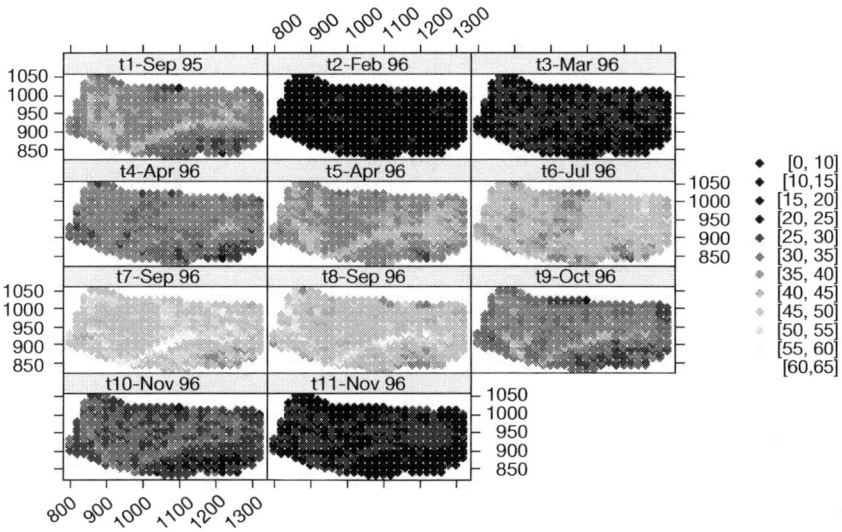

Figure 7.2   Posting maps of soil moisture values from each survey.

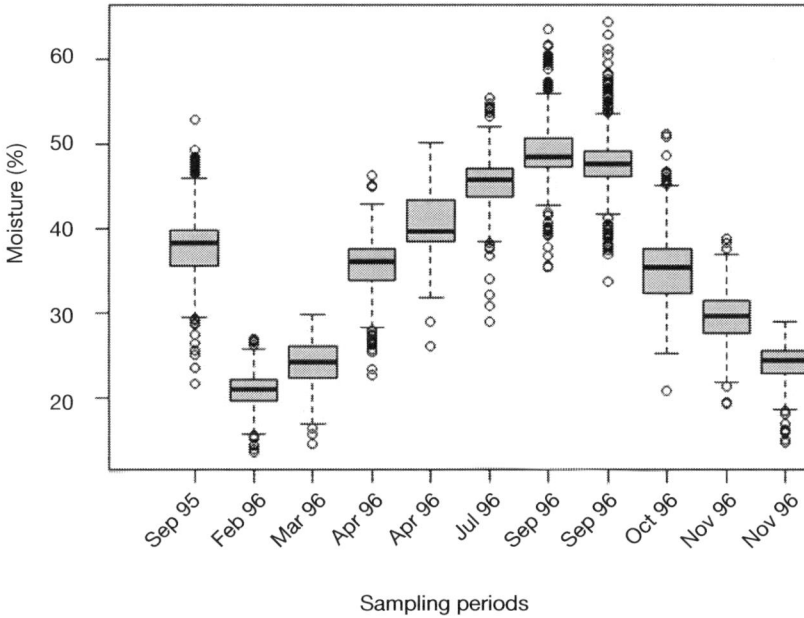

Figure 7.3   Boxplots of soil moisture measurements for each survey.

Modelling the trend

In this case study we used multiple linear regression to model the trend component of the space-time model (Equation 7.1). Our initial selection of a pool of potential covariates was based on availability and whether there was a physical basis for using them. For example, the spatial pattern of soil moisture variation has been correlated with topography (elevation and slope) and soil texture (Starr, 2005). Topography controls the flow of water and its accumulation across the landscape, whereas texture accounts for the spatial variation in infiltration rates and water-holding capacities. Clearly rainfall is related to soil moisture, and as in the case of the Tarrawarra catchment previous studies (Western *et al.*, 1999) have found that average soil moisture levels are generally high during winter and low during summer, showing a possible association with rainfall, which is greater in winter.

A digital elevation model (DEM) on a 5-m grid was available for the Tarrawarra catchment and from this slope and topographic wetness index (TWI) were calculated. The TWI combines slope and upslope area to represent where water accumulates in the landscape. For example, steep upper slopes with little upslope area will have small TWI values as water would not accumulate in these locations. On the other hand, flat regions at the base of slopes would have larger TWI values as the upslope area is large and the slope is small so water would accumulate in these locations. Moore *et al.* (1991) provide a classic review of digital terrain modelling. In addition to terrain attributes, a map of clay content was available for the study site. Therefore, the potential covariates to explain the spatial variability of soil moisture were elevation, slope, TWI and clay content (Figure 7.4). The DEM and slope maps reveal an area of low elevation and slope where water is expected to accumulate preferentially. This pattern is clearer in the TWI map where the highest values are represented. Clay content varies across the study area, reflecting a somewhat different pattern compared to those evident for the variation in elevation, slope and TWI.

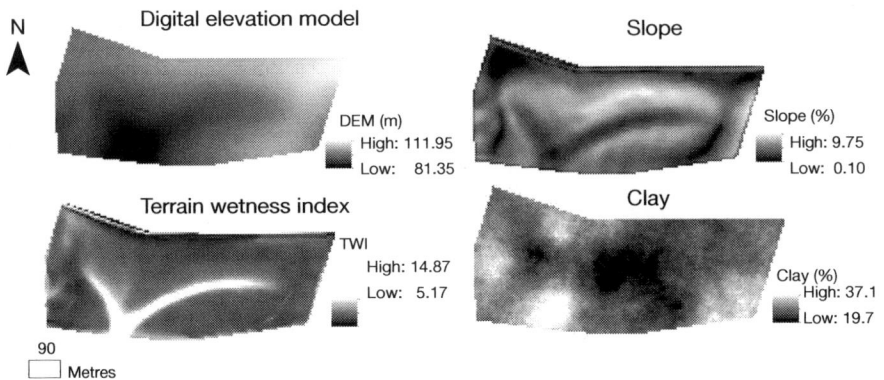

Figure 7.4   Spatially varying properties to be considered as covariates in the space-time trend model.

A time-series of daily rainfall was the only potential covariate that was available for modelling the variation in soil moisture over time. Since soil moisture at any time depends on current and previous rainfall we need to transform the daily rainfall measurements to account for this. To account for this factor we applied an idea presented by Wang *et al.* (2011) that used a discounted variable for stream flow to model suspended sediment concentrations. Similar to rainfall and soil moisture, the effect of stream flow on sediment concentration is not solely dependent on the current flow but also on the cumulative effect of prior (or antecedent) flow conditions. In this case a discounted variable means that a weighted average of rainfall from the present and past observations is calculated where the weight is decreased exponentially the further in the past the observation was made; in other words the current rainfall is allocated the largest weight. We considered different discount factors (df) of 0.5, 0.7, 0.9, 0.95, 0.99 and 0.999. The df controls the rate per day at which the weighting decays with time in the past. For example, with a df of 0.5, the rainfall of two days ago gets a weighting of $0.5^2 = 0.25$ when calculating the average rainfall. These range in interpretation from being equal to the long-term average of rainfall (df = 0.999) to being equal to the weighted average of rainfall of the past few days (df = 0.5). The values considered for rainfall and corresponding discounted rainfall variables are presented in Figure 7.5 for each sampling period (*t*1 to *t*11).

It should be noted that none of the covariates used vary in both space and time, they are either uniform in space and vary in time (rainfall) or uniform in time and vary in space (elevation). One example of a covariate that varies in space and time would be a time-series of remotely sensed images from which the greenness of the vegetation cover could be calculated as this relates to evapotranspiration, which is indirectly related to soil moisture.

The optimal combination of covariates in the multiple linear regression model used to represent the trend was determined by backward elimination. This

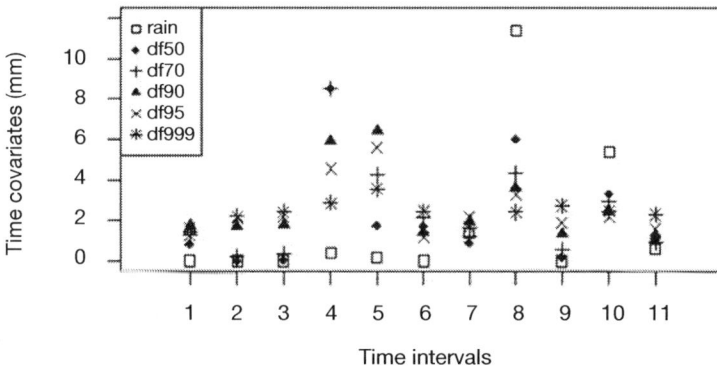

Figure 7.5 Temporally varying properties to be considered as covariates in the space-time trend model.

involves fitting an initial model with all potential covariates and sequentially removing the least statistically significant covariate one at a time. This process ends when only statistically significant covariates are left, and this becomes the final model. In this case study the final model was found to include all temporal covariates (rainfall and the discounted rainfall) together with TWI and slope. The final model is given in Table 7.1. The total variance explained by the model (given by the R^2 value) is equal to 60 per cent.

Figure 7.6 shows maps of soil moisture created using the space-time trend model. It is noticeable how the trend predictions reproduce the physical characteristics given by the covariates such as the larger soil moisture values in the area of low elevation in the catchment (given by the larger TWI and smaller slopes). Also, changes in moisture between wet and dry periods are influenced by the rainfall in the period leading up to the sampling event.

Space-time modelling of the residuals

Figure 7.7 shows post plots of the residuals from the space-time trend model. They are useful for two reasons. First, they can be used to identify where and when the trend model predicts poorly, for example the worst predictions are for November 1996. Second, they can be used to improve the space-time model predictions by adding the kriged residuals to the trend model to obtain improved predictions of soil moisture. It should be noted that the magnitude of the residuals [−3.5%, 3%] is quite small relative to the actual values of the soil moisture, which are typically 20 per cent to 40 per cent. This implies that the space-time trend model is a reasonable predictor of soil moisture.

The first step for space-time geostatistical modelling is to estimate the autocorrelation among the residuals in space, time and in space-time. Figure 7.8 shows the experimental marginal spatial and temporal variograms (dots) with the fitted models (solid lines). The spatial variogram has a large nugget effect (intercept on the ordinate), which can be explained by the fact that it involves pooling

Table 7.1 Statistics of multiple linear regression model

	Coefficients	Standard error	p-value
Intercept	22.2	1.03	< 0.001
Rain	4.8	0.07	< 0.001
df50	27.0	0.45	< 0.001
df70	42.1	0.76	< 0.001
df90	42.0	1.13	< 0.001
df95	30.7	1.00	< 0.001
df999	9.4	0.25	< 0.001
slope	0.5	0.05	< 0.001
TWI	1.6	0.08	< 0.001

Space-time trend

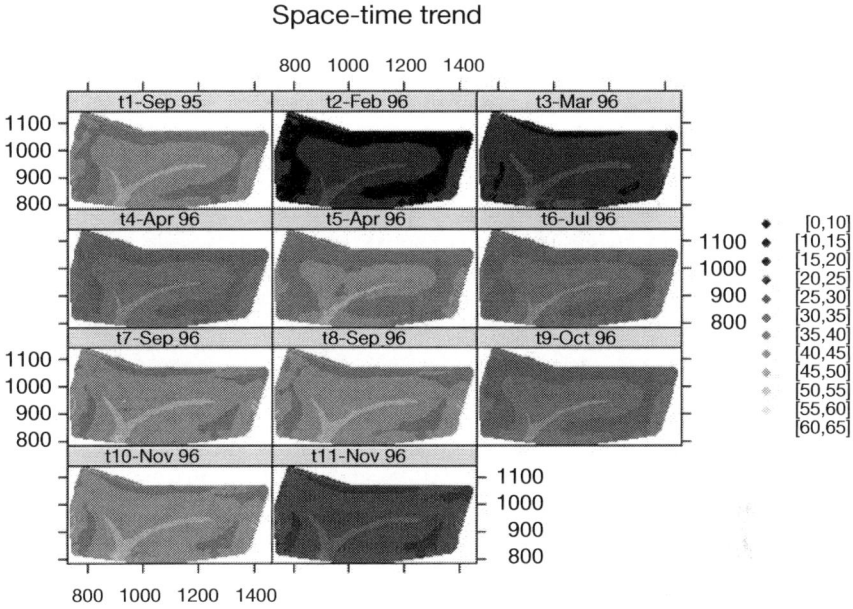

Figure 7.6 Maps of soil moisture based on the space-time trend model.

Residuals

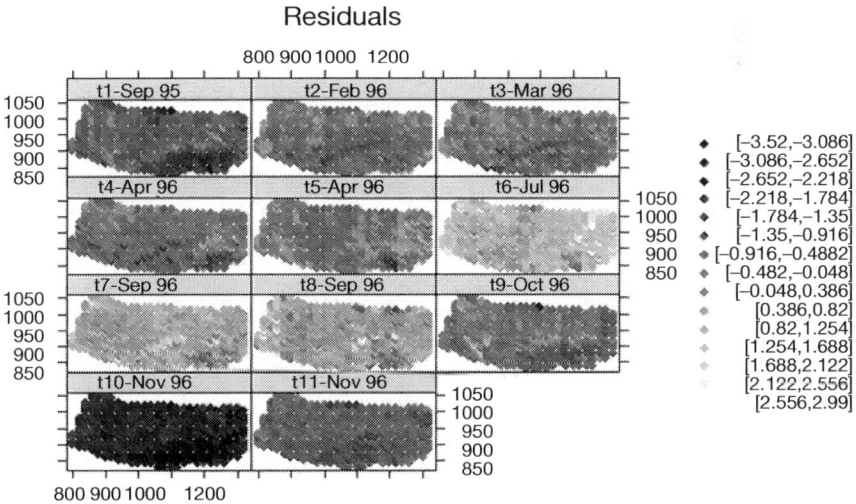

Figure 7.7 Maps of residuals from the space-time trend model.

locations at different time intervals that have very different residuals (Figure 7.7). On the other hand, the temporal variogram indicates a possible cyclic behaviour that could be explained by rainfall distribution throughout the year. The experimental space-time variogram (Figure 7.9) was obtained using the code proposed by De Cesare *et al.* (2002) to fit the space-time variogram surface. The surface obtained is consistent with the point estimates of the marginal space and time variograms in Figure 7.8. In particular, there is reasonable temporal correlation up to a range of 45 days but little spatial correlation is evident.

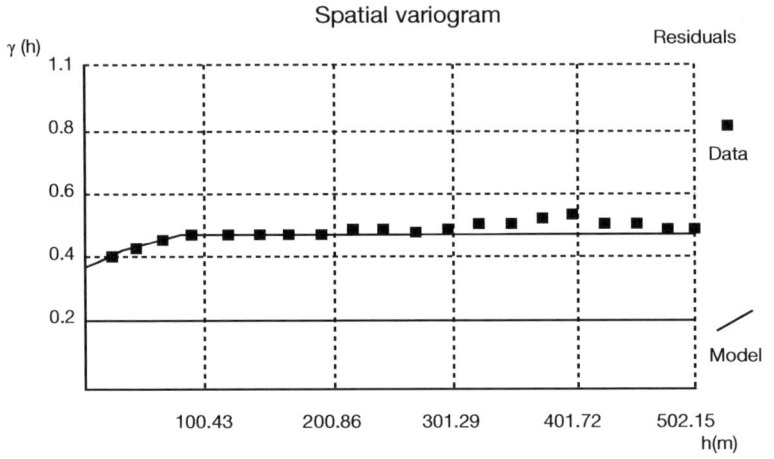

$c_0 = 0.4$, $c_1 = 0.1$, $a = 100$m, Spherical model

(a)

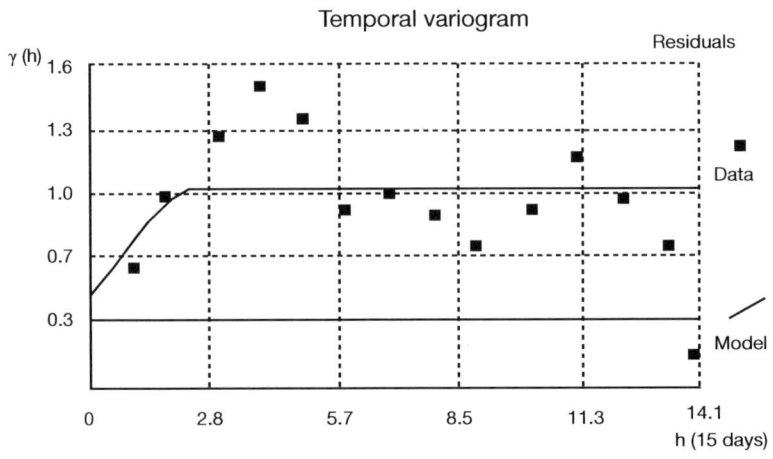

$c_0 = 0.4$, $c_1 = 0.6$, $a = 45$ days, Spherical model

(b)

Figure 7.8   Spatial (a) and temporal (b) marginal variograms of the residuals.

As mentioned above, a sum-metric model (Equation 7.3) was fitted to the experimental space-time variogram. This model was then used to predict the residuals for the entire study area by ordinary kriging (see Chapter 4). The predicted values shown in Figure 7.10 are small when compared with the trend predictions. The predictions given by the space-time trend model and the kriged residuals were summed together to produce the final maps of soil moisture (Figure 7.11). The patterns are very similar to those shown in maps based solely on the space-time trend predictions (Figure 7.6).

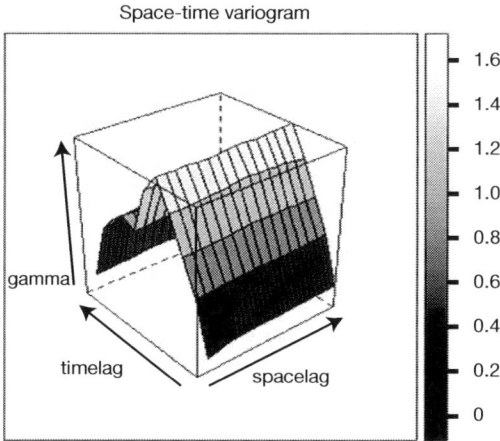

Figure 7.9 Model of space-time variogram of the residuals.

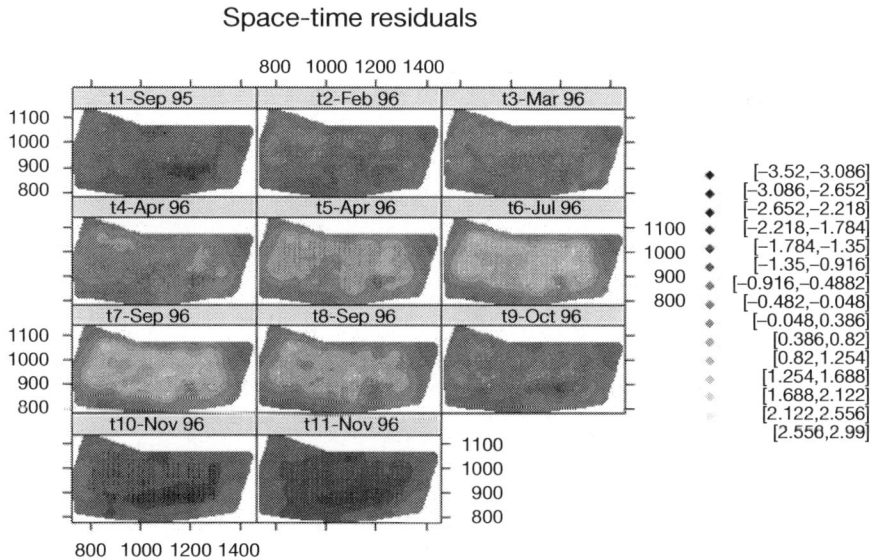

Figure 7.10 Maps of residuals produced using space-time geostatistics.

Space-time soil moisture

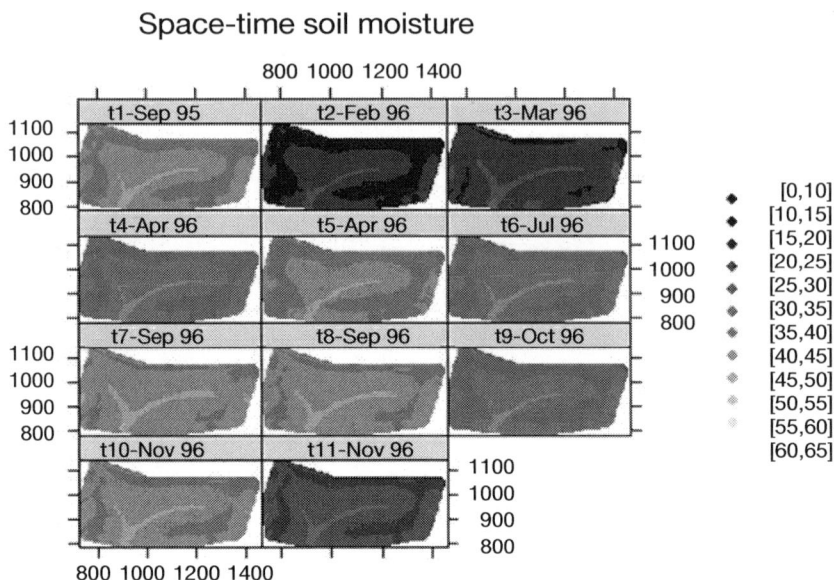

Figure 7.11 Maps of soil moisture produced using sum of trend and geostatistical model.

Discussion and conclusions

In this study we have developed a space-time model for predicting soil moisture. We found that a trend model using readily available covariates such as rainfall, elevation and associated terrain attributes explained 60 per cent of the variation in the observed data. Space-time kriging of the residuals was used to further improve the model predictions. Although space-time modelling and prediction is uncommon in research and applications of PA, it has great potential for the management of soil and crop properties that vary in time.

For example, the space-time predictions of soil moisture as presented here could be used to identify when (in time) and where (in space) irrigation water should be applied. Chapter 10 describes some of the technological limitations in precision irrigation application in terms of the distance over which irrigation rates can be varied. Aside from irrigation scheduling the analysis in this chapter could be used for management at coarser resolutions. For example, parts of a catchment that are continuously wet could have different varieties (or species) of pasture sown that are more suitable for waterlogged conditions.

One impediment to the adoption of more space-time modelling studies and its application is the lack of sensor technology to measure key soil and crop properties. Chapters 5 and 6 outline the latest developments. Clearly manual point sampling as used in this case study is not practical on a large scale, but in the near future sensor technology should enable the direct measurement of properties of interest to PA and the development of space-time models as described in the Tarrawarra case study. In addition, improvements in sensor technology should provide a rich source of useful covariates for the development of space-time trend models.

In summary, most research and applications of PA focus on how we vary management spatially according to the crop-growing environment. However, an often ignored question and perhaps the next step forward is *when* to perform a management action such as the fertilizer application or irrigation. Space-time modeling as described in this chapter can help answer this question and enable the more precise management of our agricultural resources in both space and time.

Acknowledgements

We thank Professor Gerard Heuvelink for his assistance with fitting the metric covariance function to the experimental space-time variogram.

References

Cressie, N. and Wikle, C. K. (2011) *Statistics for Spatio-Temporal Data*, Hoboken, NJ: John Wiley & Sons.

De Cesare, L., Myers, D. E. and Posa, D. (2002) 'FORTRAN programs for space-time modeling', *Computer & Geosciences*, 28, 205–12.

Dimitrakopoulos, R. and Luo, X. (1994) 'Spatiotemporal modeling: covariances and ordinary kriging systems', in R. Dimitrakopoulos (ed.) *Geostatistics for the Next Century*, Dordrecht: Kluwer Academic Publishing, pp. 88–93.

Heuvelink, G. B. M. and van Egmond, F. M. (2010) 'Space-time geostatistics for precision agriculture: a case study of NDVI mapping for a Dutch potato field', in M. A. Oliver (ed.) *Geostatistical Applications for Precision Agriculture*, Dordrecht: Springer, pp. 117–37.

Heuvelink, G. B M., Musters, P. and Pebesma, E. J. (1996) 'Spatio-temporal kriging of soil water content', in E. Y. Baafi and N. A. Schofield (eds) *Geostatistics Wollongong '98*, Dordrecht: Kluwer Academic Publishing, pp. 1020–30.

Kyriakidis, P. C. and Journel, A. G. (1999) 'Geostatistical space-time models: a review', *Mathematical Geology*, 31, 651–84.

Lark, R. M., Catt, J. A. and Stafford, J. V. (1998) 'Towards the explanation of within-field variability of yield of winter barley: soil series differences', *Journal of Agricultural Science*, 131, 409–16.

Marchant, B. P., Crawford, D. M. and Robinson, N. J. (2012) 'Spatial and temporal prediction of soil properties from legacy data', in B. Minasny, A. B. McBratney and B. P. Malone (eds) *Digital Soil Assessments and Beyond: Proceedings of the 5th Global Workshop on Digital Soil Mapping 2012*, Sydney: CRC Press, pp. 239–44.

Moore, I. D., Grayson, R. B. and Ladson, A. R. (1991) 'Digital terrain modeling: a review of hydrological, geomorphological, and biological applications'. *Hydrological Processes*, 5, 3–30.

Snepvangers, J. J. J. C., Heuvelink, G. B. M. and Huisman, J. A. (2003) 'Soil water content interpolation using spatio-temporal kriging with external drift', *Geoderma*, 112, 253–71.

Starr, G. C. (2005) 'Assessing temporal stability and spatial variability of soil water patterns with implications for precision water management', *Agricultural Water Management*, 72, 223–43.

Wang, Y.-G., Kuhnert, P. and Henderson, B. (2011) 'Load estimation with uncertainties from opportunistic sampling data: a semiparametric approach', *Journal of Hydrology*, 396, 148–57.

Western, A. W. and Grayson, R. B. (1998) 'The Tarrawarra data set: soil moisture patterns, soil characteristics and hydrological flux measurements', *Water Resources Research*, 34, 2765–8.

Western, A. W., Grayson, R. B. and Green, T. R. (1999) 'The Tarrawarra project: high resolution spatial measurement, modelling and analysis of soil moisture and hydrological response', *Hydrological Processes*, 13, 633–52.

Part 3
Management

8 Site-specific management and delineating management zones

Dennis L. Corwin

Introduction

Conventional agriculture has served humankind well. However, the limitations of conventional agricultural practices are becoming increasingly evident as the world's food producers push production to the limit with finite resources and the public closely scrutinizes the environmental ramifications of producers' efforts. Because conventional agriculture disregards the spatial and temporal complexity of the interacting factors that affect crop yield and focuses on productivity, it will not be able to meet the global economic, environmental and limited-resource challenges of the future. Barring any new technological breakthroughs, precision agriculture, or more specifically, site-specific crop management, is the next logical step to meet world food demands using state-of-the-art scientific knowledge and technology that can address these spatial and temporal complexities.

Factors responsible for within-field crop-yield variation

Ever since the classic paper by Nielson *et al.* (1973) concerning the variability of field-measured soil water properties, the significance of within-field spatial variability of soil properties has been scientifically acknowledged. However, until recently, with the introduction of global positioning systems (GPS) and yield-monitoring equipment, documentation of crop yield and soil variability at the field scale was difficult to establish. Now there is well-documented evidence that spatial variation within fields is highly significant and amounts to a factor of 3–4 or more for crops (Birrel *et al.*, 1995; Verhagen *et al.*, 1995) and up to an order of magnitude or more for soil (Corwin *et al.*, 2003a).

Spatial variation in crop yield is the result of a complex interaction of edaphic (i.e. soil-related, such as salinity, organic matter, nutrients, texture), biological (e.g. disease, pests, earthworms, microbes), anthropogenic (e.g. irrigation management, leaching efficiency, soil compaction due to farm equipment), topographic (e.g. slope, elevation, aspect) and meteorological (e.g. relative humidity, temperature, rainfall, wind) factors. All of these factors vary spatially, but some vary both temporally and spatially, resulting in complex spatial patterns that cannot be measured with fixed sensors or a single plant or soil sample.

Conventional versus site-specific management

Although it is well known that soil is spatially heterogeneous, conventional farming currently treats a field uniformly with respect to the application of fertilizer, planting density, pesticides, soil amendments, irrigation water and other inputs, which ignores the naturally inherent variation in soil and crop conditions between and within fields. Conventional agriculture, therefore, inherently under- or over-applies inputs such as irrigation water, fertilizer, pesticides and soil amendments in some parts of fields. Failure to address within-field temporal and spatial variation in edaphic properties, as well as variation in anthropogenic, biological, meteorological and topographical factors that affect crop yield, has detrimental effects on economic benefits because of reduced yield in certain areas of a field and on the environment because of over-applications of agrochemicals and water, which is a waste of finite resources. This costs the producer and the public money, depletes finite resources and degrades soil, surface-water and groundwater resources.

In contrast to conventional agriculture, site-specific crop management, or more simply site-specific management (SSM), attempts to manage soil, pests and crops based upon spatial variation within a field (Larson and Robert, 1991). Site-specific management is a form of precision agriculture where decisions on resource application and agronomic practices closely match crop requirements as they vary within a field; consequently, the collective actions are differential rather than uniform. The aims of SSM are to apply inputs (e.g. irrigation water, fertilizer, pesticides, soil amendments, etc.) *when*, *where* and in the *amount* needed to optimize crop yield.

Need for site-specific management

Although total yields continue to rise on a global basis, there is a disturbing trend with some major crops such as wheat and maize that are reaching 'yield plateaus' (World Resources Institute, 1998). The prospect of feeding a projected additional 2–3 billion people over the next 30–40 years poses greater challenges than those faced over the past 30–40 years because of finite resources. In addition there are impacts on the environment that result in unsustainability, changes in climatic conditions that threaten agriculturally productive regions and increased water scarcity. It is unlikely that conventional agriculture can solve these challenges.

In an effort to feed the world's population, conventional agriculture has affected the environment detrimentally with the loss of natural habitat, use and misuse of pesticides and fertilizers, and degradation of the soil and water resource. By 1990, poor agricultural practices had contributed to the degradation of 38 per cent of the roughly 1.5 billion ha of crop land worldwide and since 1990 the losses have continued at an annual rate of 5–6 million ha (World Resources Institute, 1998).

Sustainable agriculture is regarded as the most viable means of meeting the food demands of the world's growing population because its aim is to optimize profitability, productivity, sustainability and use of resources and reduce environmental impacts. Site-specific management is the most promising means of attaining sustainable agriculture because it addresses the weaknesses of conventional agriculture, namely spatial and temporal variation.

Components of site-specific management (SSM)

A site-specific management system consists of five fundamental components: (1) spatial referencing; (2) measurement and monitoring of crop, soil and environmental attributes; (3) attribute mapping; (4) decision support system (DSS) and (5) differential action. The technologies to bring SSM into its own fell into place in the mid 1990s, particularly with the maturation of global navigation satellite systems (GNSS), *in situ* and on-the-go sensor technologies, geographic information systems (GIS) and variable-rate technology (VRT).

Spatial referencing links data to a specific coordinate location on the Earth's surface. Global navigation satellite systems, such as the Global Positioning System (GPS), are the technology that has made geo-referencing possible with accuracy from several metres to sub-metre. Geo-referencing from GNSS made it possible to provide accurate location information that is crucial for sensor technology, GIS and VRT. In addition, geo-referencing has provided producers with (1) a navigation aid known as parallel tracking, which allows farm equipment operators to visualize their position with respect to previous passes; (2) auto-guidance to steer agricultural vehicles automatically with occasional oversight by the operator and (3) autonomous vehicles where the operator's presence is not required, allowing for the safe application of farm chemicals that are hazardous to human health.

A variety of sensors and monitors are needed for SSM to measure various crop, soil, landscape and environmental variables. These *in situ* and on-the-go sensors include sensors for crop yield and quality; crop reflectance for biomass, vigour and stress; soil attributes of apparent electrical conductivity (EC_a), reflectance, pH and natural gamma radiation emission; and elevation. With the spatial referencing capabilities of GNSS, geo-referenced attributes of crop, soil and relief are gathered from direct sensor measurements (e.g. pH, yield and elevation) or from sensor-directed soil sampling (e.g. EC_a-directed soil sampling). Yield monitoring and mapping refers to the collection of geo-referenced data on crop yield and crop characteristics, such as moisture content, while the crop is being harvested. Similarly, elevation data for a field can be recorded during planting, cultivation or harvesting using real-time kinetics (RTK) GPS equipment to provide horizontal and elevation measurements at centimetre accuracy, which are useful in auto-steering, strip tillage, drainage and digital elevation mapping (DEM). The on-the-go measurement of pH using the Veris pH Manager (Veris Technologies, Salinas, KS) is a sensor technology that is growing rapidly for variable-rate liming of field areas where pH is too low or for application of sulphur to areas that are too alkaline for the crop. Geo-referenced measurements of EC_a with electrical resistivity (e.g. Veris 3100)* or electromagnetic induction (e.g. Geonics EM38

---

* All product identification is provided solely for the benefit of the reader and does not imply endorsement by the USDA.

and DUALEM-2) have been used to map a combination of soil properties that potentially influence crop yield. Apparent soil electrical conductivity sensors are sensitive to soluble salts (i.e. salinity), clay, water content, organic matter and bulk density; consequently, the EC_a measurement is a complex combination of all these properties. Because of its complexity, geo-spatial EC_a measurements are used to direct soil sampling to reflect the range and variability of those soil properties that predominantly influence EC_a in a particular field. By applying statistical sampling designs to geo-referenced EC_a measurements, including design-based sampling (e.g. random sampling, stratified random sampling, unsupervised classification, etc.) or model-based sampling (e.g. response surface sampling), sample locations are established that will reflect the range and spatial variation of those soil properties correlated to EC_a for that field. Directed sampling using EC_a is the most widely used approach for characterizing the spatial variation of soil properties (Corwin and Lesch, 2005a). Arguably, sensors that measure EC_a provide the widest range of spatial information for SSM because EC_a is influenced by several soil-related properties. This is often the case when EC_a correlates with crop yield. However, in dryland agriculture EC_a often does not correspond to yield-limiting factors. In these cases, a strong argument can be made that crop sensors will provide the most useful information, whereas EC_a sensors will usually excel for irrigated agriculture on arid and semi-arid soils. This is clearly demonstrated by the ability of EC_a-directed sampling to provide salinity and texture maps that allow for the separation of osmotic (i.e. salinity) and matric (i.e. water content) potential effects of salt-affected soils on crop yield, which is not possible with crop sensors.

Remote sensing is also one of the more common sensor technologies used in SSM. Remote sensing is defined as the acquisition of data about an object without being in physical contact (Elachi, 1987). Satellite imagery (e.g. Landsat 5 and Landsat 7; SPOT) provides multispectral images, where the normalized differential vegetative index (NDVI) bands have been used successfully to identify soil factors that affect crops and also the nutrient status of crops. However, the remote sensing systems most common to agricultural applications are from airborne systems, which produce images with more detailed resolution than Landsat or SPOT images. There are also handheld optical sensors that belong in the remote sensing category. Handheld sensors provide the advantage of ease-of-use and of allowing measurements to be taken at a time of the user's choosing, with high resolution.

The temporal and spatial data from sensors result in extremely large datasets, therefore computer software that can compile, organize, manipulate and display attributes as maps is needed. This software is referred to as a geographic information system. A variety of commercial GISs are available; the most common of which is ESRI's ArcGIS. The current GIS software is capable of overlaying data layers (e.g. permeability, salinity, water content, clay, nitrates, pH, etc.) and of performing sophisticated spatial analysis of data that includes geostatistical techniques such as kriging or co-kriging and deterministic techniques such as inverse distance weighting and global polynomial interpolation, to mention a few. All of these capabilities of a GIS are useful in creating digital maps for SSM.

In general terms, a DSS is an information system that supports decision-making. A DSS in the context of SSM provides the means to examine the temporal and spatial variation in crop growth and yield to formulate differential actions. From a DDS, areas are established where unique treatment is needed. These areas of unique treatment are referred to as site-specific management zones or site-specific management units (SSMUs). Site-specific management units provide information on where and how much action is needed, and are regions within a field that have a relatively homogeneous combination of yield-limiting factors for which a single rate of a specific input is appropriate. Variable-rate application and variable-position technologies for the differential management of fertilizers, irrigation, pests, soil amendments and plant density use the information from SSMUs to meet optimized goals of profitability, productivity, use of resources and environmental impact.

Site-specific management units may or may not be temporally and spatially stable. An example of a temporally unstable SSMU is one that delineates an area of an irrigated field where a soil amendment, such as gypsum, is added to reduce large concentrations of Na on exchange sites as a means of improving permeability. The SSMUs will become irrelevant once site-specific applications of gypsum reduce Na levels sufficiently and increase permeability to an acceptable level. In contrast, plant available water is predominantly influenced by texture, which from an agricultural perspective is temporally stable; consequently, SSMUs for irrigation to meet plant available water needs are generally temporally and spatially stable.

To manage within-field variability site-specifically, geo-referenced information about relevant characteristics for crop production must be available. The technology for SSM is available now, but information on spatial and temporal variation is often inadequate (van Uffelen *et al.*, 1997). Yield maps provide information on the integrated effects of physical, chemical and biological processes under certain weather conditions (van Uffelen *et al.*, 1997), and provide the basis for implementing SSM by indicating where crop inputs need to be varied based upon spatial patterns of crop productivity (Long, 1998). However, the inputs necessary to optimize crop productivity and minimize environmental impacts can be derived only if the factors that gave rise to the observed spatial crop patterns are known (Long, 1998). Yield maps alone cannot provide sufficient information to distinguish between the various sources of variability, and cannot provide clear guidelines without information on the effect of variability in the weather, pests and diseases, and soil physical and chemical properties on crop yield and quality for a particular year (van Uffelen *et al.*, 1997). Each factor that affects the within-field variation in yield needs to be characterized spatially to be able to manage a crop on a site-specific basis. For this reason researchers are currently evaluating multi-sensor platforms that can provide a full spectrum of geo-referenced data for soil, crop and environment.

Delineating site-specific management units (SSMUs)

An important aspect of SSM is the delineation of site-specific management units (SSMUs). Determination of SSMUs is difficult as a result of the complex combination and interaction of factors that influence crop yield. Ideally, an SSMU will account for the spatial variation of all factors that affect the variation in crop yield, including edaphic, meteorological, biological, anthropogenic and topographic factors, and will optimize productivity, profitability, sustainability, resource utilization and environmental impacts. This has not yet been achieved completely. However, SSMUs have been defined based on edaphic and anthropogenic factors derived from EC_a-directed soil sampling (Corwin *et al.*, 2003b; Corwin and Lesch, 2005a), soil-type surveys and map overlays of topsoil depth and elevation (Kitchen *et al.*, 1998), topographic attributes and landscape position data to map productivity based on plant available water (Jones *et al.*, 1989; Jaynes *et al.*, 1994; Sudduth *et al.*, 1997) and soil fertility (Khosla *et al.*, 2002; Chang *et al.*, 2004), to mention a few.

In most instances, the delineation of SSMUs relied on crop and soil proximal sensors to establish crop inputs that are commonly applied using VRT. Common crop inputs applied using VRT include nutrients (N, P and K), manure, lime, gypsum, seeding rate, herbicides, pesticides and irrigation water. Site-specific management units have been established for each of these crop inputs using a variety of different factors. For instance, P and K management zones have been established from topography, grid or directed soil sampling, soil survey maps and EC_a maps. Manure and N management zones have been derived from soil texture, organic matter, yield patterns, bare soil photographs, NO_3-N and crop canopy reflectance. Lime management zones have come from soil pH and soil texture, while gypsum management zones have come from grower knowledge, yield patterns, EC_a maps and soil tests for pH and Na. Seeding-rate management zones have been based on historical yield maps and topsoil depth. Herbicide management zones have been derived from weed maps, soil organic matter and soil texture; and pesticide management zones have been derived from soil properties. Site-specific irrigation management units have been established from soil texture, topography, yield zones and EC_a-directed soil sampling.

Crop and soil proximal sensors for delineating SSMUs

Ground-based proximal sensors are sensors that take measurements from within a distance of 2 m from the soil surface. They may take measurements of the soil, such as electrical, electromagnetic or radiometric sensors, or of plants, such as crop yield or spectral sensors. Proximal sensors are of particular importance to site-specific management because they can obtain large volumes of reliable spatial and temporal data of soil and plant properties at relatively low cost and labour input (see Chapter 6).

Adamchuk *et al.* (2004) categorized proximal sensors into six categories: (1) electrical and electromagnetic; (2) optical and radiometric; (3) mechanical; (4) acoustic; (5) pneumatic and (6) electrochemical. Numerous reviews and technical papers have been prepared dealing with proximal sensors, with just a few of the more current ones listed in Table 8.1. Each sensor is typically affected by more than one agronomic property. Table 8.2 outlines the agronomic properties influencing each category of proximal sensor.

Table 8.1 Selected recent references using proximal soil sensors to map soil properties for applications in precision agriculture (modified from Adamchuk et al., 2004)

Category of proximal sensor	Review article	Sensor	Technical reference
Electrical and EMI	Corwin and Lesch (2005a)	ER EMI Capacitance	Corwin et al. (2003b) Corwin and Lesch (2005b, 2005c) Andrade et al. (2001)
Optical	Ben-Dor et al. (2009)[a]	Single wavelength Multispectral or Hyperspectral	Shonk et al. (1991) Maleki et al. (2008), Mouazen et al. (2007)
Radiometric	Huisman et al. (2003)	GPR Microwave	Lunt et al. (2005) Whalley and Bull (1991)
Mechanical	Hemmat and Adamchuk (2008)	Draft Load cells and penetrometers	Mouazen and Roman (2006) Chung et al. (2003), Verschoore et al. (2003)
Acoustic and pneumatic		Microphone Air pressure transducer	Liu et al. (1993) Clement and Stombaugh (2000)
Electrochemical		ISFET ISE	Birrell and Hummel (2001) Adamchuk et al. (2005), Sethuramasamyraja et al. (2008)

Notes
EMI = electromagnetic induction, ER = electrical resistivity, GPR = ground penetrating radar, ISFET = ion-selective field effect transistor, ISE = ion-selective electrode.
a   Review includes remote and proximal sensors.

Table 8.2 Soil properties influencing proximal sensors (modified from Adamchuk et al., 2004)

Category of proximal sensor	Agronomic soil property									
	Texture (sand, silt, clay content)	OM	θ	EC or Na	Cp or ρ_b	Depth of topsoil or hard pan	pH	Residual NO_3 or total N	Other macro-nutrients	CEC
Electrical and EMI	X	X	X	X	X	X	–	X	–	X
Optical and radiometric	X	X	X	–	–	–	X	X	–	X
Mechanical	–	–	–	–	X	X	–	–	–	–
Acoustic and pneumatic	X	–	–	–	X	X	–	–	–	–
Electrochemical	–	–	–	X	–	–	X	X	X	–

Notes
EMI = electromagnetic induction, OM = soil organic matter, θ = water content, EC = electrical conductivity (salinity), Na = sodium content, Cp = compaction, ρ_b = bulk density, CEC = cation exchange capacity.

To a varying extent from one field to the next, crop patterns are affected by edaphic properties. Bullock and Bullock (2000) indicated that efficient methods for measuring within-field variation accurately in soil physical and chemical properties are important for precision agriculture. No single sensor will measure all the soil properties influencing crop-yield variation; consequently, combinations of sensors have been recommended by Corwin and Lesch (2010) to add supplemental soil and plant information that can be used to better define SSMUs, resulting in a mobile multi-sensor platform. Of all of the proximal sensors, EMI and ER sensors are the most thoroughly researched and commonly used proximal sensors for the measurement of edaphic properties influencing crop yield (Corwin and Lesch, 2003, 2005a).

Case study: delineation of SSMUs with proximal sensor-directed sampling

In a strict sense, the task of delineating SSMUs is complicated because all edaphic, anthropogenic, topographic, biological and meteorological factors influencing a crop's yield must be considered, and the SSM goals of productivity, profitability, sustainability, resource use and environmental impacts must also be taken into account. One means of simplifying the complexity of delineating SSMUs is to confine the goal to crop productivity and to define SSMUs based on one or two factors, such as edaphic and anthropogenic properties, and determine the extent of the variation in crop yield due to these factors as done by Corwin *et al.* (2003b) and Corwin (2005).

Measurements of EC_a have been used to characterize spatial variation in soil salinity, nutrients (e.g. NO_3^-), water content, texture-related properties, bulk density-related properties (e.g. compaction) and leaching and organic matter related properties (Corwin and Lesch, 2005a). As pointed out by Corwin and Lesch (2003), if crop yield correlates with EC_a, then EC_a is measuring one or more soil properties that directly or indirectly influence crop yield. Corwin (2005) hypothesized that if EC_a correlates with crop yield then it can be used to develop a crop-yield response model that can delineate SSMUs. The following case study describes the EC_a-directed soil sampling methodology for delineating SSMUs based on this hypothesis.

Study site

A 32.4-ha irrigated field in the Broadview Water District of the San Joaquin Valley's west side in central California (approximately 100 km west of Fresno) was used as the study site. The soil at the site is a Panoche silty clay (thermic Xerorthents), which is slightly alkaline with good surface and subsurface drainage. The subsoil is thick, friable, calcareous and easily penetrated by roots and water. Cotton was grown at the study site in 1999. In the arid southwestern USA the primary soil properties affecting cotton yield are salinity, soil texture and structure, plant-available water, trace elements (particularly B) and ion toxicity from Na^+ and Cl^- (Tanji, 1996).

EC_a-directed soil sampling protocols

The spatial variation of properties thought to affect cotton yield was characterized following the general EC_a-directed soil sampling survey protocols developed by Corwin and Lesch (2005b,c). The basic elements of a field-scale EC_a survey specifically applied to precision agriculture include: (1) site description and EC_a survey design; (2) geo-referenced EC_a data collection; (3) soil sampling strategies based on geo-referenced EC_a data; (4) soil sample collection; (5) physical and chemical analysis of pertinent soil properties; (6) statistical and spatial analysis; (7) geographic information system (GIS) database development and (8) approaches for delineating SSMUs. The basic steps within each component are outlined in Table 8.3 and discussed in detail in Corwin and Lesch (2005b). The following describes the steps for the delineation of SSMUs at the Broadview Water District study site.

Site description, yield monitoring and EC_a survey (steps 1 and 2)

Site description is the first step and involves the recording of metadata to define site boundaries and the location of control points, which provide useful information for yield monitoring and EC_a survey. Decisions regarding EC_a measurement intensity should be based on the project objectives and scale of the project (e.g. field, landscape, basin or regional scale). For instance, a field-scale project of 30–40 ha would have measurements taken more closely together than a landscape-scale project of thousands or tens of thousands of hectares. Once this preliminary information is gathered, yield monitoring and EC_a data collection can begin.

Spatial variation of cotton yield was measured at the study site in August 1999 using a four-row cotton picker equipped with a yield sensor and global positioning system (GPS). A total of 7,706 cotton yield readings were recorded (Figure 8.1a). Each yield observation represented an area of approximately 42 m². From August 1999 to April 2000 the field was fallow.

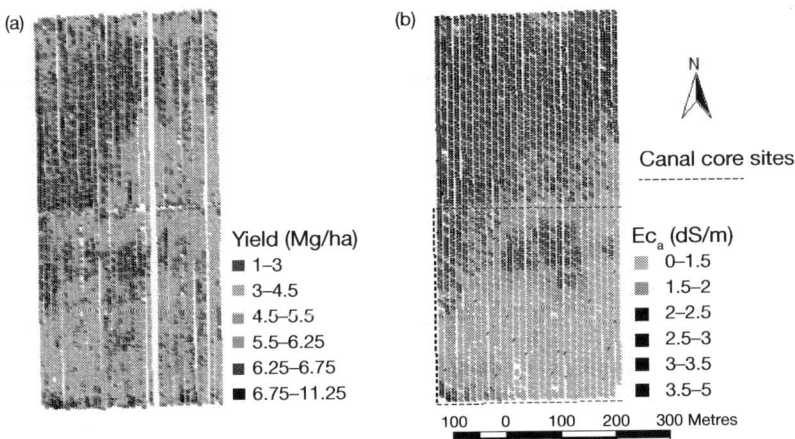

Figure 8.1 Maps of (a) cotton yield and (b) EC_a measurements including 60 soil sampling sites. Modified from Corwin *et al.* (2003b) with permission.

Table 8.3 Outline of steps for an EC$_a$ field survey for precision agriculture applications (modified from Corwin and Lesch 2005b)

---

1   *Site description and EC$_a$ survey design*
    (a) record site metadata
    (b) define project's or survey's objective
    (c) establish site boundaries
    (d) select GPS coordinate system
    (e) establish EC$_a$ measurement intensity

2   *EC$_a$ data collection with mobile GPS-based equipment*
    (a) geo-reference site boundaries and significant physical geographic features with GPS
    (b) measure geo-referenced EC$_a$ data at the pre-determined spatial intensity and record associated metadata

3   *Soil sampling strategies based on geo-referenced EC$_a$ data*
    (a) statistically analyse EC$_a$ data using an appropriate statistical sampling design to establish the soil sample site locations
    (b) establish sampling depth, sample depth increments and number of cores per site

4   *Soil core sampling at specified sites designated by the sample design*
    (a) obtain measurements of soil temperature through the profile at selected sites
    (b) at randomly selected locations obtain duplicate soil cores within a 1-m distance of one another to establish local-scale variation of soil properties
    (c) record soil core observations (e.g. mottling, horizonation, textural discontinuities, etc.)

5   *Laboratory analyses of appropriate soil physical and chemical properties defined by project objectives*

6   *Statistical and spatial analyses to determine the soil properties that affect EC$_a$ and crop yield (provided EC$_a$ correlates with crop yield)*
    (a) perform a basic statistical analysis of physical and chemical data by depth increment and by composite depths
    (b) determine the correlation between EC$_a$ and physical and chemical soil properties by depth increment and by composite depths
    (c) determine the correlation between crop yield and physical/chemical soil properties by depth and by composite depths to determine depth of concern (i.e. depth with consistently highest correlation, whether positive or negative, of soil properties to yield) and the soil properties that have a significant effect on crop yield (or crop quality)
    (d) conduct an exploratory graphical analysis to determine the relationship between the significant physical and chemical properties and crop yield (or crop quality)
    (e) formulate a spatial linear regression (SLR) model that relates soil properties (independent variables) to crop yield or crop quality (dependent variable)
    (f) adjust this model for spatial autocorrelation, if necessary, using residual maximum likelihood (REML) or some other technique
    (g) conduct a sensitivity analysis to establish dominant soil property influencing yield or quality

7   *GIS database development and graphic display of spatial distribution of soil properties*

8   *Select approach for delineating site-specific management unit*

In March 2000 an intensive geo-referenced EC$_a$ survey (Figure 8.1b) was conducted. Such surveys can be done with either mobile electromagnetic induction (EMI) or mobile fixed-array electrical resistivity (ER) equipment. Figure 8.2 shows mobile EMI (Figure 8.2a) and mobile fixed-array ER (Figure 8.2b) equipment developed by Rhoades and colleagues at the US Salinity Laboratory (Rhoades, 1992a,b; Carter *et al.*, 1993). The EC$_a$ survey conducted in March 2000 used the mobile fixed-array ER equipment shown in Figure 8.2b. The fixed-array ER electrodes were spaced to measure EC$_a$ to a depth of 1.5 m, which is roughly the depth of the root zone. Over 4,000 EC$_a$ measurements were recorded. Step 2 of the protocols outlined in Table 8.3 provides the basic procedure followed for the EC$_a$ survey. Details are provided in Corwin and Lesch (2005b).

Figure 8.2 Mobile GPS-based EC$_a$ measurement equipment: (a) electromagnetic induction (EMI) rig with a close-up of the sled holding the EMI unit and (b) electrical resistivity (ER) rig with a close-up of one of the ER electrodes.

Sample site selection, soil sampling and soil analysis (steps 3, 4 and 5)

Sample site selection is based on spatial variation of the geo-referenced EC_a measurements. Corwin and Lesch (2010) used a response surface sampling design to minimize the number of sites needed to characterize the spatial variation in this case study. This sampling design is available in ESAP software developed at the US Salinity Laboratory (Lesch *et al.*, 2000). Sites are chosen with ESAP's response surface sampling design to (1) represent about 95 per cent of the observed range in the bivariate EMI survey data; (2) represent the average of the EC_a readings for the entire field and (3) be spatially distributed across the field to minimize any clustering. In other words, ESAP creates a three-dimensional surface of the EC_a measurements and uses the variation in that surface to select sites that meet these criteria. In most instances, the number of sites selected by ESAP is not the minimum (i.e. six sampling locations per field), but is based primarily on the resources available to conduct soil sample analyses.

Following the EC_a survey at the Broadview site, soil samples were collected at 60 locations based on the ESAP-95 version 2.01 software (Lesch *et al.*, 2000) analysis of the EC_a survey data. These sample locations reflect the observed spatial variation in EC_a while simultaneously maximizing the spatial uniformity of the sampling across the study area. Figure 8.1b shows the distribution of EC_a survey data in relation to the locations of the 60 soil sampling sites. Soil core samples were taken at each site at 0.3-m increments to a depth of 1.8 m: 0–0.3, 0.3–0.6, 0.6–0.9, 0.9–1.2, 1.2–1.5 and 1.5–1.8 m. The soil samples were analysed for soil properties thought to affect cotton yield, including pH, boron (B), nitrate nitrogen (NO_3-N), Cl^-, salinity (EC_e), leaching fraction (LF; defined as the fraction of applied water at the soil surface that drains beyond the root zone), gravimetric water content (θ_g), bulk density (ρ_b), % clay and saturation percentage (SP). All samples were analysed following the methods outlined in Agronomy Monograph 9, Part 1 (Blake and Hartage, 1986) and Part 2 (Page *et al.*, 1982).

Statistical and spatial analysis (step 6)

Statistical analyses were conducted using SAS software (SAS Institute, 1999). They consisted of three stages: (1) determination of the correlation between EC_a and cotton yield with data from the 60 sites; (2) exploratory statistical analysis to identify the significant soil properties affecting cotton yield and (3) development of a crop-yield response model by ordinary least squares regression adjusted for spatial autocorrelation with restricted maximum likelihood (REML).

Correlation between crop yield and EC_a

The locations of EC_a and cotton yield measurements did not overlap exactly; therefore, ordinary kriging was used to determine the expected cotton yield at the 60 sites. The spatial correlation structure of yield was modelled by an isotropic exponential variogram. The following fitted exponential variogram was used to describe the spatial structure at the study site:

$$v(h) = (0.76)^2 + (1.08)^2 [1 - \exp(-h/109.3)], \tag{8.1}$$

where h is the lag distance.

The fitted variogram model was used with the data to estimate cotton yield at the 60 sites by kriging. The correlation of EC_a to yield at the 60 sites was 0.51. The moderate correlation between yield and EC_a suggests that one or more soil properties that influence EC_a also affect cotton yield making the EC_a-directed soil sampling strategy a viable approach at this site. The similarity of the spatial distributions of cotton yield (Figure 8.1a) and EC_a measurements (Figure 8.1b) visually confirms the reasonably close relationship of EC_a to yield.

Exploratory statistical analysis

Exploratory statistical analyses were used to reduce the number of potential soil properties influencing cotton yield and to establish the general form of the cotton yield response model. The exploratory statistical analyses comprised a preliminary multiple linear regression (MLR) analysis, a correlation analysis and scatter plots of yield versus potentially significant soil properties. The preliminary MLR and correlation analyses were used to establish the significant soil properties influencing cotton yield, whereas the scatter plots were used to formulate the general form of the cotton yield response model.

Both preliminary MLR and correlation analysis showed that the 0–1.5 m depth increment resulted in the best correlations and best fit of the data to cotton yield; consequently, the 0–1.5 m depth increment was considered to correspond to the active root zone. The correlations between cotton yield and soil properties in Table 8.4 show strong correlations of θ_g, EC_e, B, % clay, Cl^-, LF and SP with cotton yield. The preliminary MLR analysis indicated that the following soil properties were most significantly related to cotton yield: EC_e, LF, pH, % clay, θ_g and ρ_b. Chlorine, B and SP were eliminated from the MLR analysis because of multicollinearity between B and EC_e, Cl^- and LF, and SP and % clay and there were no direct cause-and-effect relationships between cotton yield and B, Cl^- and SP over the ranges of measurements found (see Corwin *et al.*, 2003b, for a detailed explanation of the issue of multicollinearity).

A scatter plot of EC_e and yield indicates a quadratic relationship where yield increases up to a salinity of 7.17 dS m^{-1} and then decreases (Figure 8.3a). The scatter plot of LF and yield shows a negative curvilinear relationship (Figure 8.3b). Yield shows a minimal response to LF below 0.4 and falls off rapidly for LF > 0.4. Clay percentage, pH, θ_g and ρ_b appear to be linearly related to yield to various degrees (Figures 8.3c,d,e and f, respectively). Even though there was clearly no correlation between yield and pH ($r = -0.01$; see Figure 8.3d), pH became significant in the presence of the other variables, which became apparent in both the preliminary MLR and in the final yield response model.

Table 8.4 Simple correlation coefficients between EC$_a$ and soil properties and between cotton yield and soil properties. Soil properties were a composite sample from 0–1.5 m (modified from Corwin *et al.*, 2003b)

Soil property[a]	Fixed-array EC$_a$[b]	Cotton yield[c]
θ_g	0.79	0.42
EC$_e$	0.87	0.53
B	0.88	0.50
pH	0.33	−0.01
% clay	0.76	0.36
ρ_b	−0.38	−0.29
NO$_3$-N	0.22	−0.03
Cl$^-$	0.61	0.25
LF	−0.50	−0.49
SP	0.77	0.38

Notes
a   Properties averaged over 0–1.5 m.
b   Pearson correlation coefficients based on 60 observations.
c   Pearson correlation coefficients based on 59 observations.
θ_g = gravimetric water content; EC$_e$ = electrical conductivity of the saturation extract (dS m^{-1}); LF = leaching fraction; SP = saturation percentage.

Based on the exploratory statistical analysis it became evident that the general form of the cotton yield response model was:

$$Y = \beta_0 + \beta_1(EC_e) + \beta_2(EC_e)^2 + \beta_3(LF)^2 + \beta_4(pH) + \beta_5(\% \text{ clay}) + \beta_6(\theta_g) + \beta_7(\rho_b) + \varepsilon, \tag{8.2}$$

where, based on the scatter plots of Figure 8.3, the relationships between cotton yield (Y) and pH, % clay, θ_g and ρ_b are assumed linear; the relationship between yield and EC$_e$ is assumed to be quadratic; the relationship between yield and LF is assumed to be curvilinear; $\beta_0, \beta_1, \beta_2, ..., \beta_7$ are the regression model parameters and ε represents the random error component.

Crop-yield response model development

The purpose of the crop-yield response model is to identify those edaphic and anthropogenic properties that have a statistically significant influence on cotton yield and to develop a quantitative relationship between cotton yield and the edaphic and anthropogenic properties. Applying ordinary least squares regression to Equation (8.2) results in the following crop-yield response model:

$$Y = 20.90 + 0.38(EC_e) - 0.02(EC_e)^2 - 3.51(LF)^2 - 2.22(pH) + 9.27(\theta_g) + \varepsilon, \tag{8.3}$$

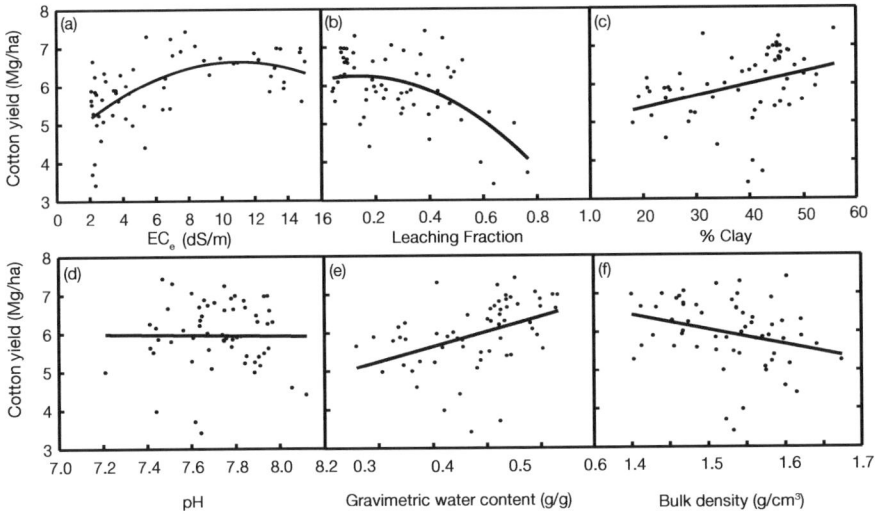

Figure 8.3 Scatter plots of soil properties and cotton yield: (a) electrical conductivity of the saturation extract (EC$_e$, dS m^{-1}), (b) leaching fraction, (c) % clay, (d) pH, (e) gravimetric water content and (f) bulk density (Mg m^{-3}).

Source: Corwin *et al.* (2003b) with permission.

where the non-significant *t*-test for % clay and ρ_b indicated that these soil properties did not contribute to the yield predictions in a statistically meaningful manner and dropped out of the regression model, whereas all other parameters were significant near or below the 0.05 level. The R^2 value for Equation (8.3) is 0.61 indicating that 61 per cent of the estimated spatial yield variation is successfully described by Equation (8.3). However, the variogram of the residuals indicates that the errors are spatially correlated, which implies that Equation (8.3) must be adjusted for spatial autocorrelation.

Restricted maximum likelihood was used to adjust Equation (8.3) for spatial autocorrelation, resulting in Equation (8.4), which is the most robust and parsimonious yield response model for cotton:

$$Y = 19.28 + 0.22(EC_e) - 0.02(EC_e)^2 - 4.42(LF)^2 - 1.99(pH) + 6.93(\theta_g) + \varepsilon. \quad (8.4)$$

A comparison of measured and simulated cotton yields at the soil sampling locations showed close agreement, with a slope of 1.13, *y*-intercept of –0.70 and R^2 value of 0.57. A visual comparison of the measured (Figure 8.4b) and predicted (Figure 8.4c) cotton yield shows a spatial association between them.

Sensitivity analysis is a means of establishing how the uncertainty in the output of a model can be apportioned to different sources of uncertainty in the model input and thereby can identify the most significant model input affecting the

Measured vs. predicted cotton yield
(interpolated data)

| Measured (based on 7706 sites) | Measured (based on 59 sites) | Predicted (based on 59 sites) | N |

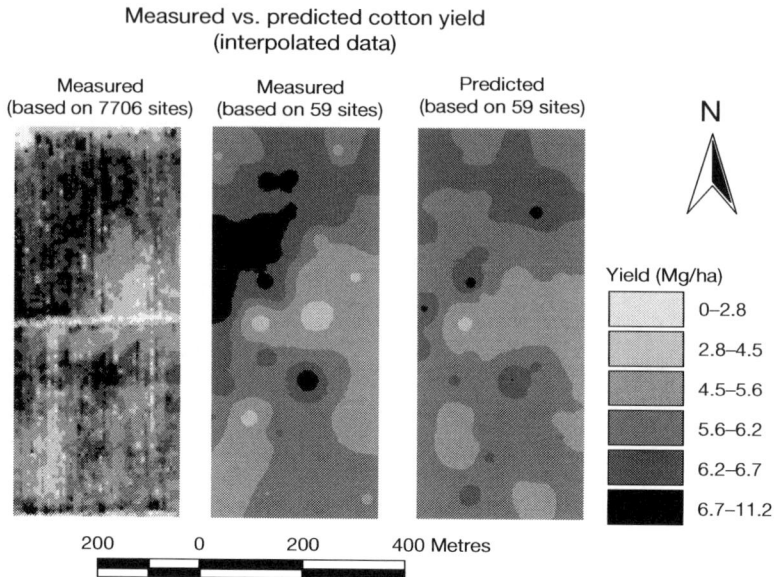

Yield (Mg/ha)

0–2.8
2.8–4.5
4.5–5.6
5.6–6.2
6.2–6.7
6.7–11.2

200   0   200   400 Metres

Figure 8.4   Comparison of (a) measured cotton yield based on 7,706 yield measurements, (b) kriged data at 60 sites for measured cotton yield and (c) kriged data at 60 sites for predicted cotton yields based on Equation (8.4).

Source: Corwin *et al.* (2003b) with permission.

output. Sensitivity analysis of Equation (8.4) reveals that LF is the single most significant factor affecting cotton yield with the degree of predicted yield sensitivity to one standard deviation change resulting in a percentage yield reduction for EC_e, LF, pH and θ_g of 4.6%, 9.6%, 5.8% and 5.1%, respectively.

GIS considerations and site-specific management units (steps 7 and 8)

All spatial data were compiled, organized, manipulated and displayed within a geographic information system (GIS). Kriging was selected as the preferred method of interpolation because in all cases it outperformed inverse distance weighting based on comparisons using jackknifing. Figure 8.5 shows maps of the four properties that affect cotton yield significantly at the Broadview site.

With several variables, the most straightforward and practical means of defining SSMUs is to use a GIS overlay approach, which overlays criteria for each of the significant properties influencing crop yield. This approach was used to delineate the SSMUs in Figure 8.6. Based on Equation (8.4), Figures 8.3 and 8.5 and knowledge of the interaction of the significant properties influencing cotton yield in the Broadview Water District, a DSS was developed where four recommendations were established to improve cotton productivity at the study site:

Factors influencing cotton yield

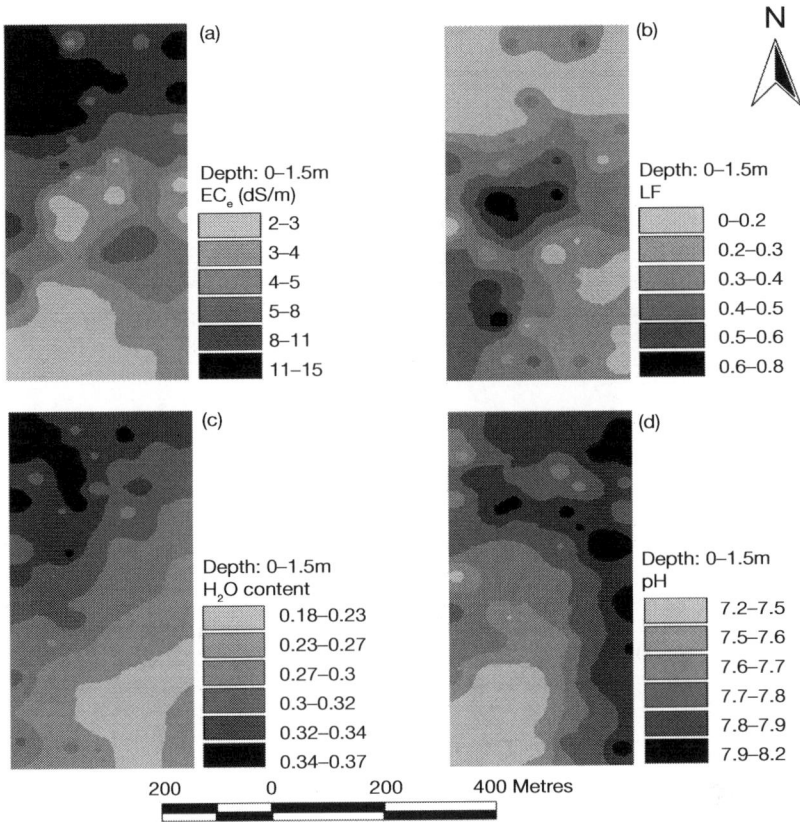

Figure 8.5 Maps of the four most significant factors (0–1.5 m) influencing cotton yield: (a) electrical conductivity of the saturation extract (EC_e, dS m^{-1}), (b) leaching fraction (LF), (c) gravimetric water content (θ_g, kg kg^{-1}) and (d) pH.

Source: Corwin *et al.* (2003b) with permission.

1 reduce the LF in strongly leached areas (i.e. areas where LF > 0.4),
2 reduce salinity by increased leaching in areas where the average root zone (0–1.5 m) salinity is > 7.17 dS m^{-1},
3 increase the plant-available water in coarse-textured areas by more frequent irrigation and
4 reduce the pH where pH > 7.9.

An overlay based on the four above recommendations produces the SSMUs of Figure 8.6.

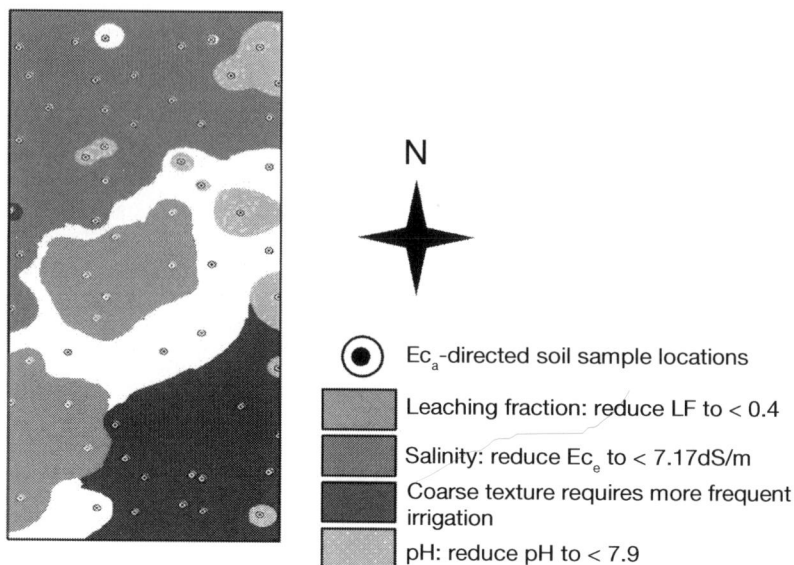

Figure 8.6   Site-specific management units (SSMUs) for a 32.4-ha cotton field in the Broadview Water District of central California's San Joaquin Valley. Recommendations associated with the SSMUs are for leaching fraction, salinity, texture and pH.

Source: Corwin and Lesch (2005a) with permission.

Equation (8.4) identified the main properties that affect cotton yield and sensitivity analysis identified the most significant of these to be LF. The three-dimensional scatter plot of yield, LF and NO_3^- indicates that LF > 0.4 leached nutrients such as NO_3^-, which resulted in a decrease in yield. The salinity level of 7.17 dS m^{-1} was established by differentiating Equation (8.4) with respect to EC_e and setting it equal to zero, which provided the salinity threshold (i.e. salinity level beyond which crop yield would decrease because of osmotic effects on plant water uptake). High salinity increases cotton production by producing stress in the plant and causing it to expend energy on growth of the reproductive portion of the plant (i.e. cotton bolls) at the expense of vegetative growth. However, there is a limit beyond which the osmotic effect of accumulated salinity causes the cotton plant to die. A salinity level of 7.17 dS m^{-1} is the limit for the conditions in this field. The recommendation to increase the frequency of irrigation on coarse-textured areas is based on the fact that irrigation schedules were roughly every two weeks as a result of water availability and after two weeks water content was near the wilting point. Setting a limit of 7.9 for the pH is a result of the effect that high pH has on the availability of micro- and macro-nutrients to the plant.

All four recommendations can be accomplished by improving water-application scheduling and distribution and by site-specific application of soil amendments.

The use of variable-rate irrigation technology at this site would enable the site-specific application of irrigation water at the times and locations needed to optimize yield.

Beneficial impacts of SSMUS to producers and the environment

Defining SSMUs is not a trivial process, but the benefits of doing so can result in substantial rewards to the producer. Ideally, SSMUs provide producers with a means of attaining sustainable agriculture by balancing profit, depletion of finite resources, detrimental environmental impacts and crop productivity. If inputs such as water, fertilizer and pesticides are applied site-specifically, that is, *when* they are needed, *where* they are needed and in the *amounts* they are needed, then crop productivity, profit, resource utilization and environmental impacts can be optimized. The application of excessive inputs or insufficient amounts, in locations where they are not needed and or at times when they cannot be used, does not benefit crop productivity and in fact may reduce it; it also reduces profit, wastes valuable resources and is deleterious to the environment.

For example, with conventional (uniform) applications of NO_3-N fertilizer there are areas of the field where excessive applications will occur. The excess N is unused by the plant, which wastes fertilizer and the producer's money, and can be readily leached through the soil or washed away in runoff water contaminating groundwater and surface water supplies, respectively. In some parts of fields insufficient inputs may be applied, which reduces productivity and profit. For example, in areas where insufficient NO_3-N is applied, crop yield is less than it could potentially be, which reduces the producer's profit potential. Site-specific management units identify areas within a field where NO_3-N fertilizer is applied in the amount and at a time and or frequency that optimizes productivity, profit, resource utilization and limits environmental degradation.

Another example concerns irrigated agriculture. With conventional applications of irrigation water (see Chapter 10), there are areas within a field that are under- and over-irrigated. Areas of the field that are over-irrigated leach valuable nutrients from the root zone into groundwater supplies. Over-irrigation wastes water, which is a valuable resource in arid and semi-arid agricultural regions. In contrast, areas of the field that are under-irrigated cannot meet the consumptive water needs of the plant, which lowers crop yield. Furthermore, under-irrigation causes salinity to accumulate in the root zone because evapotranspiration (ET) exceeds irrigation amounts, leaving salts in the irrigation water behind, which lowers yield as a result of osmotic and specific ion toxicity effects. The site-specific application of irrigation water will minimize the leaching of salts and nutrients into the groundwater, yet adequately remove salts from the root zone to prevent a decrease in yield and provide plants with sufficient water to meet ET demands.

Growing demand for food worldwide as a result of increased calorie intake by a growing population, greater global competition for limited resources and worldwide awareness of humankind's impact on the environment make agriculture a highly competitive and crucial sector of the world's economy. To remain competitive in a global economy, producers know that they must take advantage of technology and the precision agriculture approach. Even though conventional farming practices have made incredible technological progress thanks to the green revolution, they are not sustainable because of the fundamental flaw that conventional farming treats heterogeneous soil in a homogeneous manner. Site-specific management provides producers with the means of taking current technology of GPS, GIS, remote sensing, variable-rate application, yield monitoring and spatial statistics to manage the spatial heterogeneity of soil to optimize productivity, profit, environmental protection and resource utilization.

References

Adamchuk, V. I., Hummel, J. W., Morgan, M. T. and Upadhyaya, S. K. (2004) 'On-the-go soil sensors for precision agriculture', *Computers and Electronics in Agriculture*, 44, 71–91.

Adamchuk, V. I., Lund, E. D., Sethuramasamyraja, B., Morgan, M. T., Dobermann, A. and Marx, D. B. (2005) 'Direct measurement of soil chemical properties on-the-go using ion-selective electrodes', *Computers and Electronics in Agriculture*, 48, 272–94.

Andrade, P., Aguera, J., Upadhyaya, S. K., Jenkins, B. M., Rosa, U. A. and Josiah, M. (2001) 'Evaluation of a dielectric based moisture and salinity sensor for in-situ applications', ASAE Paper 01-1010, St. Joseph, MI: American Society of Agricultural Engineers.

Ben-Dor, E., Chabrillat, S., Demattê, J. A. M., Taylor, G. R., Hill, J., Whiting, M. L. and Sommer, S. (2009) 'Using imaging spectroscopy to study soil properties', *Remote Sensing of Environment*, 113, S38–S55.

Birrel, S. J. and Hummel, J. W. (2001) 'Real-time multi-ISFET/FIA soil analysis system with automatic sample extraction', *Computers and Electronics in Agriculture*, 32, 45–67.

Birrel, S. J., Borgelt, S. C. and Sudduth, K. A. (1995) 'Crop yield mapping: comparison of yield monitors and mapping techniques', in P. C. Robert, R. H. Rust and W. E. Larson (eds) *Proceedings of the Second International Conference on Site-Specific Management for Agricultural Systems*, Madison, WI: American Society of Agronomy, Crop Science Society of America and Soil Science Society of America, pp. 15–32.

Blake, G. R. and Hartge, K. H. (1986) 'Bulk density', in A. Klute (ed.) *Methods of Soil Analysis, Part 1, Physical and Mineralogical Methods*, 2nd edn, Agronomy Monograph 9, Madison, WI: American Society of Agronomy, Crop Science Society of America and Soil Science Society of America, pp. 363–75.

Bullock, D. S. and Bullock, D. G. (2000) 'Economic optimality of input application rates in precision farming', *Precision Agriculture*, 2, 71–101.

Carter, L. M., Rhoades, J. D. and Chesson, J. H. (1993) 'Mechanization of soil salinity assessment for mapping', ASAE Paper 93-1557, St. Joseph, MI: American Society of Agricultural Engineers.

Chang, J., Clay, D. E., Carlson, C. G., Reese, C. L., Clay, S. A. and Ellsbury, M. M. (2004) 'Defining yield goals and management zones to minimize yield and nitrogen and phosphorus fertilizer recommendation errors', *Agronomy Journal*, 96, 825–31.

Chung, S., Sudduth, K. A. and Hummel, J. W. (2003) 'On-the-go soil strength profile sensor using a load cell array', ASAE Paper 03-1071, St. Joseph, MI: American Society of Agricultural Engineers.

Clement, B. R. and Stombaugh, T. S. (2000) 'Continuously-measuring soil compaction sensor development', ASAE Paper 00-1041, St. Joseph, MI: American Society of Agricultural Engineers.

Corwin, D. L. (2005) 'Geospatial measurements of apparent soil electrical conductivity for characterizing soil spatial variability', in J. Álvarez-Benedí and R. Muñoz-Carpena (eds) *Soil-Water-Solute Process Characterization: An Integrated Approach*, Boca Raton, FL: CRC Press, pp. 640–72.

Corwin, D. L. and Lesch, S. M. (2003) 'Application of soil electrical conductivity to precision agriculture: Theory, principles, and guidelines', *Agronomy Journal*, 95, 455–71.

Corwin, D. L. and Lesch, S. M. (2005a) 'Apparent soil electrical conductivity measurements in agriculture', *Computers and Electronics in Agriculture*, 46, 11–43.

Corwin, D. L. and Lesch, S. M. (2005b) 'Characterizing soil spatial variability with apparent soil electrical conductivity: I. Survey protocols', *Computers and Electronics in Agriculture*, 46, 103–33.

Corwin, D. L. and Lesch, S. M. (2005c) 'Characterizing soil spatial variability with apparent soil electrical conductivity: II. Case study', *Computers and Electronics in Agriculture*, 46, 135–52.

Corwin, D. L. and Lesch, S. M. (2010) 'Delineating site-specific management units with proximal sensors', in M. A. Oliver (ed.) *Geostatistical Applications for Precision Agriculture*, New York: Springer, pp. 139–65.

Corwin, D. L., Kaffka, S. R., Hopmans, J. W., Mori, Y., Lesch, S. M. and Oster, J. D. (2003a) 'Assessment and field-scale mapping of soil quality properties of a saline-sodic soil', *Geoderma*, 114, 231–59.

Corwin, D. L., Lesch, S. M., Shouse, P. J., Soppe, R. and Ayars, J. E. (2003b) 'Identifying soil properties that influence cotton yield using soil sampling directed by apparent soil electrical conductivity', *Agronomy Journal*, 95, 352–64.

Elachi, C. (1987) *Introduction to the Physics and Techniques of Remote Sensing*, New York: John Wiley & Sons.

Hemmat, A. and Adamchuk, V. I. (2008) 'Sensor systems for measuring spatial variation in soil compaction', *Computers and Electronics in Agriculture*, 63, 89–103.

Huisman, J. A., Hubbard, S. S., Redman, J. D. and Annan, A. P. (2003) 'Measuring soil water content with ground penetrating radar: a review', *Vadose Zone Journal*, 2, 476–91.

Jaynes, D. B., Colvin, T. S. and Ambuel, J. (1994) 'Yield mapping by electromagnetic induction', in P. C. Robert, R. H. Rust and W. E. Larson (eds) *Proceedings of the Second International Conference on Site-Specific Management for Agricultural Systems*, Madison, WI: American Society of Agronomy, Crop Science Society of America and Soil Science Society of America, pp. 383–94.

Jones, A. J., Mielke, L. N., Bartles, C. A. and Miller, C. A. (1989) 'Relationship of landscape position and properties to crop production', *Journal of Soil and Water Conservation*, 44, 328–32.

Khosla, R., Fleming, K., Delgado, J. A., Shaver, T. M. and Westfall, D. G. (2002) 'Use of site-specific management zones to improve nitrogen management for precision agriculture', *Journal of Soil and Water Conservation*, 57, 513–18.

Kitchen, N. R., Sudduth, K. A. and Drummond, S. T. (1998) 'An evaluation of methods for determining site-specific management zones', in *Proceedings of the*

North Central Extension-Industry Soil Fertility Conference, St. Louis, MO: Potash and Phosphate Institute, pp. 133–9.

Larson, W. E. and Robert, P. C. (1991) 'Farming by soil', in R. Lal and F. J. Pierce (eds) *Soil Management for Sustainability*, Ankeny, IA: Soil and Water Conservation Society, pp. 103–12.

Lesch, S. M., Rhoades, J. D. and Corwin, D. L. (2000) *ESAP-95 Version 2.01R: User Manual and Tutorial Guide*, Research Report 146, USDA-ARS, Riverside, CA: US Salinity Laboratory.

Liu, W., Gaultney, L. D. and Morgan, M. T. (1993) 'Soil texture detection using acoustic methods', ASAE Paper 93-1015, St. Joseph, MI: American Society of Agricultural Engineers.

Long, D. S. (1998) 'Spatial autoregression modeling of site-specific wheat yield', *Geoderma*, 85, 181–97.

Lunt, I. A., Hubbard, S. S. and Rubin, Y. (2005) 'Soil moisture content estimation using ground-penetrating radar reflection data', *Journal of Hydrology*, 307, 254–69.

Maleki, M. R., Mouazen, A. M., De Ketelaere, B., Ramon, H. and de Baerdemaeker, J. (2008) 'On-the-go variable-rate phosphorus fertilization based on a visible and near-infrared soil sensor', *Biosystems Engineering*, 99, 35–46.

Mouazen, A. M. and Roman, H. (2006) 'Development of on-line measurement system of bulk density based on on-line measured draught, depth and soil moisture', *Soil & Tillage Research*, 86, 218–29.

Mouazen, A. M., Maleki, M. R., de Baerdemaeker, J. and Ramon, H. (2007) 'On-line measurement of some selected soil properties using a VIS-NIR sensor', *Soil & Tillage Research*, 93, 13–17.

Nielson, D. R., Biggar, J. W. and Erh, K. T. (1973) 'Spatial variability of field-measured soil-water properties', *Hilgardia*, 42, 215–59.

Page, A. L., Miller, R. H. and Kenney, D. R. (eds) (1982) *Methods of Soil Analysis, Part 2, Chemical and Microbiological Properties*, 2nd edn, Agronomy Monograph 9, Madison, WI: American Society of Agronomy, Crop Science Society of America and Soil Science Society of America.

Rhoades, J. D. (1992a) 'Instrumental field methods of salinity appraisal', in G. C. Topp, W. D. Reynolds and R. E. Green (eds) *Advances in Measurement of Soil Physical Properties: Bringing Theory into Practice*, SSSA Special Publication 30, Madison, WI: Soil Science Society of America, pp. 231–48.

Rhoades, J. D. (1992b) 'Recent advances in the methodology for measuring and mapping soil salinity', in L. Moncharoen (ed.) *Proceedings of the International. Symposium on Strategies for Utilizing Salt-Affected Lands*, Bangkok: Funny Publishing, pp. 39–58.

SAS Institute (1999) *SAS Software*, Version 8.2, Cary, NC: SAS Institute.

Sethuramasamyraja, B., Adamchuk, V. I., Dobermann, A., Marx, D. B., Jones, D. D. and Meyer, G. E. (2008) 'Agitated soil measurement method for integrated on-the-go mapping of soil pH, potassium and nitrate contents', *Computers and Electronics in Agriculture*, 60, 212–25.

Shonk, J. L., Gaultney, L. D., Schulze, D. G. and Van Scoyoc, G. E. (1991) 'Spectroscopic sensing of soil organic matter content', *Transactions of the American Society of Agricultural Engineers*, 34, 1978–84.

Sudduth, K. A., Drummond, S. T., Birrell, S. J. and Kitchen, N. R. (1997) 'Spatial modeling of crop yield using soil and topographic data', in J. V. Stafford (ed.)

Proceedings of the First European Conference on Precision Agriculture, Oxford: BIOS Scientific Publishers, 1, pp. 439–47.

Tanji, K. K. (ed.) (1996) *Agricultural Salinity Assessment and Management*, New York: ASCE.

van Uffelen, C. G. R., Verhagen, J. and Bouma, J. (1997) 'Comparison of simulated crop yield patterns for site-specific management', *Agricultural Systems*, 54, 207–22.

Verhagen, A., Booltink, H. W. G. and Bouma, J. (1995) 'Site-specific management: balancing production and environmental requirements at farm level', *Agricultural Systems*, 49, 369–84.

Verschoore, R., Pieters, J. G., Seps, T., Spriet, Y. and Vangeyte, J. (2003) 'Development of a sensor for continuous soil resistance measurement', in J. Stafford and A. Werner (eds) *Precision Agriculture*, Wageningen, The Netherlands: Wageningen Academic Publishers, pp. 689–95.

Whalley, W. R. and Bull, C. R. (1991) 'An assessment of microwave reflectance as a technique for estimating the volumetric water content of soil', *Journal of Agricultural Engineering Research*, 50, 315–26.

World Resources Institute (1998) *1998–99 World Resources: A Guide to the Global Environment*, New York: Oxford University Press.

9 Precision weed management

Roland Gerhards

Introduction

Weed populations in arable fields have often been found to be distributed heterogeneously in time and space (Marshall, 1988; Thornton *et al.*, 1990; Wiles *et al.*, 1992; Johnson *et al.*, 1996; Gerhards *et al.*, 1997; Christensen and Heisel, 1998; Nordmeyer *et al.*, 2003; Gerhards and Christensen, 2003). They often occur in aggregated patches of varying size or in stripes along the direction of cultivation. The spatial distribution of weeds, however, has mostly been disregarded in weed-management decisions, and herbicides or mechanical weed-control methods are applied uniformly across the whole field. The use of field-scale mean density estimates in spatially heterogeneous weed populations results in under-prediction of yield loss at locations where weed density is high and over-prediction in parts of the field where weed densities are low or weeds are absent (Lindquist *et al.*, 1998; Brain and Cousens, 1990). Spatial variation in weed density must, therefore, be considered in weed-management strategies. When there is large within-field variation in weed density, variable-rate patch spraying based on the need for weed control may reduce the area treated with herbicides and hence the economic and environmental costs.

A major step towards a practical solution for site-specific weed management was the development of precise and powerful sampling techniques to determine automatically and continuously within-field variation of crop cover and weed seedling populations (Lamb and Brown, 2001; Vrindts and de Baerdemaeker, 1997; Biller, 1998; Gerhards and Oebel, 2006).

There are two approaches for site-specific weed management in arable crops. In the offline or map-based approach, the weed distribution is measured at geo-referenced points and then interpolation methods are used to create weed distribution maps. The weed distribution maps are converted into application maps by selecting a threshold of the weed density above which weed controls are required. The application maps are then used to direct a patch sprayer or vehicles for mechanical weed control. In many studies, this map-based approach of site-specific weed management has been applied successfully resulting in examples with 20–90 per cent of field areas left untreated (Timmermann *et al.*, 2003; Gerhards and Oebel, 2006). In the real-time (online) approach, weed detection and patch spraying are performed simultaneously. Sensors are mounted on the spray boom to detect weed infestations, and thresholds are set to turn the spray boom on when the weed threshold is exceeded.

Many authors have measured the effectiveness of offline site-specific weed control methods. Gerhards and Oebel (2006) grouped the weed species into three classes: namely grass-weeds, annual broad-leaves and perennial weed species. They used simple weed density thresholds for all three classes to vary the herbicide dose between 0 and 100 per cent of the recommended uniform rate dose. Perennials such as *Convolvulus arvensis* L. and *Cirsium arvense* L. were found to be highly aggregated in annual grains, maize and sugar beet with less than 20 per cent of the field being infested. Grass-weeds covered on average 30–40 per cent of the fields at infestation levels greater than the economic thresholds and annual broad-leaves were observed on 20–90 per cent of fields (Timmermann *et al.*, 2003; Gerhards and Oebel, 2006) (Table 9.1).

Studies at the University of Bonn (Timmermann *et al.*, 2003), where site-specific weed control has been applied in five fields since 1997, showed that weed patches were relatively stable in location and size. Weed species' distributions were found to be heterogeneous in all crops studied. High density patches required application of the full herbicide dose, but in other areas herbicide application was not warranted. The average weed density in the experiments varied from 19 to 58 weeds m^{-2}. Herbicide savings ranged from 6% in maize up to 77% in winter wheat for herbicides against broad-leaves and from 22% in winter rape to 65% in winter barley for herbicides against grass-weeds. In 2005, herbicide savings varied from 19% in winter rape to 81% in winter wheat for herbicides against broad-leaves and from 20% in winter rape to 79% in winter wheat for herbicides against grass-weeds. The greatest savings were realized in cereals followed by sugar beet, maize and winter rape. Efficacy of site-specific

Table 9.1 Herbicide savings using site-specific weed control (Gerhards and Oebel, 2006)

Crop (field size)	Herbicide savings against broad-leaves (%)	Herbicide savings against grass-weeds (%)
Spring barley (17.5 ha)	18	42
Winter rape (11.5 ha)	20	22
Winter barley (8.1 ha)	38	34
Maize (4.6 ha)	6	46
Winter wheat (5.3 ha)	77	69
Sugar beet (5.8 ha)	57	46
Winter barley (8.5 ha)	39	56
Spring barley (8.4 ha)	26	71
Winter rape (6.6 ha)	19	20
Winter wheat (20.0 ha)	58	65
Spring barley (2.4 ha)	40	76
Winter wheat (5.3 ha)	81	79

weed control in this study was measured by counting the weed density before and approximately three weeks after weed control in the sprayed and unsprayed areas. The relative reduction in weed density between these dates was defined as efficacy. In all the experiments within this study, site-specific, post-emergent herbicide application resulted in an efficacy of 85% to 98%. Despite the additional costs of the weed sensing and application technology, site-specific weed control resulted in higher economic returns compared to conventional uniform applications even over periods of several years (Oerke *et al.*, 2010).

Christensen *et al.* (2003) determined an economic optimal herbicide dose with respect to the local weed density, the weed competition and the population dynamics. This strategy was tested in a five-year experiment and resulted in higher crop yields, lower soil seed banks and weed control costs similar to those for conventional decision models.

The results of these experiments show that site-specific weed management can reduce weed control costs and reduce the impact of weed control on the environment. However, the offline mapping process is time-consuming and a broader acceptance of site-specific weed management in practical agriculture will require an online system that combines weed species detection and herbicide application in one operation.

Dynamics of weed populations

There are many reasons for the complex spatial and temporal variation observed in weed populations. The dynamics of weed populations are affected by the biological characteristics of weed species, the competitiveness of the crop, soil properties, direct weed-control methods and various farming practices such as tillage, crop rotation, time of seeding and harvesting (Mortensen *et al.*, 1998; Nordmeyer and Niemann, 1992; Timmermann *et al.*, 2002). The major weed species have developed specific adaption and survival strategies to persist in cropping systems (Radosevich *et al.*, 1997). Those strategies include the production of a large number of seeds over a long period of time and seed dormancy (e.g. *C. album*). In addition, successful weed species have the capacity to survive under variable environments based on high phenotypic and genetic plasticity to invade new sites (e.g. *Abutilon theophrasti*). Many weeds are able to compete strongly for space, light, water and nutrients with the crops through rapid growth rates and efficiency in using water and nutrients. Several weeds produce mature seeds in a much shorter time than crops so that the seeds are spread long before a dense crop stand has been established (e.g. *Galingsoga parviflora*). Other weed species, such as *Cirsium arvense* and *Agropyron repens*, have the ability to persist and spread via seeds and vegetative reproduction tissues. Those perennial weeds can emerge much faster than annual plants.

Nordmeyer and Niemann (1992) found that blackgrass (*Alopecurus myosuroides*) populations mostly occurred at locations in the field where the clay content was relatively large. Timmermann *et al.* (2002) reported that the crop rotation had a long-term effect on weed density and weed species' composition. In fields that had been planted with 50 per cent maize in the rotation more than

20 years ago, the density of *C. album* was still much higher than in fields with a large percentage of winter annual grains in the rotation. The crop rotation also had a considerable effect on the organic matter content. Fields that had been planted with potatoes had less organic matter content than fields where mostly grains were planted. In turn, the organic matter content strongly influenced the weed species' composition. For example, *Galium aparine* predominantly occurred in fields with high organic matter contents (Timmermann *et al.*, 2002).

Weed mapping

The distribution of weed seedlings in the field has often been assessed using discrete weed mapping or continuous-area sampling (Rew and Cousens, 2001). In most of these studies, discrete weed mapping is applied on a regular sampling grid that is established in the field. The grid size varies from a few metres to approximately 50 m and depends on the width of the spray boom used for site-specific herbicide application. The density and or coverage of emerged weed seedlings are counted in sampling frames placed at the grid intersection points before and after the post-emergence herbicide application.

A major step towards a practical system for site-specific weed management was the development of precise and powerful sampling techniques to determine automatically and continuously within-field variation of weed seedling populations. Airborne remote sensing can be applied to identify *Avena fatua* L. and *Avena sterilis* ssp. *ludoviciana* (Durieu) Nyman populations in wheat, although Lamb and Brown (2001) could not detect densities of less than 19 plants m^{-2} because of the resolution of the sensor. Better resolution can be achieved by mounting proximal optoelectronic sensors on a tractor. Felton and McCloy (1992), Vrindts and de Baerdemaeker (1997) and Biller (1998) have used optoelectronic sensors to measure the reflectance in the green, red and near-infrared light wave bands. Green leaves are characterized by a high reflectance in the green and near-infrared and a low reflectance in the red spectrum compared with the reflectance curve of bare soil.

Two different methods have been used to record within-field variation of weed populations continuously. The first is to surround and record the borders of aggregated patches of weed species such as *Avena fatua* using a data logger connected to a differential global positioning system (DGPS) (Colliver *et al.*, 1996). The second is to map weed patches during harvest operations (Barroso *et al.*, 2005). However, Weis *et al.* (2008) suggest that the most promising approach for weed detection is a continuous ground-based detection method based on image analysis. They developed and mounted digital bi-spectral cameras (VIS/IR) on a vehicle. With each bi-spectral camera, two images were taken at the same time in the near-infrared spectrum (770–1150 nm) and in the red spectrum (610–670 nm), Figure 9.1b. The images of both cameras were normalized and subtracted (IR–VIS) in real time (Figure 9.1a). With this camera, a strong contrast between green plants and soil, mulch and stones was achieved even under variable illumination and soil moisture conditions. Artificial lightning was not necessary. A grey level threshold was set automatically and the contours of all white objects in the picture were extracted. Objects that were smaller or bigger than plants

(a)

IR-image >700 nm

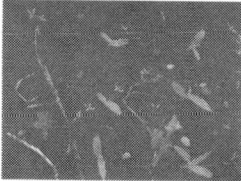

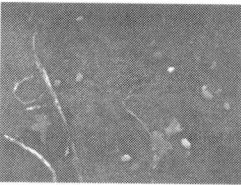

VIS-image 610–690 nm (res)

IR-VIS difference image

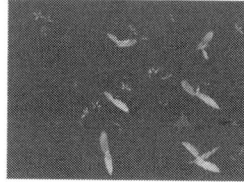

Binary image

(b)

Structure of the bi-spectral camera system

ultrasonic sensor   bi-spectral camera   artificial light source

Figure 9.1   Principle of a bi-spectral camera for weed identification (after Gerhards and Oebel, 2006).

were automatically removed from the image. Geometrical, shape and structural features of the contours of approximately 30 weed species, maize, winter wheat, winter barley and sugar beet were calculated and stored in a database. A different set of plants was used to determine the rate of identification based on different classification algorithms. Average identification rates were between 73 and 89 per cent when plant species were grouped into three different classes. Weed distribution maps were created using manual sampling and automatic camera weed identification (Figure 9.2). Cameras were mounted on a spray boom at a distance of 3 m and images were taken every second resulting in approximately 10 per cent of the field area being sampled. This is a much larger proportion than that achieved by regular grid sampling.

Interpolated weed maps produced manually or automatically can be reclassified into weed infestation levels (Gerhards *et al.*, 1997). Then a weed treatment map can be produced to provide a decision rule for the patch sprayer (Figure 9.3).

Spatial and temporal stability of weed distributions

The mapping techniques described above are vital for any site-specific weed control system. However, they have also allowed researchers to learn more about the spatial variation and the temporal persistence of weed populations. Such understanding can help to improve further weed control strategies. Krohmann *et al.* (2002) studied the dynamics of weed seedling distributions over five years in a rotation of maize, sugar beet, winter wheat and winter barley, and in continuous maize. They found that weed-distribution maps obtained in maize and sugar beet were suitable for site-specific weed control in winter wheat and winter barley (Figure 9.4).

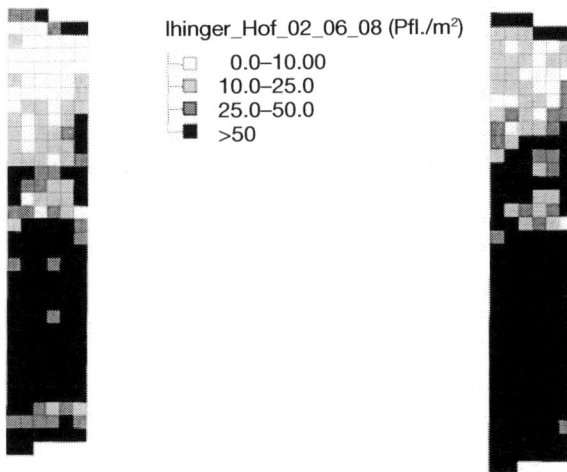

Ihinger_Hof_02_06_08 (Pfl./m²)

- 0.0–10.00
- 10.0–25.0
- 25.0–50.0
- >50

Figure 9.2 Weed distribution maps in a sugar beet field of 5.3 ha at Ihinger Hof Research Station, 2008, using visual weed sampling (left) and automatic imaging classification (right).

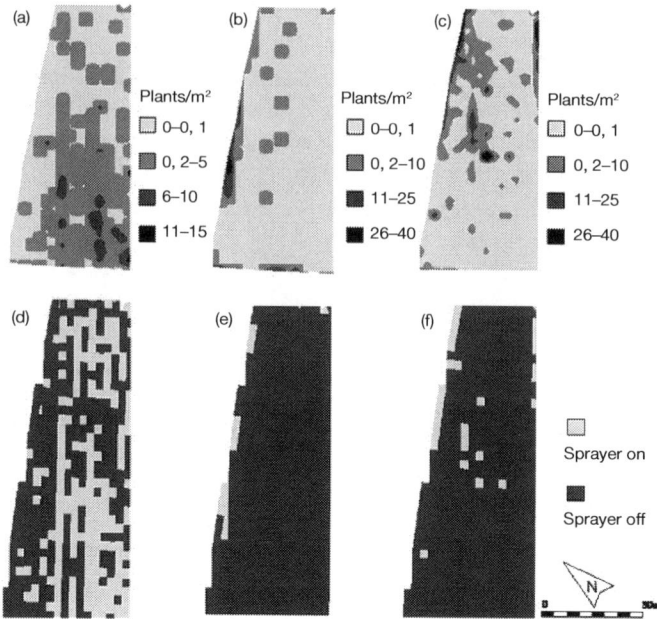

Figure 9.3 Distribution of different weed species (a–c) in a 3-ha spring barley field 2003 and application maps as a decision rule for the patch sprayer (d–f). Maps were created according to economic weed thresholds for all three weed species classes (Gerhards *et al.*, 1997).

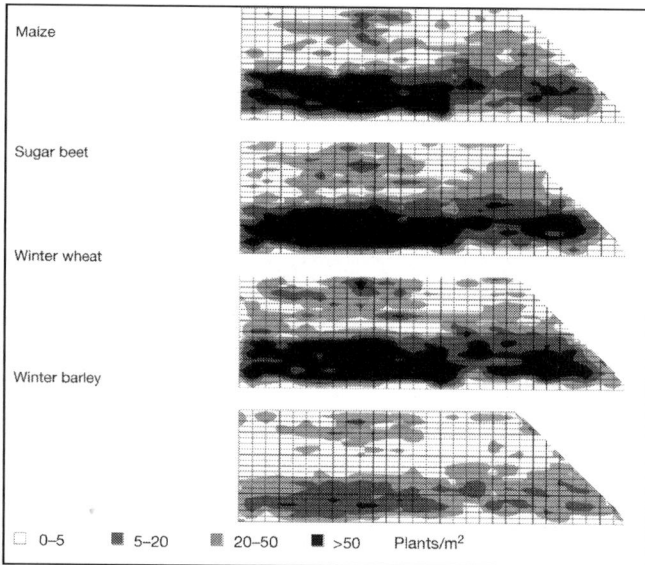

Figure 9.4 Distribution of *Viola arvensis* Murr. in maize, sugar beet, winter wheat and winter barley in a 5-ha arable field at Dikopshof Research Station near Bonn, Germany (modified after Krohmann *et al.* 2002).

Ritter and Gerhards (2008) reported that density, location and size of *Alopecurus myosuroides* (Huds.) did not change significantly when site-specific weed control methods were applied over a period of eight years in a rotation of winter annual cereals, maize and sugar beet. In each of the three fields studied, weed seedling distributions were heterogeneous. Density was higher in maize and sugar beet than in winter cereals. Patches with densities of more than 25 plants per m² consistently recurred each year at the same areas in the fields. Weed density reduction from the use of herbicides and other weed control methods was satisfactory each year indicating that site-specific weed control methods are sustainable for long-term weed suppression. Herbicide savings against *A. myosuroides* ranged from 50 per cent in sugar beet to 75 per cent in winter barley.

Ritter and Gerhards (2008) also studied weed population dynamics of *Galium aparine* (L). and *A. myosuroides* with the use of site-specific weed management. They found that most of the population properties tested were weed density dependent. It had previously been thought that individual weeds without competition evolve more successfully and produce more seeds, but this study provided evidence to the contrary. With increasing weed density, weed biomass and fecundity increased in this study (Figures 9.5 and 9.6). The authors suggested that individual weeds benefit from growing in patches, perhaps because of reduced competition by the crop or lower herbicide efficacy.

An understanding of fundamental weed population biology would improve our ability to develop site-specific management decisions. Weed population models have been applied to quantify the effects of site-specific weed management practices (Paice *et al.*, 1997). However, the mechanism of weed patch stability is not well understood. A few studies have reported that the efficacy of weed control methods is less in weed patches than at low density

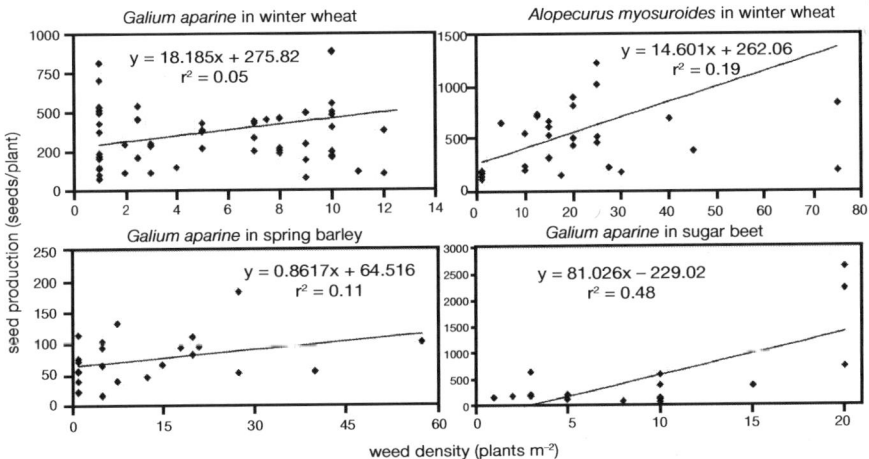

Figure 9.5 Weed density and seed production of *Galium aparine* and *Alopecurus myosuroides* in various crops (Ritter and Gerhards, 2008).

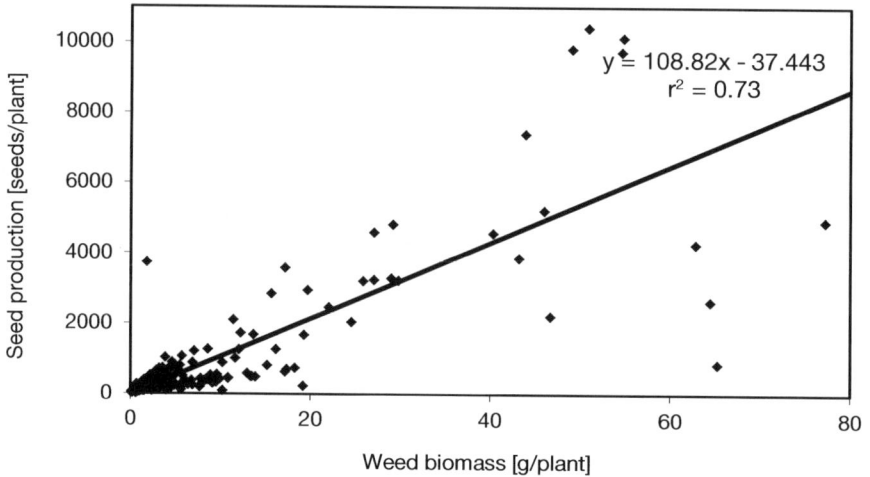

Figure 9.6 Correlation of weed biomass and seed production of *Galium aparine* and *Alopecurus myosuroides* over all crops (Ritter and Gerhards, 2008).

locations (Mortensen *et al.*, 1998). Krohmann *et al.* (2002) found that the persistence of weed populations was also attributed to weed seedlings that emerged after weed control methods had been applied. Those individuals were able to produce viable seeds in maize and sugar beet, but not in winter wheat and winter barley. The authors assume that competition with the crop was greater in winter annual grains, therefore late-emerging weed seedlings were suppressed.

Few studies have attempted to quantify spatial stability of weed patches in agricultural fields. If weed patches are consistent in density and location over many years, maps from one year could be used to direct sampling and to regulate weed control methods in subsequent years. Wilson and Brain (1991) found that the pattern of blackgrass (*A. myosuroides* Huds.) patches persisted during a 10-year study. The persistence of patches was attributed to the poor ability of blackgrass to colonize new locations when effective herbicides had been applied. The pattern of patches was most stable in fields planted to cereals. Pester *et al.* (1995) observed significant stability for velvetleaf populations using Pearson's, Spearman's rank and chi-square correlation analysis to quantify year-by-year relationships between weed density at individual X, Y-coordinates of the sampling grid in four fields. Walter (1996) also used the chi-square correlation method and found that field violet (*V. arvensis*), common lambsquarters (*C. album* L.) and prostrate knotweed (*Polygonum aviculare* L.) distributions were stable in cereal grain fields over three years.

Gerhards *et al.* (1996) studied the spatial stability of velvetleaf (*A. theophrasti* Medik.), hemp dogbane (*Apocynum cannabinum* L.), common sunflower (*Helianthus annuus* L.), yellow foxtail (*Setaria glauca* L.) and green foxtail (*Setaria viridis* L.) over four years (1992–5) in two fields in eastern Nebraska.

The first field was planted to soya beans in 1992 and maize in 1993, 1994 and 1995. The second field was planted to maize in 1992 and 1994 and soya beans in 1993 and 1995. Weed density was sampled prior to post-emergence herbicide application at approximately 800 locations per year in each field on a regular 7-m grid. The same locations were sampled every year. Weed density at locations between the sample sites was determined by linear triangulation interpolation. The weed seedlings were significantly aggregated and large areas free of weeds were observed in both fields. Common sunflower, velvetleaf and hemp dogbane patches were very persistent in the east–west and north–south directions, and in location and area over the four years in the first field. Foxtail distribution and density increased continuously in each of the four years in the first field and decreased in the second field. A geographic information system (GIS) was used to overlay maps from each year for each species. This showed that 36 per cent of the sampled area was free from common sunflower, 62.5 per cent from hemp dogbane and 11.5 per cent from velvetleaf in the first field, but only 1 per cent was free from velvetleaf in the second field. The persistence of broadleaf weed patches observed in this study suggests that weed seedling distributions mapped in one year are good predictors of future seedling distributions.

Heijting *et al.* (2007) found strong spatial correlations for *Echinochloa crusgalli*, *C. album*, *C. polyspermum* and *Solanum nigrum* populations over three years of continuous maize cultivation. They attributed spatial and temporal stability of weed populations to their high recruitment.

Application technologies for site-specific weed control

Specialist technology is required to apply herbicide according to a site-specific strategy. Variable-rate sprayers have been developed that separately turn on and off sections of the spray boom to vary herbicide dosage according to the size and density of patches of specified weed species in the field (Stafford and Miller, 1993; Paice *et al.*, 1997). Herbicide dosage is mainly regulated by the pressure in the hydraulic system. Application maps for offline systems are created based on interpolated maps of weed distribution and economic weed thresholds. With the technology developed at the University of Bonn, three different application maps can be acted upon at the same time using a multiple sprayer with three separated hydraulic circuits (Figure 9.7). The sprayer allows the herbicide mixture to be varied on-the-go. Each of the sprayer circuits has a boom width of 21 m divided into seven sections of 3 m. Each sprayer circuit and each section are turned on and off separately by a control unit via solenoid valves. The main hydraulic circuit of each of the three sprayer circuits is similar to that on a conventional sprayer, with an output from the main pump fed to a pressure control valve which ensures the concentration matches that set by the spray computer. During the herbicide application, the spray control system is linked to an onboard computer with the weed treatment maps. A differential mode GPS is used for real-time location of the patch sprayer. The onboard computer compares the actual position of the sprayer with the information in the weed

Figure 9.7   GPS-controlled patch sprayer with three separated hydraulic circuits.

treatment maps, and signals are transmitted to the control unit via a data bus to open each individual solenoid valve when herbicide application is warranted. In the same way, the herbicide dose is adjusted to the recommended rate in the treatment map.

An alternative approach is to use sprayers with an integrated direct injection system (Figure 9.8). In injection sprayers, herbicides and their carrier (water) are stored separately. According to the recommendations of the weed treatment map (offline application) or the weed analysis camera system (online application), the herbicides are metered into the carrier and mixed immediately before entering the nozzles.

In current direct injection systems there are two crucial factors that affect the accuracy of real-time patch application. The first is lag time, that is, the time it takes for the mixed solution to flow from the injection point to the spray nozzles. Commercial direct injection sprayers operate with one central injection point close to the main pump, thus lag times are rather large. Depending on the working width of the sprayer and the position of the injection point, lag times of up to 40 s have been measured. The second factor is the occurrence of non-uniform mixtures caused by inadequate mixing of herbicides and carrier in the boom as a result of varying physical properties of different herbicides such as formulation, viscosity and solubility.

Figure 9.8 shows an experimental direct injection system with one injection point for each boom section (3 m each). The lag time for this configuration is between 4 s and 7 s. An even shorter lag time of between 0.8 s and 1.4 s can be achieved by the injection of the herbicide at each nozzle.

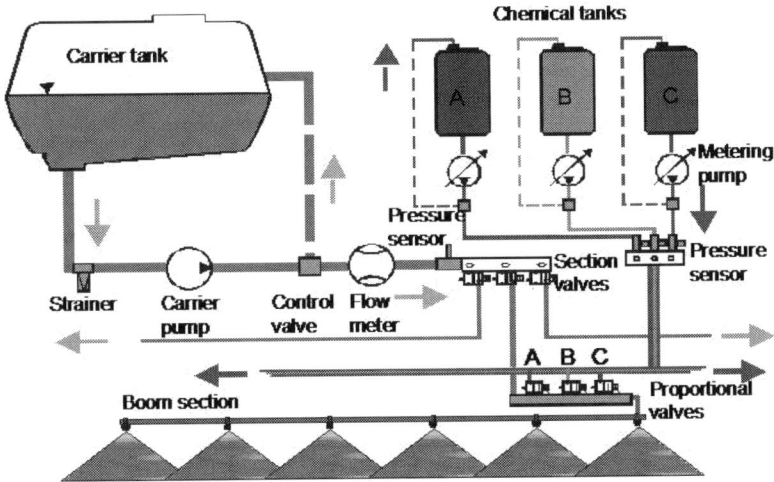

Figure 9.8 Schematic view of the hydraulic system for the direct injection of herbicides (after Oerke *et al.*, 2010).

Conclusions

With accurate knowledge of the spatial and temporal variability of weed populations there is great potential for precise weed control methods that use less herbicide and result in less herbicide residue in the environment and food chain. Such systems have been successfully tested for all major arable crops across fields of different sizes. Further improvements to the site-specific weed control systems are required. The greatest challenge is the development of robust real-time sensor technologies for the detection of weed species. Those sensors need to be implemented into patch spraying systems. Precise decision rules to determine the optimal patch spraying strategy for observed weed distributions also require further development.

Weed mapping has helped us to understand weed–crop interactions and population dynamics of weed species. It enables the effects on yield of different weed infestations to be quantified and used in models of the spatial and temporal variability of weed populations under different crop-management systems. By continuing to monitor weed maps and yield maps over several years it is possible to assess whether the weed-control strategy is sustainable.

References

Barroso, J., Ruiz, D., Fernandez-Quintanilla, C., Leguizamon, E. S., Hernaz, P., Ribeiro, A., Dias, B., Maxwell, B. D. and Rew, L. J. (2005) 'Comparison of different sampling methodologies for site-specific management of *Avena sterilis*', *Weed Research*, 45, 165–74.

Biller, R. H. (1998) 'Pflanzenunterscheidung und gezielter Einsatz von Herbiziden' (Differentiating between plants and targeted application of herbicides) (in German), *Forschungs-Report*, 1, 34–6.

Brain, P. and Cousens, R. (1990) 'The effect of weed distribution on predictions of yield loss', *Journal of Applied Ecology*, 27, 735–42.

Christensen, S. and Heisel, T. (1998) 'Patch spraying using historical, manual and real-time monitoring of weeds in cereals', *Journal of Plant Diseases and Protection*, Special issue 16, 257–63.

Christensen, S., Heisel, T., Walter, A. M. and Graglia, E. (2003) 'A decision algorithm for patch spraying', *Weed Research*, 43, 276–84.

Colliver, C. T., Maxwell, B. D., Tyler, D. A., Roberts, D. W. and Long, D. S. (1996) 'Georeferencing wild oat infestations in small grains: accuracy and efficiency of three survey techniques', in P. C. Robert, R. H. Rust and W. E. Larson (eds) *Proceedings of the Third International Conference on Precision Agriculture*, Madison, WI: American Society of Agronomy, Crop Science Society of America and Soil Science Society of America, pp. 453–63.

Felton, W. L. and McCloy, K. R. (1992) 'Spot spraying', *Agricultural Engineering*, 11, 9–12.

Gerhards, R. and Christensen, S. (2003) 'Real-time weed detection, decision making and patch spraying in maize, sugar-beet, winter wheat and winter barley', *Weed Research*, 43, 1–8.

Gerhards, R. and Oebel, H. (2006) 'Practical experiences with a system for site-specific weed control in arable crops using real-time image analysis and GPS-controlled patch spraying', *Weed Research*, 46, 185–93.

Gerhards, R., Pester, D. Y. and Mortensen, D. A. (1996) 'Characterizing spatial stability of weed populations using interpolated maps', *Weed Science*, 45, 108–19.

Gerhards, R., Sökefeld, M., Schulze-Lohne, K., Mortensen, D. A. and Kühbauch, W. (1997) 'Site-specific weed control in winter wheat', *Journal of Agronomy and Crop Science*, 178, 219–25.

Heijting, S., van der Werf, W., Stein, A. and Kropff, M. J. (2007) 'Are weed patches stable in locations? Application of an explicitly two-dimensional methodology', *Weed Research*, 47, 381–95.

Johnson, G. A., Mortensen, D. A. and Martin, A. R. (1996) 'A simulation of herbicide use based on weed spatial distribution', *Weed Research*, 35, 197–205.

Krohmann, P., Timmermann, C., Gerhards, R. and Kühbauch, W. (2002) 'Ursachen für die Persistenz von Unkrautpopulationen', *Journal of Plant Diseases and Protection*, Special issue 18, 261–8.

Lamb, D. W. and Brown, R. B. (2001) 'Remote sensing and mapping of weeds in crops', *Journal of Agricultural Engineering Research*, 78, 117–25.

Lindquist, J. L., Dieleman, J. A., Mortensen, D. A., Johnson, G. A. and Wyse-Pester, D. Y. (1998) 'Economic importance of managing spatially heterogeneous weed populations', *Weed Technology*, 12, 7–13.

Marshall, E. J. P. (1988) 'Field-scale estimates of grass populations in arable land', *Weed Research*, 28, 191–8.

Mortensen, D. A., Dieleman, J. A. and Johnson, G. A. (1998) 'Weed spatial variation and weed management', in J. L. Hatfield, D. D. Buhler and B. A. Stewart (eds) *Integrated Weed and Soil Management*, Chelsea, MI: Ann Arbor Press, pp. 293–309.

Nordmeyer, H. and Niemann, P. (1992) 'Möglichkeiten der gezielten Teilflächenbehandlung mit Herbiziden auf Grundlage von Unkrautverteilung und Bodenvariabilität', *Journal of Plant Diseases and Protection*, Special issue 13, 539–47.

Nordmeyer, H., Zuk, A. and Häusler, A. (2003) 'Experiences of site-specific weed control in winter weeds', in J. Stafford and A. Werner (eds) *Precision Agriculture*, Wageningen, The Netherlands: Wageningen Academic Publishers, pp. 457–62.

Oerke, E.-C., Gerhards, R., Menz, G. and Sikora, R. A. (eds) (2010) *Precision Crop Protection: The Challenge and Use of Heterogeneity*, 1st edn, Dordrecht: Springer.

Paice, M. E. R., Day, W., Rew, L. J. and Howard, A. (1997) 'A simulation model for evaluating the concept of patch spraying', *Weed Research*, 36, 373–88.

Pester, D. Y., Mortensen, D. A. and Gotway, C. A. (1995) 'Statistical methods to quantify spatial stability of weed populations', *North Central Weed Science Society*, 50, 52.

Radosevich, S., Holt, J. and Ghersa, C. (1997) *Weed Ecology: Implications for Weed Management*, 2nd edn, New York: John Wiley & Sons.

Rew, L. J. and Cousens, R. D. (2001) 'Spatial distribution of weeds in arable crops: are current sampling methods appropriate?', *Weed Research*, 41, 1–18.

Ritter, C. and Gerhards, R. (2008) 'Population dynamics of *Galium aparine* L. and *A. myosuroides* (Huds.) under the influence of site-specific weed management', *Journal of Plant Diseases and Protection*, Special issue 21, 209–14.

Stafford, J. V. and Miller, P. C. H. (1993) 'Spatially selective application of herbicide to cereal crops', *Computers and Electronics in Agriculture*, 9, 217–29.

Thornton, P. K., Fawcett, R. H., Dent, J. B. and Perkins, T. J. (1990) 'Spatial weed distribution and economic thresholds for weed control', *Crop Protection*, 9, 337–42.

Timmermann, C., Gerhards, R. and Kühbauch, W. (2002) 'Ursachen für Ertragsunterschiede in Ackerschlägen', *Journal of Agronomy and Crop Science*, 187, 1–9.

Timmermann, C., Gerhards, R. and Kühbauch, W. (2003) 'The economic impact of the site specific weed control', *Precision Agriculture*, 4, 249–60.

Vrindts, E. and de Baerdemaeker, J. (1997) 'Optical discrimination of crop, weed and soil for on-line weed detection', in J. V. Stafford (ed.) *Precision Agriculture 1997: Proceedings of the First European Conference on Precision Agriculture*, Oxford: BIOS Scientific Publishers, pp. 537–44.

Walter, W. (1996) 'Temporal and spatial stability of weeds', in H. Brown, G. W. Cussans, M. D. Devine, S. O. Duke, C. Fernandez-Quintanilla, A. Helweg, R. E. Labrada, M. Landes, P. Kudsk and J. C. Streibig (eds) *Proceedings of the 2nd International Weed Control Congress, Flakkebjerg, Denmark*, pp. 125–30.

Weis, M., Ritter, C., Gutjahr, C., Rueda-Ayala, V., Gerhards, R. and Schölderle, R. (2008) 'Precision farming for weed management: techniques', *Gesunde Pflanzen*, 60, 171–81.

Wiles, L. J., Oliver, G. W., York, A. C., Gold, H. J. and Wilkerson, G. G. (1992) 'Spatial distribution of broadleaf weeds in North Carolina soybean (*Glycine max*) fields', *Weed Science*, 40, 554–7.

Wilson, B. J. and Brain, P. (1991) 'Long-term stability of distribution of *Alopecurus myosuroides* Huds. within cereal fields', *Weed Research*, 31, 367–73.

10 Site-specific irrigation water management

Robert G. Evans and E. John Sadler

Introduction

Irrigation can be fundamentally characterized as a temporal adaptation to seasonal and annual variation in rainfall (Turral *et al.*, 2010). As variation in rainfall in both time frames is very common, irrigation is widely practised, and it is considered important from almost all perspectives. Many reasons support the importance of irrigation. Irrigation of lands accounts for over 70 per cent of fresh water consumed in the world, and is also a major user of energy for farming operations and pumping. Irrigated land constitutes approximately 18 per cent of the world's total cultivated farmland, but produces more than 40 per cent of its food and fibres. Irrigated agricultural activities support diverse components of the world's food chain and provide much of the fruit, vegetables and cereals consumed by humans plus the grain fed to animals that are used eventually as human food. It also provides much of the feed to sustain animals used for work in many parts of the world, and these lands provide considerable sources of food and foraging areas for migratory and local birds, as well as for other wildlife.

Irrigation has shaped the economies of many semi-arid and arid areas, permanently colouring the social fabric of numerous regions around the world over many centuries. It has stabilized rural communities, increased income and provided many new opportunities for economic growth. These practices have enabled human habitation, at times with quite dense populations, where it otherwise could not exist (Postel, 1999). In short, irrigated crop production underpins large sections of current society and lifestyles throughout the world. Consequently, irrigation will necessarily continue to be a major part of the world's future agricultural production systems, even though agriculture's access to one of its most critical resources (i.e. water) will be restricted.

The importance of water and its uneven geographical distribution has made its acquisition a matter of great contention between various users and nations. Where agricultural water resources were previously of little concern, they are coming under increasing pressure from other issues, including declining water tables in many areas, industrial and municipal pollution and escalating energy costs. In addition, there is also a persistent loss of arable land to soil salination, soil erosion and urbanization, which collectively emphasize the need to maximize irrigated food productivity wherever and whenever possible. Population growth and competing uses are also forcing major changes in how water is allocated. Thus, current water use patterns and practices may not be sustainable in many regions of the world. Nevertheless, output from irrigated agriculture will have to increase dramatically in the next few decades to help meet the projected world food requirements and to maintain food security.

These compelling and competing physical, societal and economic demands are forcing major, fundamental transformations in how crops are grown, the distribution of where and which crops are planted, and how much and what quality of water is made available for irrigation in many locations. Water and energy conservation measures and the implementation of advanced irrigation technologies will be required (Clemmens and Allen, 2005). These changes make it very likely that a future limited by water and energy will be the catalyst that finally causes the integrated use of several precision agriculture (PA) technologies in irrigated agriculture, including site-specific irrigation strategies and systems. Site-specific irrigation covers a wide range of technologies and practices, each with many variations.

Site-specific variable-rate irrigation (SS-VRI) can be defined as the ability to vary water application depths spatially across a field to match the specific soil, crop and or other conditions in unique zones within a field. It is included in the spectrum of PA technologies because advanced SS-VRI methods use many of the same management tools, and they can potentially impose treatments in ways that optimize plant responses for each unit of water applied in different areas of the same field. Site-specific water applications can be accomplished with most existing irrigation technologies, and the farmer has many options for how these goals can be accomplished. These technologies will not be appropriate for all situations.

This chapter is intended to enhance understanding of different irrigation methods, system limitations and management options as well as various sources of within-field variability. The challenges and opportunities to adapt site-specific irrigation technologies for sustainable crop production are introduced and discussed briefly.

Irrigation technology

Site-specific variable-rate irrigation is the high-technology evolution of current irrigation technology, and to understand it well one must view SS-VRI in the context of these technologies. Existing irrigation technology is well advanced and if implemented to its full potential would conserve significant amounts of water. However, acquisition of an advanced irrigation technology does not always result in improved management, and anticipated water savings are often unrealized for many reasons, including time constraints on growers and limited economic incentives. Consequently, the greatest impacts on productivity and conservation will be gained by advances in irrigation management in ways that minimize economic risk to growers. This implies a process of continuous, sequential changes that physically improves existing systems and refines management to levels where irrigators can then implement the spectrum of advanced irrigation technologies, supporting technologies and management options incrementally.

There are basically three types of irrigation systems: surface (which flows by gravity), and pressurized micro and sprinkler systems. Surface (also called gravity) irrigation introduces water at the highest elevation point in a field and the water flows over the soil surface by gravity. Pressurized irrigation systems include several types of microirrigation and sprinkler systems where water is conveyed by pressurized pipes or tubing to a point in the field where it is applied by various devices. In addition, pressurized systems commonly function as site-specific application methods for fertilizers and other agrochemicals (chemigation), which require appropriate management to avoid environmental contamination. Technology levels in all three types of systems vary widely, and they all have advantages and disadvantages (Hoffman *et al.*, 2007). Each irrigation method is more or less amenable to achieving at least some site-specific goals under certain conditions.

Surface or gravity irrigation methods

The oldest and the most widely practised method of water application is surface irrigation. It is estimated that surface systems collectively irrigate about 95 per cent of the world's irrigated land (FAO, AQUASTAT, 2011). Surface irrigation methods typically have the lowest capital requirements and the lowest efficiencies, but unit area labour costs may be much greater than pressurized irrigation methods. Soil erosion may be a problem on steeper slopes.

Because the irrigation water is distributed across the field on the soil surface, inherent variation in soil properties, slope, roughness, length of run and inflow rates can cause non-uniform infiltration across a field, which greatly constrains the ability to meet highly efficient irrigation and uniformity goals. These types of variability are difficult to manage even if the whole field and irrigation system are designed for uniform applications.

There are highly efficient and uniform methods of surface irrigation such as precision-graded level basins, but their use on large fields is generally limited by topography. However, if the design is such that small parcels can be irrigated independently (i.e. small basins), site-specific irrigation that accounts for relatively small levels of variability can be achieved indirectly through design rather than better management of these small zones. In fact, this is often the default case in many fields and paddies of small landholders worldwide.

Microirrigation methods

Microirrigation includes drip and microsprinklers. It is a flexible set of technologies that can potentially be used on almost every crop, soil type and climatic zone if justified economically. It is characterized by the applications of water in small amounts using frequent irrigations (i.e. daily). It is also sometimes referred to as localized irrigation and can provide numerous crop production and water conservation benefits that address many of the water quality and supply challenges facing modern irrigated agriculture. Microirrigation acreage is increasing steadily worldwide and is fully expected to continue its rapid growth in the foreseeable future (Lamm *et al.*, 2007). Novel applications for microirrigation systems, such as the reuse of municipal wastewater in turf areas, are providing new opportunities for growth.

These systems can be laid on the soil surface or buried, and can be permanent or temporary installations. They can be used on gardens as well as in fields from 0.5 ha to more than 50 ha. Because of the relatively high capital cost of these advanced systems and the need for high levels of management, modern microirrigation technologies are typically permanent installations on small blocks of land (e.g. < 10 ha) of intensively managed, high-value specialty crops (e.g. vegetables, grapes, nuts, fruits and berries). Microirrigation could also become more common in perennial forage crops (e.g. alfalfa) and annual row crops such as maize using widely spaced, buried drip lines in areas with limited water supplies or high water costs.

Greenhouse culture is largely irrigated by microirrigation methods. Year-round greenhouse production of fresh vegetables, cut flowers and other high-value crops is rapidly expanding, with some of the largest increases occurring around the Mediterranean Basin. All the major factors affecting plant growth can be controlled and managed precisely in these conditions, including the environment, water, nutrients, pests and diseases. Thus, these growing systems provide the ultimate opportunities for the application of many PA concepts.

Microirrigation systems are well suited for automated adaptive control with real-time sensor feedback and decision-making capabilities. However, site-specific water management often occurs by default as a function of the scales of underlying environmental variability. Microirrigation systems are generally designed and installed in small blocks to match the needs of particular crop varieties or variations in soil texture and topography across a farm. Thus, additional site-specific irrigation within blocks may not be warranted, although it is technically possible.

A low-cost microirrigation method called pitcher or clay-pot irrigation (Siyal *et al.*, 2009) is used on some small fields (e.g. < 0.5 ha) in several areas in northern Africa and the Middle East. This labour-intensive method uses unglazed, porous clay pots which are buried in the ground up to their neck to irrigate individual plants (i.e. fruit and nut trees) that are typically spaced 1 m or more apart. There may be more than one pitcher per plant, filled manually with water on at least a daily basis. Water seeps out through the unglazed walls of the pitcher to irrigate the crop.

Sprinkler irrigation methods

Sprinkler irrigation can be accomplished with a number of different systems currently in use. These include hand-move laterals, wheel-move laterals and continuous-move systems such as centre pivots and solid-set systems (movable or permanent) that encompass a diverse set of water application devices called sprinklers. However, most of this discussion will concern self-propelled (continuous-move) sprinkler systems because they are by far the most widely used method.

Self-propelled sprinkler irrigation systems such as centre pivots and linear (also called lateral) move systems have allowed large-scale agricultural development of marginal lands that were unsuitable for surface irrigation, including areas with light, sandy soil and a large variation in topography within the same field. These adaptable systems have experienced tremendous growth around the world over the past 50 years as a result of: (1) their potential for efficient and uniform water application; (2) their high degree of automation, requiring less labour than most other irrigation methods; (3) their coverage of large areas and (4) their ability to apply water and labelled agrichemicals (chemigation) economically and safely over a wide range of soil, crop and topographic conditions. As in the case of microirrigation, these systems apply limited amounts of water during each irrigation event and require frequent applications. A single machine can irrigate fields ranging from about 5 ha to over 200 ha.

These large machines basically consist of a pipeline (lateral) mounted on motorized structures (towers) with large rubber tyres. The section between the two towers is called a span, which can vary from about 30 m to 70 m in length. Sprinkling devices are mounted on or below the pipe. Maximum application depths of water applied are controlled by varying the speed of machine travel. Field-scale water applications by these systems can be quite uniform.

A centre pivot machine basically rotates around a 'pivot' at one end, usually in the centre of the field. These continuously moving systems can irrigate areas ranging from part circle segments to whole circles. Various optional accessories are also available to irrigate portions of field corners. Self-propelled linear move sprinkler systems look and perform almost identically to centre pivot systems except that they move in straight lines to irrigate square- or rectangular-shaped fields. Linear move systems require a global positioning system (GPS) or other methods for physical guidance, and they require considerably greater management and manual oversight than centre pivot systems.

Field variability in irrigated agriculture

Traditionally, irrigation system design and management has strived for maximum uniformity of water applications over the entire irrigated field. However, agricultural fields are never physically or biologically uniform, and substantial agronomic and environmental differences can occur. The effects of different spatial and temporal sources of variation on management can be additive, and the stochastic variation of several interrelated factors across a field can affect crop growth, yields and crop quality. Often, the underlying causes of crop performance variation are not well understood, and they can vary substantially from field to field and year to year. Some sources of variation such as pest problems may raise both temporal and spatial considerations.

Spatial and temporal variation can influence irrigation management decisions, and site-specific managers attempt to characterize the major sources of variation throughout the growing season. However, growers cannot practically manage for all of the many sources of variability. Therefore, they tend to group the most critical properties into relatively homogeneous management zones within a field, and those which respond well to similar management practices (see Chapter 8). Thus, different PA technologies may be managed simultaneously at different scales in the same field. For example, the smallest zone that a large-scale farmer (e.g. > 20-ha fields) prefers to manage for irrigation is probably in the order of about 5 ha, even though other PA technologies (e.g. planting and spraying) will often be managed at much smaller scales (e.g. 0.1 ha).

These factors can also accelerate localized leaching of soil nutrients and other agrichemicals past the root zone. Field variability can result in excessive energy use for pumping, can affect irrigation system design initially and, later, can affect management decisions. Furthermore, some management decisions may introduce additional sources of within-field variability.

Spatial variation

Spatial variation of soil physical properties (e.g. texture with depth, topography and aspect) and many soil chemical properties (e.g. pH, salinity and organic matter) are for practical purposes permanent. Whereas, soil nutrition, infiltration characteristics, soil compaction and pest issues are more dynamic and can vary significantly throughout the season. Soil topographic differences can play an important role in infiltration, erosion and surface runoff, resulting in redistribution of applied water on and off the field. All of this can profoundly affect variation in crop stands, weed growth, and crop yield and quality. Water infiltration rates can vary among locations within the field and may also be affected by tillage practices, weeds, soil compaction and multi-year cropping sequences.

Temporal variation

Temporal availability of water or labour can profoundly affect the timing and depth of irrigations. Other short-term and long-term temporal variation can occur because of dynamic processes that include plant nutritional status, pest management programmes, changing crop growth stages and crop water stresses. Variation in micrometeorological conditions and the impact of weather events such as spatially variable rainfall distributions, frost, high winds or hail can likewise contribute to temporal variation in irrigation requirements or soil water status. Abnormal short-term heat waves or cool periods can introduce additional complications and variability. Temporal variation can also be affected by row spacing, aspect, slope, runoff and temporary ponding, non-uniformity of irrigation and fertilizer applications, and external causes such as damage from herbicide drift. Distributions of undesirable insects, weeds and diseases are seldom uniformly distributed in either time or space, but they can also cause large temporal differences in water requirements and crop growth across a field.

Infiltration rates have long been known to vary between irrigation events. Temporal variation in soil water content affects decisions on irrigation frequency and amounts and also how available water is distributed throughout the season.

Variation in soil water content at planting can also result in large differences in emergence across the field, which can change localized plant densities and the timing of water use throughout the season, which ultimately affects total yield. Recognition of the sources of temporal variation through feedback systems, remote sensing and field sampling can lead to the dynamic adjustment of management zones that alter throughout the season as conditions change, which contributes to optimal economic crop production.

Variation related to irrigation management

Irrigation management decisions can contribute to within-field variation in soil moisture content on both spatial and temporal scales. Despite good designs and best intentions, irrigations are generally not uniform, which induces further variation that is superimposed on the natural spatial variation. Management can sometimes purposefully account for the variation across a field, but it can also have unintended consequences by increasing the variability of other factors. For example, some portions of the field may be over-irrigated and others under-irrigated. Ponding or persistent wet areas may occur in low areas of a field affecting yield and farming operations, and results in unintended excesses of applied fertilizers or pesticides.

Suboptimal management decisions that affect crop production include over- or under-estimating crop evapotranspiration (water use), which can cause excessive or under-applications of water and nutrients. For example, when dry areas appear in a centre pivot irrigated field as a result of either spatial or temporal variation, irrigators tend to run the irrigation system for longer and more often across the entire field to ensure adequate water is supplied to the affected areas. These practices result in much of the area being over-irrigated, and large amounts of water and energy are being wasted to compensate for limited drought in relatively small areas of a field.

Irrigation system design and maintenance also contributes to non-uniform water application as a result of the spacing and flow characteristics of the installed sprinkler nozzles. Elevation differences across a field can change pressure distributions in a centre pivot system, which affects application rates and may cause dry areas or excessive accumulation of water in low areas. Effects of elevation can be regulated to some degree with pressure regulating valves or flow control nozzles on individual sprinkler heads.

Irrigation equipment breakdowns can leave parts of a field unwatered for extended periods. Damaged or improper nozzle selection, broken pressure regulators, leaking gaskets and other issues resulting from improper maintenance will all contribute to non-uniform applications along the length of the machine. These system factors not only affect the amount of water applied to a given area within the field, but they also compound the problems introduced by other sources of variation. This issue becomes particularly acute when applying nutrients or pesticides through the irrigation system.

Other management decisions, such as tillage, fertilizer application practices and pesticide spray programmes, seeding rates, cultivar selection, herbicide persistence and long-term crop rotations, may also affect irrigation decisions. Efforts to compensate for different sources of non-uniformity may have unintended consequences by creating unwanted conditions in other areas of a field (e.g. ponding), which must be balanced in some way.

Irrigation management

Good irrigation management strives to optimize total yield and quality for maximum economic returns across a field, but approaches vary in response to specific field conditions and available capital and labour resources. Irrigation management can range from relatively simple for small fields to extremely complex for large, highly variable fields.

Management is directly affected by water quality, crop rotation, water application patterns of the various systems and irrigation equipment limits. Decisions on irrigation timing and frequency are greatly affected by irrigation system characteristics, changing environmental conditions, cultural practices, pest management, water supply, economic considerations and field size.

Managers must be able to address the many constraints imposed by the field site and integrate them with management options on varying temporal and spatial scales before and during irrigations. The field site constraints are often the most difficult to address because they can be numerous, with some varying in extent during the season and from year to year.

Thus, good irrigation decision-making requires a highly skilled and competent manager. The required management skills are developed as a result of strong personal commitments to improved farm profitability and environmental benefits. These skills must be gained through education and technical assistance, and are maintained by the irrigator's continued commitment and are encouraged by economic incentives.

The design of a water application system will determine the maximum potential performance level, whereas management dictates the actual benefits realized and the magnitude of any positive or negative ecological impacts. Thus, the amount of water that can be conserved by advanced irrigation systems and practices depends on the ability of a particular type of irrigation system to implement various improved management strategies economically. Higher levels of irrigation management and systems that enable greater efficiency of water use can potentially reduce operating costs as well as energy, nutrient and water use.

'Site-specific' inherently implies that management of the various technologies will be climate-, crop- and region-specific, and that transferability of research results to different regions will not always be appropriate. In particular, transfer from highly automated, high technology systems in the USA and some other countries to the labour-intensive farming systems in Asia, for example, may not always be successful.

Water management strategies for humid areas will be quite different from approaches to SS-VRI management in arid areas. This is because humid areas annually receive precipitation in excess of maximum crop water use, whereas arid areas receive less than the required amounts for optimal crop production. Thus, in humid areas, SS-VRI management is basically focused on minimizing yield reductions caused by the timing and duration of short-term drought (i.e. two to three weeks) and accounting for the variation in incident rainfall across a field for maximum crop yields. Irrigation totals are usually less than 20 per cent of the total annual crop water use. For fully watered crop production in humid areas, the greatest water savings from site-specific irrigation may be achieved by spatially maximizing the use of non-uniform incident precipitation over the growing season. In contrast, management in arid areas is focused on managing season-long drought stresses using a whole-systems approach, and maximum crop water use is not always an option.

Irrigation may supply more than 80 per cent of total seasonal crop water use in arid regions. This dependence on irrigation means that the greatest potential water savings could come from appropriately managed deficit-irrigation strategies (managed drought stress) in arid regions. However, excessive soil salinity is a common occurrence in semi-arid and arid areas under irrigation, and managing to control this major problem requires extraordinary efforts.

Salinity control

High levels of irrigation management can particularly benefit one specific irrigation-related issue that deserves separate mention, and that is salinity control. Compared with rain, irrigation water often contains much larger amounts of dissolved substances (salts) and other contaminants, many of which accumulate on the soil surface because of evaporation and transpiration by the crops. Soil salinity can affect crop growth to various degrees depending on the water and soil salt concentrations and plant tolerances to salinity (Hillel and Vlek, 2005). Soil salinity problems may be exacerbated by rising groundwater tables associated with the introduction of inefficient irrigation or with restrictive subsurface conditions, both of which can result in evaporation of groundwater from the soil surface leaving the salts behind and greatly increasing salt concentrations at the surface.

In many locations, irrigated crop production is physically sustainable only if both soil salts and groundwater elevations are adequately managed. Fortunately, unlike irrigation among early civilizations, modern irrigation and drainage technologies in conjunction with advanced irrigation management options can control soil salinity and maintain sustainable crop production. This requires the use of well-managed, efficient irrigation systems and may also necessitate the installation of expensive field drainage systems to control groundwater levels and to facilitate leaching of salts through the soil profile (Smedema *et al.*, 2004). Field drainage systems can be buried perforated pipelines or large open channels to intercept and divert groundwater flows to off-site locations.

Irrigation management for salinity control generally focuses on minimizing deep water percolation losses that can cause the water table to rise. However, when excessive salts accumulate on or near the soil surface, it is usually necessary to leach salts periodically (e.g. once or twice a season) from the soil profile by the deliberate over-application of irrigation water or higher rainfall amounts.

The frequency of managed leaching and water amounts for adequate leaching must be increased as the salinity of the applied water rises, but the total amount of water needed is relatively small (e.g. 2–10 per cent of total applied seasonally). At times, salts can accumulate in localized, small areas within a field, and site-specific applications of water for leaching purposes to those areas may be required to maintain crop production. However, this use of SS-VRI has not been widely adopted, perhaps because of the concomitant need for subsurface drainage.

Site-specific irrigation

Irrigation management can amplify the negative effects of variation or it can help to minimize these effects depending on management. Site-specific irrigation management strategies can be justified to account for variation under non-uniform growing conditions. Reducing areas of excess water applications within a field will decrease the potential for runoff, limit the movement of nutrients and agrochemicals below the plant root zone and create conditions for improving crop yield and quality.

In China and India, the average farm size is about 0.5 ha and 1.5 ha, respectively (FAO, AQUASTAT, 2011) and most use some type of surface irrigation. The small farm size and smaller field sizes mean that many site-specific agriculture practices (e.g. hand-weeding, hand-planting) are commonly used by default in some of the poorest areas of the world. Thus, much of surface irrigation already uses some degree of site-specific management, and managing adjacent paddies or fields to adapt to variation is certainly possible. For example, nutrient content in irrigation water is almost certainly going to change as it descends a series of paddies on a hillside with water cascading from one paddy to the next. Thus, fertility management of each paddy could be adapted to account for the changes, which is a realization of site-specific management at a scale that is appropriate for the situation. However, potential site-specific benefits are not usually evident in these situations because of the low levels of technology, unreliable water deliveries, high labour requirements and degraded soils (e.g. salinity, erosion, waterlogging), plus the limited availability of energy, major and minor nutrients, pesticides, improved seed and other basic resources. Consequently, the greatest opportunity for increased food production in these small fields as well as in many larger fields in more developed areas may be to focus on raising existing systems and their management to higher levels, rather than injecting advanced aspects of site-specific irrigation at this time.

Most PA technologies generally address individual components of the cropping system such as fertilizer management, planting, pest management and harvesting. However, SS-VRI is much more complex over a season because of intricate spatial and temporal effects on soil–water–plant relationships and the ability to control more parameters. In addition, SS-VRI can be strongly influenced by decisions and characteristics covered by other PA technologies. Irrigations are also repeated numerous times during the growing season, which is often not the case with some other PA technologies that are only applied once a season.

Success under these conditions is likely to be achieved only with a true whole-system approach. Thus, a significant challenge is that implementation of SS-VRI generally has the most taxing requirements and the most complicated and costly control systems of all PA technologies. The addition of site-specific application of fertilizers and other agrochemicals by the irrigation system, which is quite common, adds another layer of complexity.

The above discussion suggests that purposely designed pressurized microir-rigation and sprinkler methods are the most suitable for advanced forms of site-specific irrigation. Of these, most of the following discussion will be directed towards self-propelled, continuous-move sprinkler irrigation systems, which cover much larger areas than microirrigation methods and whose use is rapidly expanding (Sadler *et al.*, 2005; Evans and King, 2012).

Site-specific variable-rate sprinkler irrigation

Self-propelled centre pivot and linear move sprinkler irrigation systems are particularly amenable to site-specific approaches because of their current level of automation and ability to cover large areas with a single lateral pipe. Site-specific variable-rate sprinkler systems can apply water at different rates to various discrete areas as the machines move across the field, as shown in Figure 10.1.

Figure 10.1 An example of an SS-VRI system being used for agronomic research to apply water site-specifically to different crops within the same field.

These technologies can be used to treat predefined areas in a whole field or small areas within a field with on/off sprinkler nozzle controls. Small-area (e.g. 0.5–5 ha) SS-VRI systems can address well-defined, localized problem areas where the cost of a full-sized SS-VRI system may not be justified. For example, SS-VRI may be used to manage water applications under only the first span from the pivot point to avoid over-irrigation or to minimize foliar disease incidence in those areas.

Over the past few years, some companies have begun to market centre pivot control panels with an option to change centre pivot travel speed in small increments ranging from 1 to 10 degrees as the machine rotates around the field. This tactic effectively changes application depths in each defined radial sector of the field, and no additional hardware (other than perhaps a GPS) is needed beyond a standard centre pivot. This practice is commonly referred to as speed or sector control. However, variation within fields seldom occurs in long, narrow triangular- or pie-slice-shaped parcels as soil and crop conditions often vary substantially in the radial direction. Thus, adjusting machine speed may not always provide a sufficient level of control for efficient SS-VRI. Despite these limitations, the most common site-specific sprinkler irrigation systems in use today are speed control systems.

In contrast, zone control involves defining irregularly shaped management areas or zones following specific guidelines and applying water differentially to each zone. Special electronic control panels are required plus the addition of a considerable number of valves, supplemental wiring, GPS and other specialized equipment.

Most zone-control SS-VRI systems vary water application depths by various forms of pulse modulation (on/off cycling of spray-type sprinkler heads) that may vary with machine speed. Valves are located on every sprinkler head or group of heads. Water is then applied to each management area or zone by controlling water output from each group of heads along the length of the machine depending on their location in the field.

Figure 10.2 illustrates one example of the conceptual operation of zone-control SS-VRI systems that apply water at different rates as the machines move across the field to adjust for temporal and spatial variation in soil and plant conditions in real time. Zone control has the largest potential for achieving more efficient management of water and energy, and will receive most attention in the remainder of this chapter.

Supporting technologies

Self-propelled sprinkler irrigation systems provide a unique control and sensor platform for optimum management because they pass over the entire irrigated area and can potentially provide real-time feedback every time they are used. Recent innovations in low-voltage sensor and wireless radio frequency (RF) data communications combined with advances in Internet technologies offer tremendous opportunities for advancement of SS-VRI. However, developing the full potential of SS-VRI systems will require integrated development and the use of advanced feedback systems and decision-support systems, which must be integrated with the system controls.

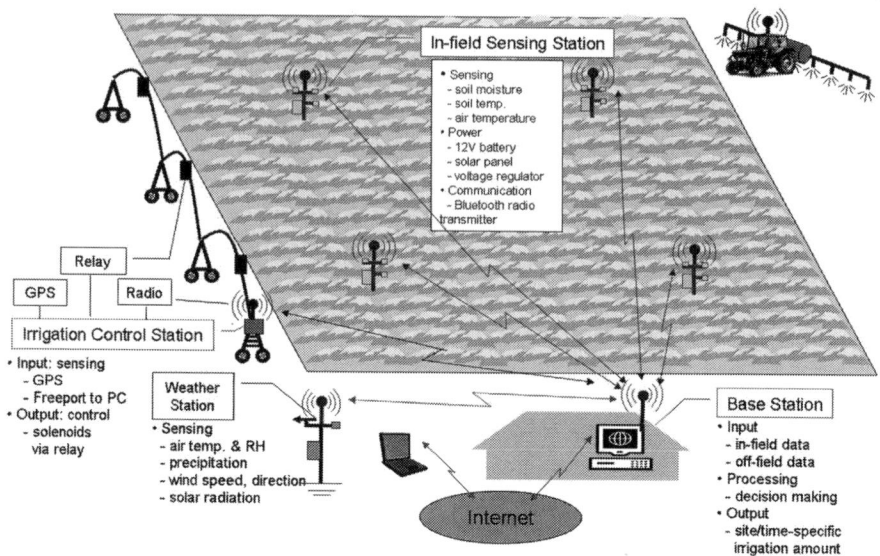

Figure 10.2 Layout of conceptual system of a within-field wireless sensor network for real-time site-specific irrigation (Kim *et al.*, 2008).

Feedback systems

In the early days of precision agriculture, it was often thought that input maps (e.g. yield or soil maps) were sufficient for defining management zones, especially for dryland production. However, for irrigation, within-season temporal variation in rainfall, runoff, infiltration and water use (all of which will affect management) require periodic feedback of plant and soil status. Temporal feedback can be used for real-time irrigation management and improved decision support.

Spatially distributed, within-field plant and soil sensors in combination with agro-weather stations are potentially more accurate for controlling VRI than historical or static map-based input projections. Sensor systems can be used to measure climatic, soil water, plant density, canopy temperature differences, irrigation application amounts and other types of variability. Remote sensing by satellites can also provide synoptic crop feedback in both space and time. Various estimating procedures in combination with predefined management zones can help to account for variation based on real-time feedback from the field. Real-time feedback from multiple sources, in combination with modern wireless communication and computerized control systems, is fundamental to the development and implementation of optimal site-specific irrigation management strategies.

Decision-support systems

Most current irrigation decision-support software (often called scientific irrigation scheduling programs) calculates the timing and duration of water applications
using algorithms that forecast irrigation based on historical weather patterns and
predicted crop water use over a relatively short period (e.g. 3–14 days). Feedback
to estimate crop water use is usually made by spot measurements (e.g. soil water)
and other data after the irrigation is completed and adjustments are made for the
following irrigation event.

Specialized software programs commonly called decision-support systems
(DSS) need to be developed for SS-VRI so that crop-status data obtained
from multiple sources can be used to reduce irrigation use and costs and to
increase water productivity compared to conventional irrigation management
based on uniform crop and field conditions. Decision-support systems must
also consider the general management philosophy and operational preferences
of the owner and or operator. In addition, DSS must be tailored to each field
and region. However, advanced DSSs have not been developed to optimize
production under the expected range of crops and conditions where SS-VRI
will be applied.

Advanced DSSs need to identify those variable-rate strategies that will optimize desired crop production responses. Theoretically, the DSS process should
integrate real-time monitoring of field conditions with predictive plant and pest
models. The integrated DSS, in turn, should interface seamlessly with the irrigation system controls, relays and a GPS to optimally manage SS-VRI as use of this
practice moves forward (McCarthy *et al.*, 2010).

Advanced DSSs will require development of holistic approaches to irrigated
crop management that include optimized irrigation scheduling of whole farms
(multiple fields) over extended time periods and consider yield potentials, crop
and water prices, maintenance and other economic factors to maximize net
returns. The various components of a DSS for site-specific applications should
also have some capability to be self-calibrating and self-learning so that they can
adjust automatically to changing environmental conditions. This process is also
referred to as adaptive control. However, the development of real-time DSSs for
SS-VRI lags substantially behind hardware development.

State of the art of SS-VRI

Computer simulation studies comparing conventional and optimized advanced
site-specific zone control sprinkler SS-VRI have reported water savings of 0 to
26 per cent, which also boosted simulated production (Evans and King, 2012).
However, the benefits of zone or speed control SS-VRI have not been verified
independently by field-scale research. Potential benefits of speed control have
also not been evaluated with either simulation models or field research.

Current uses of SS-VRI zone control technologies on agricultural fields are
generally on a fairly coarse scale. Probably the most common use is for site-
specific non-applications of water to complex-shaped, non-cropped areas such

as waterways, ponds, roads or rocky outcrops where selected interior sprinkler heads are turned off as the machine moves over these areas. The use of SS-VRI for general crop production is still quite limited and is mostly directed towards adjusting for soil textural differences and treating symptoms such as runoff, ponding, limited well capacities, fluctuating water supplies, maintenance issues, nutrient management and related concerns for maximum crop production. These often do not produce measureable savings in water or energy use, although total yields may increase.

Several innovative technologies have been developed by researchers and industry to apply irrigation water variably to meet anticipated whole field management needs in SS-VRI systems (Buchleiter *et al.*, 2000; Evans *et al.*, 2000; Sadler *et al.*, 2000; Evans and King, 2012). Almost all SS-VRI research to date has been directed towards the development and improvement of hardware and basic control software to improve irrigation performance aimed at maximizing total yields.

Case study: site-specific sprinkler irrigation

This case study describes the use of SS-VRI on a large irrigated farm in the state of Washington (USA) for one year (1998) on one 40-ha centre pivot irrigated potato field. This field was planted to potatoes in early spring and managed in the same way as other potato fields on the farm except for irrigation. About 80 per cent of growing season nitrogen fertilizer (32 per cent urea-ammonium nitrate solution) was also applied through the irrigation system, but done uniformly with small irrigation amounts.

Custom-built SS-VRI controls and wiring panels were developed, and the controls, valves, solenoids and wiring were installed on an existing electrically powered centre pivot machine in the autumn of 1997 (Evans and Harting, 1999; Harting, 1999). Because machine speed sets the maximum depths of applied water, a variable-frequency drive motor was installed at the end tower so as to ensure a relatively constant travel speed for more uniform applications. The position of each tower was calculated every minute, based on real-time data from a beacon-corrected GPS located at the end tower. The GPS was connected by wire to a microprocessor at the pivot for data processing and control. The control program compared the known position of every sprinkler group with an internal table specifying the amounts to apply to each area for that irrigation event and sent instructions to each control valve along the length of the machine as warranted.

Irrigation amounts were varied by pulsing sprinkler nozzles on and off over a 61-second time interval to obtain the desired depths of application. Sprinkler heads were spaced about 3.3 m apart, and banks of four sprinkler heads (~13-m widths) were cycled on and off as a group to apply the required amounts of water as the machine moved. Some of the sprinkler groupings are shown in Figure 10.3 as dashed white lines as examples. Each sprinkler group was controlled and managed to apply water site-specifically within each management area covered by that particular bank of nozzles as the machine moved over

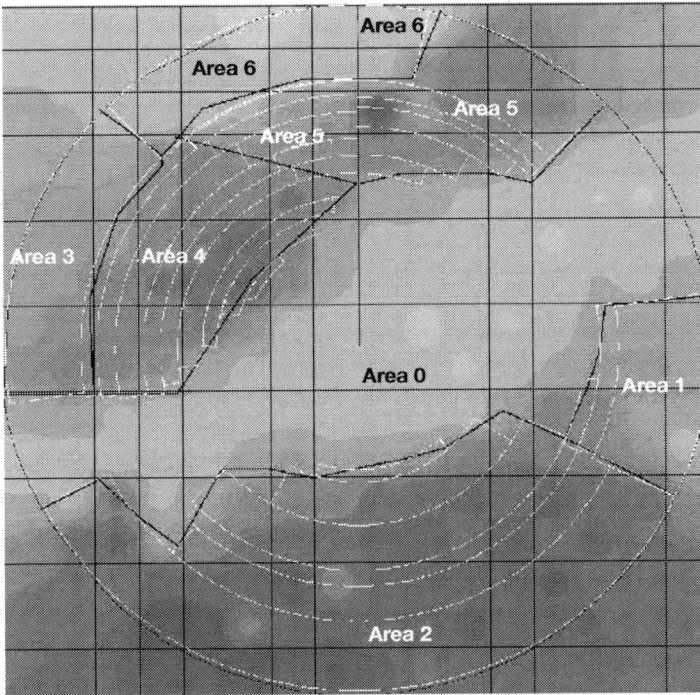

Figure 10.3 Management areas based on the fraction of sand content and topography on a centre pivot irrigated potato field in central Washington, USA, 1998. The lighter the colour, the greater the percentage of sand, which ranged from 37 per cent to 89 per cent in the top 30 cm of soil.

the field. Maximum application amounts were determined by the preset machine travel speed, which did not vary during an irrigation event. Reduced applications were a percentage of the maximum depths applied.

Knowledge of soil and topographic variation is fundamental to the development of site-specific management areas and for prescribing irrigation management in them, primarily because different soil textures will have different soil water-holding capacities. In this case study, the soil varies with depth and directly overlies basalt bedrock, which has its own variable topography. Soil samples were collected on a 0.2-ha grid for the top 60 cm and analysed for percentage sand, silt and clay. Figure 10.3 shows the static management zones or areas that were based on sand-fraction analysis of the top 30 cm, interpolated using geostatistical techniques (see Chapter 4), and then adjusted slightly depending on topography and the farm manager's agronomic judgement. The areas with the largest percentage sand are Areas 0 and 6. The deepest soil is in Areas 0, 1 and 2, and the shallowest soil is in Areas 4 and 5. The highest surface elevation in the field is at the top of Area 6 (~30 m higher than the pivot point) and the lowest point is in the lower left portion of Area 2 (~5 m lower than the pivot point) in Figure 10.3. The field had been irrigated uniformly in the past, and was selected for this study because portions of Areas 0, 1 and 2 in Figure 10.3 often became saturated because of subsurface bedrock conditions even though the soil is considered sandy. The excessively wet areas were primarily due to runoff from the steep slopes in Areas 3, 4 and 5 that collected at the bottom of the hill in Area 0 and also moved laterally to Areas 1 and 2. The excess water resulted in decreased yields and traction problems during cultural activities such as spraying and harvest, which caused considerable expense and loss of income for the farmer, as well as an increased potential of soil disease.

The SS-VRI treatments were done with only small adjustments in water application rates, but the effects were cumulative. Irrigations were more frequent (e.g. 5 times wk^{-1} vs. 3 times wk^{-1}) with smaller applications of water (faster machine travel) than used on other fields on the farm. Applications in Area 0 were not reduced (100 per cent), whereas there were small reductions to amounts applied to Areas 1, 2, 3 and 6 (95 per cent of maximum). Application amounts to Areas 4 and 5 were at the 90 per cent level. Maintaining adequate soil water levels in Area 0, which had the lowest water-holding capacity, was the primary determinant for scheduling irrigations.

Soil water levels were monitored by weekly readings of neutron scatter soil moisture gauges in each of the six areas to about 1.2 m. Aerial infrared photography was also used as a visual check during summer to identify excessively dry areas, emerging pest issues and other problems.

The farm had at least 20 years of good historical records on production, water use, weather and yields for each field. After harvest, the irrigation manager compared yield and water use for the test year by correlation analysis between potato production and applied water on the same field and adjacent fields. Yields were slightly adjusted based on past records on the SS-VRI field and the nearby fields to account somewhat for year-to-year climatic variation. The results from 1998 were estimated to be about 0.75 metric tons ha^{-1} average increase in harvested yield over the entire field, which was an increase of about 8 per cent. Quality of the tubers from the entire field was rated as above average both by market criteria and in comparison with uniform management. It was also estimated that seasonal water applications were reduced by about 12 per cent on the SS-VRI field compared to similar fields in 1998, which resulted in energy savings and less leaching. Observed soil erosion on the steeper slopes was also substantially reduced.

These advantages were not considered sufficient to offset the high cost of retrofit, management and maintenance for SS-VRI. In addition, the low cost of both water and energy helped to make the economics less favourable. Nevertheless, the case study showed the feasibility of SS-VRI approaches and provided data to suggest where price points for these inputs might tip the economic balance.

Economics of SS-VRI

A standard 45-ha centre pivot system will often cost between US$40,000 and US$45,000. A linear move sprinkler system can cost about twice as much per irrigated area as a centre pivot. Land, energy and water supply development can cost much more than the machine. Water development expenses depend on the source of water and power (i.e. electricity, diesel or natural gas) and on pipelines to get the water and power to the pivot point or various water delivery points for lateral move machines. Generally, the largest annual expense for these machines is for power or fuel to pump water.

Economic risk to growers is increased under advanced SS-VRI because of reduced water applications. In addition, it is quite likely that production costs will not decrease in proportion to the decreased inputs because management costs will probably increase.

Zone control adds about US$200–US$550 ha^{-1} over standard systems depending on size and options, without field sensor systems, decision support or basic field mapping services (about US$15–US$20 ha^{-1} depending on the type and scale). Speed or sector control adds US$25–US$125 ha^{-1}, depending on the options.

The use of SS-VRI to avoid irrigating non-cropped areas will clearly be economical in terms of water and nutrient savings, as well as the avoidance of environmental and regulatory penalties. However, in more than 20 years of public and private research pertaining to SS-VRI, demonstrated proof of any economic benefits for agronomic production has failed to materialize. Much of this is probably because the marginal costs for relatively small water savings (e.g. 5–15 per cent) are relatively high, which often makes purchasing and managing SS-VRI for maximum profit difficult to justify economically and the impacts difficult to measure.

Challenges to adoption of SS-VRI

Farmers are entrepreneurs and need to manage crops effectively and efficiently in ways that position them to make the most profit with the least negative impact to the environment. Site-specific irrigation technologies that can spatially direct the amount and frequency of water and appropriate agrochemical applications could potentially be very powerful tools to increase water productivity, reduce inputs and minimize adverse water quality impacts.

The significant under-utilization of zone-control SS-VRI technologies is likely to continue until: (1) cost-effectiveness is increased by higher water and energy costs; (2) regulatory limits on water application amounts are implemented broadly and (3) suitable economic incentives for compliance with environmental and other regulations are implemented and enforced. In addition, information on how to manage these systems with demonstrated increased economic returns using irrigation-VRI technologies must be illustrated with compelling regional research results.

References

Buchleiter, G. W., Camp, C., Evans, R. G. and King, B. A. (2000) 'Technologies for variable water application with sprinklers', in R. G. Evans, B. L. Benham and T. P. Trooien (eds) *Proceedings of 4th Decennial National Irrigation Symposium*, St. Joseph, MI: American Society of Agricultural Engineers, pp. 316–21.

Clemmens, A. J. and Allen, R. G. (2005) 'Impact of agricultural water conservation on water availability', in R. Walton (ed.) *Proceedings of the World Water and Environmental Resources Congress*, Reston, VA: ASCE (CD-ROM).

Evans, R. G. and Harting, G. B. (1999) 'Precision irrigation with center pivot systems on potatoes', in R. Walton and R. E. Nece (eds) *Proceedings of ASCE 1999 International Water Resources Engineering Conference*, Reston, VA: ASCE (CD-ROM).

Evans, R. G. and King, B. A. (2012) 'Site-specific sprinkler irrigation in a water limited future', *Transactions of the ASABE*, 55, 493–504.

Evans, R. G., Buchleiter, G. W., Sadler, E. J., King, B. A. and Harting, G. B. (2000) 'Controls for precision irrigation with self-propelled systems', in R. G. Evans, B. L. Benham, and T. P. Trooien (eds) *Proceedings of 4th Decennial National Irrigation Symposium*, St. Joseph, MI: American Society of Agricultural Engineers, pp. 322–31.

FAO, AQUASTAT (2011) *Global Map of Irrigation Areas*. Available online at <http://www.fao.org/nr/water/aquastat/irrigationmap/index.stm> (accessed 30 November 2011).

Harting, G. B. (1999) 'As the pivot turns', *Resource*, 6, 13–14.

Hillel, D. and Vlek, P. L. G. (2005) 'The sustainability of irrigation', *Advances in Agronomy*, 87, 55–84.

Hoffman, G. J., Evans, R. G., Jensen, M. E., Martin, D. L. and Elliott, R. L. (eds) (2007) *Design and Operation of Farm Irrigation Systems*, 2nd edn, St. Joseph, MI: American Society of Agricultural and Biological Engineers.

Kim, Y., Evans, R. G. and Iversen, W. M. (2008) 'Remote sensing and control of an irrigation system using a wireless sensor network', *IEEE Transactions on Instrumentation and Measurement*, 57, 1379–387.

Lamm, F. R., Ayars, J. E. and Nakayama, F. S. (eds) (2007) *Microirrigation for Crop Production*, Developments in Agriculture 13, Amsterdam: Elsevier.

McCarthy, A. C., Hancock, N. H. and Raine, S. R. (2010) 'VARIwise: a general-purpose adaptive control simulation framework for spatially and temporally varied irrigation at sub-field scale', *Computers and Electronics in Agriculture*, 70, 117–28.

Postel, S. (1999) *Pillar of Sand: Can the Irrigation Miracle Last?*, New York: Worldwatch Books, W. W. Norton & Co.

Sadler, E. J., Evans, R. G., Buchleiter, G. W., King, B. A. and Camp, C. R. (2000) 'Site-specific irrigation: management and control', in R. G. Evans, B. L. Benham and T. P. Trooien (eds) *Proceedings of 4th Decennial National Irrigation Symposium*, St. Joseph, MI: American Society of Agricultural Engineers, pp. 304–15.

Sadler, E. J., Evans, R. G., Stone, K. C. and Camp, C. R. (2005) 'Opportunities for conservation with precision irrigation', *Journal of Soil and Water Conservation*, 60, 371–9.

Siyal, A. A., van Genuchten, M. T. and Skaggs, T. H. (2009) 'Performance of a pitcher irrigation system', *Soil Science*, 174, 312–20.

Smedema, L. K., Vlotman, W. F. and Rycroft, D. W. (2004) *Modern Land Drainage: Planning, Design and Management of Agricultural Drainage Systems*, London: Taylor & Francis.

Turral, H., Svendsen, M. and Faures, J. M. (2010) 'Investing in irrigation: reviewing the past and looking to the future', *Agricultural Water Management*, 97, 551–60.

11 The economics of precision agriculture

Ben P. Marchant, Margaret A. Oliver,
Thomas F. A. Bishop and Brett M. Whelan

Introduction

Precision agriculture (PA) in its various forms can have many benefits for the profitability and sustainability of agriculture and of the environment. However, farmers are yet to be fully convinced that these benefits generally outweigh the costs of implementing PA technology and this has hindered the rate of adoption. The profitability of PA is location-specific because of differences in the soil, management and climate (Lowenberg-DeBoer, 2000). It also varies according to the value of the crop. Therefore, economic analyses are required to determine when PA is worthwhile.

Economics studies how limited resources are allocated (Howard, 2010); in agriculture these resources may include seeds, fertilizers, soil ameliorants such as lime and gypsum, water, pesticides and herbicides, purchase of remotely sensed imagery or proximally sensed surveys, machinery, farm maintenance of such things as fencing, employees, management time, computer facilities and software. The questions that farmers pose are: how will PA help us to allocate these resources and what will the costs and benefits be? A farmer's first concern might be whether the use of PA will increase the profit they expect to achieve each year, but there are other potential benefits. Farming is inherently risky and in any one year unpredictable threats such as pests, crop diseases and adverse weather conditions can reduce yields and lead to monetary losses that threaten the solvency of the farm business. Such risks can be managed through PA. Indeed, as described in Chapter 1, farmers have always applied some form of PA in the management of their land to reduce risk. The idea of farmers being risk-averse is thus not a new concept. Risk-aversion aims to decrease the probability of bad outcomes, that is, low profit and income, but this often reduces the maximum profit that can be achieved. Dillon *et al.* (2007) suggest that PA has as much, if not more, potential to reduce production risk as to increase profit. However, firm economic evidence will be required, since many risk-averse farmers will treat recommendations to reduce fertilizer rates with scepticism. Farmers have always preferred to avoid risk by applying more rather than less fertilizer; one example of this is Stewart and McBratney (2001), who found standard recommendations for N application in cotton fields to be too large and in one field actually caused a decline in yield at higher application rates of N.

Farmers are also aware of sustainability issues and will think beyond the short term to maintain the quality of their land, in particular soil, and the potential for good yields in future. They are conscious of the need to avoid damage to the environment that may invoke penalties from the environmental agencies in their country, and PA enables them to monitor their activities on the land and show that they have followed 'best practice'. The European Union, for example, has designated land where nitrate could drain easily into ground- and surface waters as nitrate vulnerable zones (NVZs) where the application of nitrogen fertilizers is limited. Precision agriculture can ensure that the fertilizer that is permitted is used as efficiently as possible.

The adoption of PA does incur costs. For low-technology solutions these could be limited to the time required to reorganize farming systems. More substantial costs will arise if farmers invest in expensive technology such as yield monitors, sensors (proximal or remote), variable-rate technology (VRT), precision irrigation systems and geographical information systems (GIS). The farmer might also have to spend time learning about and operating this technology or spend money to hire consultants to provide assistance.

Many studies have analysed different types of PA systems within economic frameworks to quantify the costs and benefits of such technologies and to determine if they should be adopted. For example, Robertson *et al.* (2007) quantified the use of autoguidance on Australian grain farms and showed that it reduced overlap in the application of fertilizers and pesticides, resulting in a 10 per cent saving on spraying, fuel use, soil compaction, amount of hired labour, working hours and has increased the timeliness of sowing. Later in this book (Case study 2), Molin describes how autopilot technology in the sugarcane industry of Brazil has increased yields. Maine *et al.* (2010) compared VRT and uniform applications of nitrogen for maize in South Africa and showed that there is some advantage in VRT, but that the benefits vary from year to year. Tenkorang and Lowenberg-DeBoer (2008) discuss the potential benefits of using remote sensing methods to identify injury and disease in crops, the effects of pests, nutrient deficiencies and water stress. They conclude that the evidence of profitability remains 'fuzzy'. Lowenberg-DeBoer (2003) reported that 63 per cent of the 108 economic studies of PA systems conducted at Purdue University in 2000 found the technology to be profitable. Profitability was correlated with crop value, and systems that could manage multiple inputs were more profitable.

Economic studies should be completely objective and lead to reliable conclusions if they are to provide appropriate guidance to farmers about the adoption of PA. However, Bullock and Lowenberg-DeBoer (2007) express concern that subjective decisions about the costs that are included in the analyses and other assumptions mean that the results can be unreliable. They highlight studies of VRT which assume unrealistically that the farmer decides upon fertilizer application rates with perfect knowledge of the forthcoming weather, disease incidence rates and other unpredictable temporal variables. They also note examples where inappropriate statistical methods are applied and some costs are neglected, particularly the costs of gathering data to provide information about the within-field variation of yield. They suggest that these studies are overly optimistic about the profitability of VRT. Therefore, Bullock and Lowenberg-DeBoer (2007) and Tenkorang and Lowenberg-DeBoer (2008) indicate the need for a common framework to assess the profitability of PA systems.

In this chapter we describe how economic assessments of PA are made. We focus particularly on studies of the profitability of VRT, but the same basic principles apply to other PA technologies such as remote sensing or autoguidance systems. First, we present a partial budget assessment of the cost-effectiveness of VRT on a field in Australia. Then we discuss the shortcomings of this approach and how they can be overcome. Finally, we consider the implications of economic assessments for the future adoption of PA.

A case study using partial budgets

Partial budgeting

The profitability of adopting PA systems can be addressed through partial budgeting (Howard, 2010). This approach compares the cost-effectiveness of a few discrete management options and focuses only on the costs and revenues that vary between the options. The first step in the approach is to define the management options clearly. For example, these could be to continue to apply fertilizer with uniform-rate technology (URT) or to switch to VRT. Then the differences between the costs and revenues from the different options are estimated. Less tangible costs and benefits should also be considered. For example, the farmer might have regrets if the use of technology means that they are providing employment for fewer staff or they might enjoy a boost to their reputation if the new technology means that the farm is perceived to be environmentally friendly. Finally, the differences in costs and benefits should be compared so that a rational decision can be made about which management option to adopt. The farmer must decide on their attitude to risk and whether they favour an option that maximizes the profit they expect to achieve on average or whether they are more risk-averse and prefer to ensure that a very large loss in any one year is unlikely.

The example here is from an experiment performed in 2004 to decide whether management practices should be changed on Rosewood, a 75-ha field in the dryland cropping region of northern New South Wales, Australia. Previously fertilizer had been applied at a uniform rate across the field. It had been suggested that the field could be split into two classes with different production potentials and that profits could be increased by using a different application rate in each class. The classes had been delineated based on maps of soil EC_a, elevation and yield in the previous year using the management class delineation approach proposed by Taylor *et al.* (2007). A major challenge in this case study was to estimate the effect that the change in management practice would have on yield and fertilizer costs. It was necessary to know the rates at which the farmer would apply fertilizer in each class and the yields that would result. Therefore, a field trial was established.

The design of the field trial is shown in Figure 11.1a. In each class, six plots were randomly selected to ensure that they did not overlap with one another and were fully contained within the field and class boundaries. They were 100 m long and four harvester-widths apart. Two replicates of three rates of nitrogen fertilizer were used for each class (0, 27, 83 kg N ha⁻¹). The remainder of each class received 60 kg N ha⁻¹, which is the rate normally applied uniformly across the field that was believed necessary to achieve the field's target yield of 4.5 t ha⁻¹.

The yield map that resulted can be seen in Figure 11.1b. This yield map was analysed to determine the effects of the different fertilizer treatments. It was noted that there were patterns of variation in the yield map that were unrelated to the fertilizer treatments. These could potentially distort the relationship between treatment and yield. Therefore, a linear mixed model (Pringle *et al.*, 2010) was fitted to the yield data by residual maximum likelihood (REML; Lark and Cullis, 2004). The linear mixed model divides the yield variation between effects as a result of treatment and other spatially correlated effects. The expected yields under each treatment in each class can be extracted from the fitted model and are shown in Table 11.1.

If the farmer were to use VRT, the decision as to which of the four application rates would be used in each class could be addressed by partial budgeting. The only factors that vary between the different treatments are the fertilizer costs and the yields. In 2004 the price of wheat was AU$150 t⁻¹ and the price of N fertilizer was AU$0.9 kg⁻¹. Table 11.1 shows the costs and revenues under the four treatments in each class. The most profitable variable-rate strategy would

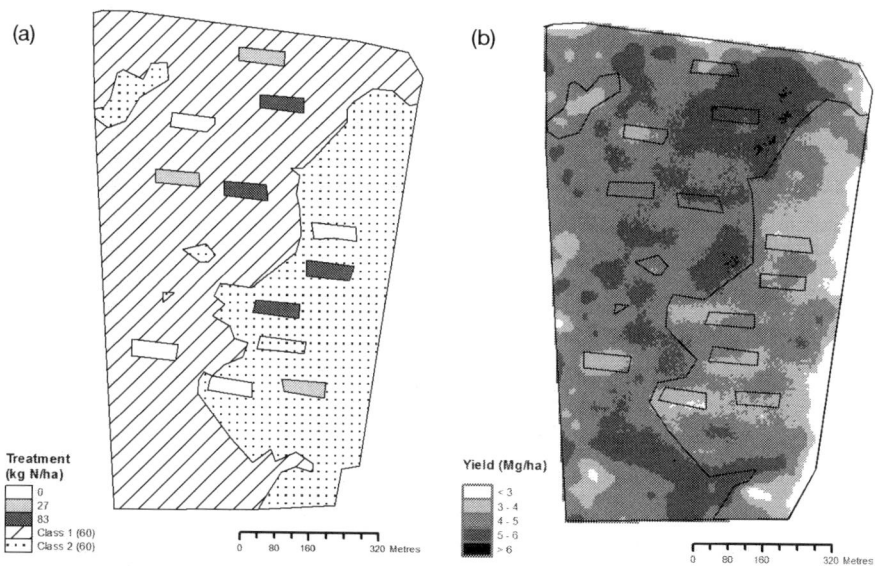

Figure 11.1 Maps of (a) nitrogen fertilizer inputs (kg N ha⁻¹) and (b) yield in the Rosewood field trial.

Table 11.1 Partial budget analysis of the Rosewood field trial

Class	Rate kg N ha⁻¹	Fertilizer cost $AU	Yield t ha⁻¹	Revenue $AU	Profit $AU
1	0	0	3.59	23,156	23,156
1	27	1,045	4.49	28,961	27,916
1	60	2,322	4.76	30,702	28,380
1	83	3,212	5.28	34,121	30,909
2	0	0	3.02	14,496	14,496
2	27	778	3.37	16,176	15,398
2	60	1,728	3.71	17,808	16,080
2	83	2,390	3.57	17,136	14,746

be to apply 83 kg ha⁻¹ in Class 1 and 60 kg ha⁻¹ in Class 2. Across the field this leads to a marginal profit (crop revenue minus fertilizer costs) of AU$46,989.

If we return to the initial question of whether VRT should be adopted, we note that the most profitable uniform rate is 83 kg ha⁻¹. This leads to a marginal profit of AU$45,655, which is AU$1,334 less than the marginal profit of VRT. Therefore, VRT should be adopted if any additional costs beyond those discussed here are less than AU$1,334.

The above illustrates how partial budgeting can be used to make rational management decisions. However, there are several shortcomings with the analysis. Only four different fertilizer rates were considered, whereas it would have been preferable to know the yield for any fertilizer rate. The experiment was based on a single growing season. Therefore, it disregards the variation that can occur between different seasons because of the weather or the presence of pests and disease. Indeed, the analysis assumes that when fertilizer rates are decided the farmer knows the yields that will be achieved under different treatments. Neither is the uncertainty in the relationships between treatment and yield quantified. Such issues were of concern to Bullock and Lowenberg-DeBoer (2007) when they reviewed economic assessments of PA. In the next section we describe their concerns with reference to a model of maize yield.

Economic assessments of PA

Bullock and Lowenberg-DeBoer (2007) suggest that a good understanding of the yield response of a crop to managed inputs is vital to quantify revenue changes with the adoption of PA. Although researchers have been estimating yield response functions for some time Bullock and Lowenberg-DeBoer (2007) indicate that many farmers in the USA know little about how much N fertilizer, for example, would maximize expected profits in a given field and would be unlikely to know how application rates might vary site-specifically to achieve maximization. Yield response in most fields is not uniform because of heterogeneity in the soil properties, drainage conditions, physiography, aspect, microclimate and so on. Therefore, it is vital to conduct field experiments to determine local relationships between yield, managed inputs and environmental variables. As far as is practical, these experiments should observe the yield response to management across the full range of soil conditions on the farm and across the full range of plausible weather conditions. Therefore, long-term experiments are required.

Bullock and Lowenberg-DeBoer (2007) assumed that any field could be divided into a number of spatial units such that the yield response and the results of field trials could be expressed as a model of the form

$$y(s,t) = f[\mathbf{x}(s,t), \mathbf{c}(s), \mathbf{z}(t)]. \tag{11.1}$$

Here $y(s,t)$ is the yield in space unit s in year t, $\mathbf{x}(s,t)$ are the managed inputs such as fertilizers and seed rates which vary between space units and years, $\mathbf{c}(s)$ are the non-managed components that vary between space units but not years (e.g. soil depth), $\mathbf{z}(t)$ are factors such as weather and disease that vary between years but not space units and $f[...]$ is the response function.

Such a model must be based on well-designed experiments. They should include sufficient repetition such that the model can be estimated with certainty. They should gather data from several years so that the effects of different weather conditions can be quantified. Measurements should be made across the full range of soil conditions within the farm. They must also use statistical methods that are suitable for the data collected.

Having established a model, the farmer will choose the managed inputs that lead to the optimal yield response. The $\mathbf{x}(s,t)$ might consist of a small number of discrete input options such as the fertilizer rates in the Rosewood experiment above. However, it might be preferable if the inputs are continuous variables that can take any value. Then the choice of inputs becomes a continuous optimization problem.

Once the variable and uniform rate inputs have been decided, the mathematical model can be used to calculate the crop yields. Assumptions about the price of fertilizer and grain are required so that the differences in costs between the URT and VRT options can be compared and the partial budgeting can be completed.

We illustrate the shortcomings of some economic assessments of PA with reference to a model of maize yield devised by Bullock and Bullock (2000). The model had the same form as Equation (11.1). In a simplified scenario, Bullock and Bullock (2000) assumed that there was only one managed input, one unmanaged spatially variable input and one unmanaged temporally variable input. These were a generic and continuous input (nominal units), soil depth (cm) and effective rainfall (cm), respectively. It was assumed that the spatial units were all of equal area and that half of them had shallow soil depth (20 cm) and the remainder had deep soil depth (150 cm). The aim of the scenario was to show how partial budgets could be used to compare the relative merits of URT and VRT and to highlight potential shortcomings in such analyses which might lead to misleading results.

An ex post *analysis*

Initially, Bullock and Bullock (2000) considered the situation where the input was selected with hindsight or *ex post*. This means that the model contained no uncertain temporal variables. The rainfall for the year was known when the input was selected. We assume it was 76.2 cm. Of course, this situation is unrealistic and long-range forecasts of weather variables are highly uncertain. It was also assumed that the farmer had perfect knowledge of the yield response function and of the soil depth in each spatial unit. The yield responses to inputs where soil is deep and shallow are shown in Figure 11.2a. In this discussion we are primarily interested in marginal profits, by which we mean the differences between input (fertilizer) costs and yields, for the management options. The price of maize (US\$98.33 kg^{-1}) and cost of fertilizer (US\$0.8125 unit^{-1}) are known exactly, therefore the relationship between input and marginal profit and hence the optimal managed inputs can be calculated without uncertainty (Figure 11.2b). The optimal inputs are those that maximize the marginal profits in Figure 11.2b.

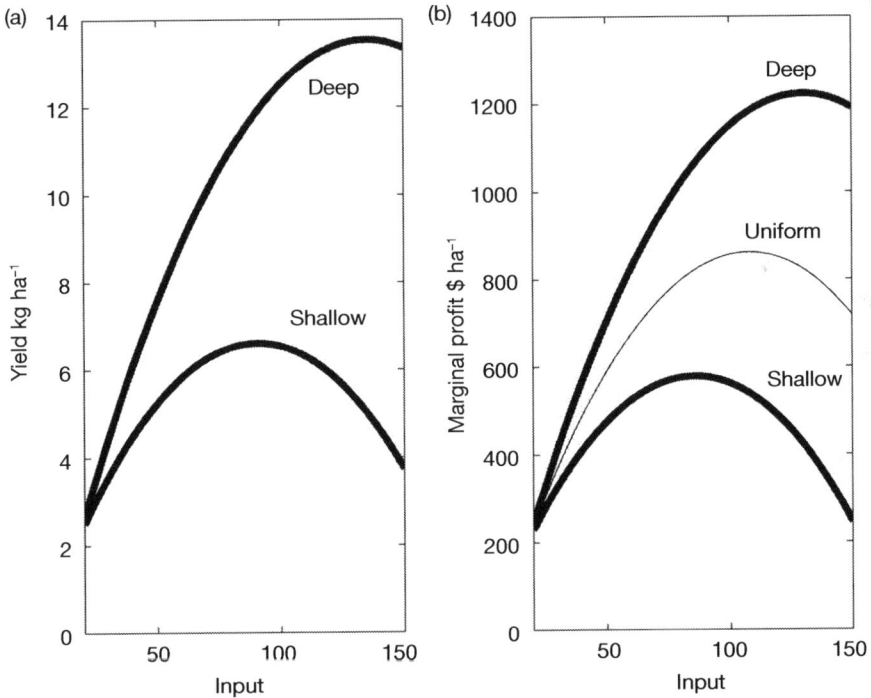

Figure 11.2 (a) Yield response to input on deep and shallow soil units when rainfall is 76.2 cm and (b) marginal profit response to input on deep and shallow soil (bold lines) and for whole field under uniform rate input when rainfall is 76.2 cm (fine line). Calculations are based upon the model of Bullock and Bullock (2000).

If VRT is adopted, we see from Figure 11.2b that the optimal inputs are 86.2 units ha^{-1} on shallow soil and 130.1 units ha^{-1} on deep soil, an average of 108.15 units ha^{-1}. The corresponding marginal profits are US\$576.50 ha^{-1} and US\$1223.34 ha^{-1}, respectively, and the average marginal profit is US\$899.92 ha^{-1}. The fine line in Figure 11.2b is the average of the profit functions on both deep and shallow soil, and it corresponds to the profitability of URT across the field. The optimal uniform rate would be US\$108.40 units ha^{-1} and this would lead to a marginal profit of US\$861.32 ha^{-1} across the field or US\$538.78 ha^{-1} on shallow soil and US\$1183.86 ha^{-1} on deep soil. Thus the use of VRT leads to an increase in marginal profits of US\$38.60 ha^{-1}. Note that in this example only a small proportion (USD\$0.08 ha^{-1}) of this additional profit results from a reduction in total fertilizer costs. The key is that VRT ensures that the fertilizer is better targeted. The farmer also has to account for the purchase, hire and running costs of the VRT before choosing the preferred strategy, but we will assume they amount to less than US\$38.60 ha^{-1} year^{-1} and that VRT appears to be cost-effective. We note that the farmer's attitude to risk is irrelevant in this example since the yield response is known with certainty.

An ex ante *analysis*

In their second scenario Bullock and Bullock (2000) considered an *ex ante* analysis. They acknowledged the uncertainty of rainfall when inputs were selected, and used historical data to form a probability distribution. For example, we assume a Gaussian distribution with mean rainfall of 76.2 cm and standard deviation of 12.7 cm. Bullock and Bullock (2000) also acknowledged that the yield response function could vary across the field and the sources of this variation might not be included in the model. The sources of this variation for each spatial unit are unknown and are drawn from independent and identical Gaussian distributions with zero mean and a standard deviation of 0.5 kg ha^{-1}.

The yield response now becomes a random variable. For any specified management input the yield is no longer known with certainty. This uncertainty can be explored by simulating multiple realizations of the rainfall and the spatial random function. Each of these realizations can be substituted into the mathematical model (Equation 11.1) for the specified input value to produce a distribution of possible yields and hence a distribution of marginal profits. The choice of management input controls the expected value and uncertainty of the marginal profits. In Figure 11.3 we see how the expected marginal profit and the bounds of the 95 per cent confidence interval vary with management input on deep and shallow soil. The confidence interval bounds are extreme marginal profits which the farmer would only expect to see once every 40 years. The maximum expected marginal profit on shallow soil is US\$550.82 ha^{-1} and this occurs when the input is 87 units ha^{-1}. Similarly the maximum expected marginal profit on deep soil is US\$1208.37 when the input is 130 units ha^{-1}. These are the variable-rate inputs that would be chosen by a farmer whose aim is to maximize expected profit, and across the whole field they would lead to an expected marginal profit of US\$879.60. A more risk-averse farmer would be more interested in limiting adverse outcomes. They could select inputs of 80 units ha⁻¹ on shallow soil and 128 units ha⁻¹ on deep soil since these inputs maximize the lower 95 per cent confidence interval on the marginal profits. These inputs are only slightly different from those that maximize the expected profit. However, the differences could be larger in a region where the variation in weather is more severe and might lead to larger yield losses, for example Australia. Across the whole field this risk-averse farmer could expect to see a marginal profit as small as US$704.65 ha⁻¹ once every 40 years.

The relationship between uniform-rate inputs and marginal profit is shown in Figure 11.4. The maximum expected profit of US$839.74 ha⁻¹ (US$39.86 ha⁻¹ less than the VRT profit) occurs when the input is 107 units ha⁻¹. If the uniform input is 103 units ha⁻¹ then the lower 95 per cent confidence bound is maximized and the once-every-40-years low marginal profit is US$666.55 ha⁻¹. This is US$38.10 ha⁻¹ less than the corresponding VRT profit, so it appears that both a profit-maximizing and a risk-averse farmer will favour VRT provided that the start-up costs do not exceed these benefits.

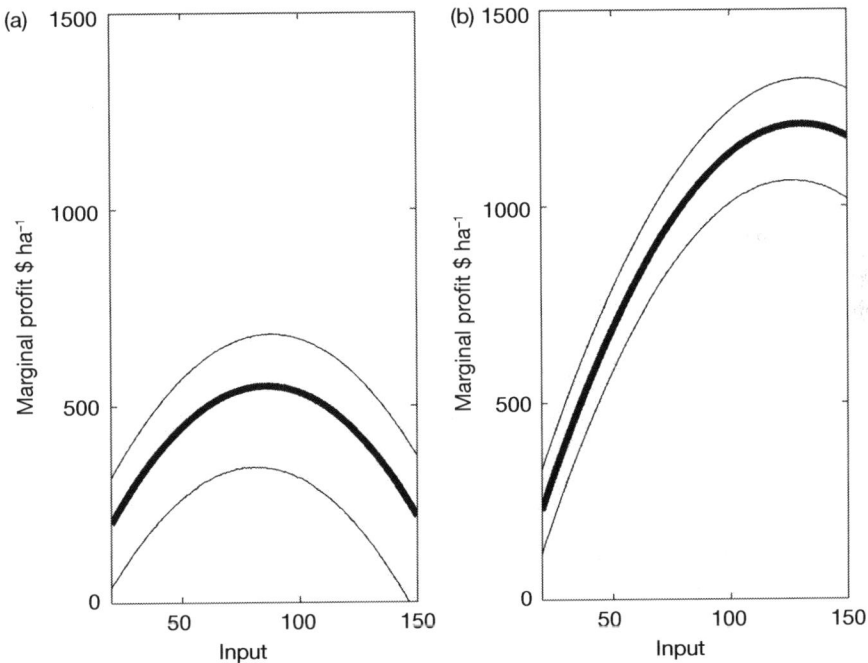

Figure 11.3 Marginal profit response to input on (a) shallow soil and (b) deep soil when rainfall is drawn from a Gaussian distribution with a mean of 76.2 cm and standard deviation of 12.5 cm. Thick lines are the expected values and fine lines are 95% confidence intervals. Calculations are based on the model of Bullock and Bullock (2000).

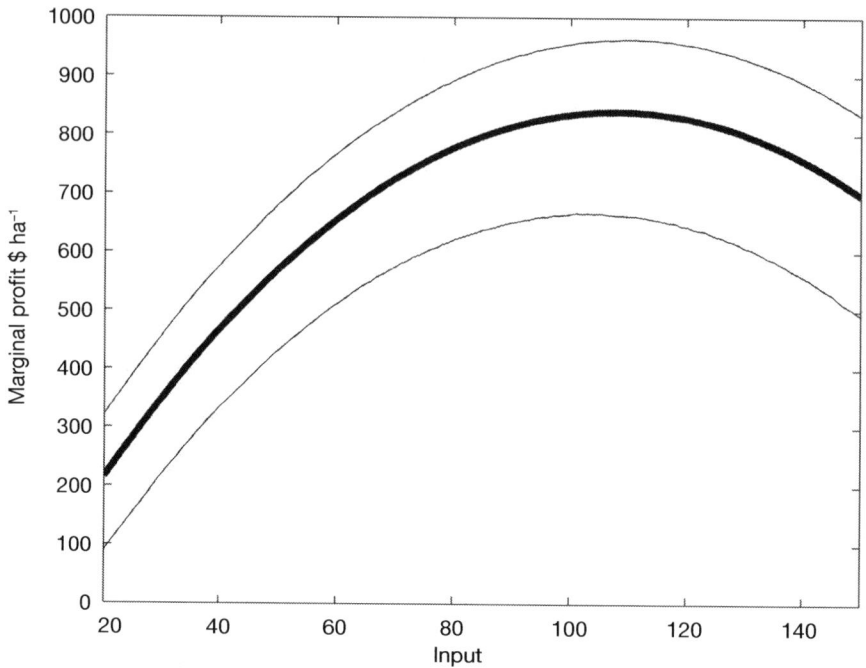

Figure 11.4 Marginal profit response to input using URT when rainfall is drawn
from a Gaussian distribution with a mean of 76.2 cm and a standard
deviation of 12.5 cm. The thick line shows the expected values and fine
lines are 95 per cent confidence intervals. Calculations are based on the
model of Bullock and Bullock (2000).

The value of information

In their final scenario, Bullock and Bullock (2000) consider the value of information. So far we have assumed that soil depth is known exactly for each spatial
unit, but in reality it will be estimated from some form of survey. Bullock and
Bullock (2002) considered how the precision of this survey affected the profitability of VRT. They used spatial analysis methods similar to those in Chapter 4
of this book to determine how errors in estimating soil depth would lead to
sub-optimal management decisions and hence to smaller profits. They assumed
that soil depth would be measured on a regular grid and interpolated by kriging.
They simulated this process using sampling grids of different intensities and
found that as the sampling became less dense VRT could not target the inputs
correctly and hence the benefits in comparison to URT started to decrease. In
addition, the cost of the survey of soil depth had to be included in the economic
analysis and this led to a further reduction in the benefit of VRT compared to
URT. This third scenario illustrates that VRT and information are economic
complements. Information about soil depth becomes valuable because it is
required by farmers to apply VRT. Equally, VRT becomes more valuable if the
information regarding soil depth is available.

A review of economic assessments of PA

Bullock and Lowenberg-DeBoer (2007) reviewed recent attempts to estimate yield response functions and to compare the profitability of URTs and VRTs. They indicated the need for a common framework to compare the different studies effectively. They classified the studies according to three factors: (1) whether management decisions would be made with the benefit of hindsight (*ex post*) or based on the farmers' uncertain knowledge of unmanaged temporal variables $z(t)$ such as weather and disease incidence (*ex ante*); (2) the experiments used to determine the yield response to different inputs and in particular crop-type, experimental design, number of seasons and input rates and (3) the statistical method used to analyse the experimental data and whether it accounted for the spatial correlation. They had realized that spatial correlation meant standard regression methods were inadequate, and that approaches such as spatially correlated linear mixed models (Lark and Wheeler, 2003) or spatial auto-regression (Anselin *et al.*, 2004) should be applied.

Anselin *et al.* (2004) and Lambert *et al.* (2004) suggested that economic assessment of VRT depends on the specification of the model used. When they used models that did not account for spatial correlation between yield data, VRT nitrogen application appeared to be unprofitable. However, it proved to be profitable once the spatial correlation had been modelled appropriately.

Similar themes were evident in the studies reviewed by Bullock and Lowenberg-DeBoer (2007). In an analysis of a one-year, one-field agronomic experiment on maize yields in Argentina, Anselin *et al.* (2004) compared the results when the yield response functions were determined *ex post* by standard regression methods and spatial auto-regression models. They found that the spatial-regression models led to more accurate information and benefits over the standard regression models of US$2.62 ha^{-1} with URT and US$4.78 ha^{-1} with VRT. This was consistent with the point made by Bullock and Bullock (2000) that information and VRT are economic complements because VRT increases the value of information and vice versa. The study demonstrated an *ex post* benefit from VRT of greater than their threshold to cover application costs (US$6 ha^{-1}). However, they did not include the costs of obtaining spatial information and running their field trial. Similarly, Lambert *et al.* (2006) also saw benefits of VRT when compared to URT in *ex post* analyses. Liu *et al.* (2006) and Ruffo *et al.* (2006) both conducted *ex ante* analyses of the benefits of VRT. Liu *et al.* (2006) included between-year variation in their yield models by assuming that the model coefficients were drawn from a probability model each year. Ruffo *et al.* (2006) included weather variables in their yield response function in the manner of Bullock and Bullock (2000). Liu *et al.* (2006) did not see a significant benefit of VRT, but Ruffo *et al.* (2006) saw a benefit of US$2.16 ha^{-1} from combining site-specific information with VRT. The benefit of site-specific information for URT was only US$0.28 ha^{-1}. The review by Bullock and Lowenberg-DeBoer (2007) concluded that although shortcomings remained in agronomic experimentation, these had been recognized by agricultural economists and agronomists, who were making rapid progress in addressing them.

Discussion

We have illustrated that results on the profitability of PA systems have been mixed largely because of the heterogeneity of agricultural systems. The viability of VRT will depend on the crop grown, the extent of within-field variability, crop prices, the available information, the costs of VRT and many other factors. However, we have also seen that some discrepancies between studies arise because of differences in assumptions in the economic analyses. Some assume that perfect information about spatial variation is available and disregard the cost of obtaining it. Others ignore the effects of temporal variation and in effect assume that the farmer knows what the weather and crop prices will be in the forthcoming season when deciding on the managed inputs. When temporal variation is included it is often modelled from a few seasons' data, which might not be representative of the long-term variation. The uncertainty of economic predictions can be underestimated if the effects of spatial correlation between experimental plots are disregarded. Clearly, economic analyses will only lead to reliable results if the primary sources of uncertainty and costs associated with VRT are accounted for.

Bullock and Lowenberg-DeBoer (2007) are encouraged by the progress that is being made in research into the economics of VRT. They note that many authors now accept the need for appropriate statistical methods and that they realize that management decisions must be made *ex ante* rather than *ex post*. They emphasize that information and VRT are economic complements. Variable-rate technology and PA are only of value if detailed information on the spatial variation of the agricultural systems is available. Equally, such information might only be of value if PA is applied. Average values of properties across fields or management zones might well suffice for uniform applications. They note that given that PA is relatively new, it is not surprising that in some cases there is insufficient spatial information to make it viable. However, now that PA exists it may provide the motivation to obtain more information, which in turn will make PA more profitable.

The costs of obtaining information must be taken into account in assessing economic viability. Soil sampling and analysis remain costly, but once some kind of 'on-the-go' analysis becomes available to assess the soil condition for the nutrients the crop requires, then it is likely we shall see a surge of interest in PA on a large scale. Equipment to measure pH on-the-go is already available and this is often done at the same time as measuring the soil's apparent electrical conductivity (Schirrmann *et al.*, 2011). The cost of information might also be reduced if machinery is developed that can combine information gathering and management into one computer-controlled application. The automated weed control system described by Gerhards in Chapter 9 is one example of this. An improvement in long-range seasonal weather forecasting would also benefit PA. For example, if a farmer knew whether the growing season was going to be wetter than normal, fertilizers could be applied accordingly in a PA context to fields that suffer poor drainage in parts and drought in others. Football Field in Bedfordshire, England, is an example of where yields in parts of the field vary according to rainfall years (see Oliver and Carroll, 2004). In wet years the northern part of the field, which has a low elevation, yields poorly, whereas the southern higher part yields well. This situation is reversed in years that are drier than normal. Therefore, we could imagine a simplistic case where we have two response functions for each part of the field, one for wet years and one for dry years.

It is worth noting that the quality of information does not correspond directly to the quantity of data. Basso *et al.* (2009) indicate that the challenge will be to identify the usefulness, importance and relevance of the data recorded to optimize farm efficiency. Blackmore (2009) supports this view in saying that the biggest challenge for farmers will be to manage information both on and off farms effectively to improve the economic viability of agricultural production and to reduce environmental impact. A benefit of this will be increased transparency and traceability in farm management.

The complementary relationship between PA and information is one reason to feel positive about the future viability of PA. It is also important to note that there are important intangible benefits of VRT that are often excluded from economic analyses. These include better soil management and maintenance of soil quality through the targeting of inputs, a better understanding of the variability within fields and opportunities to control the environmental impact of agriculture. These are all important in ensuring food security for the world's increasing population.

References

Anselin, L, Bongiovanni, R. and Lowenberg-DeBoer, J. (2004) 'A spatial econometric approach to the economics of site-specific nitrogen management in corn production', *American Journal of Agricultural Economics*, 86, 675–87.

Basso, B., Fountas, S., Sartori, L., Cafiero, G., Pedersen, S. M., Sørensen, C., Pesonen, L., Werner, A. and Blackmore, B. S. (2009) 'Farmer's risk in decision making: the case of nitrogen', in E. J. van Henten, D. Goense and C. Lokhorst (eds) *Precision Agriculture '09*, Wageningen, The Netherlands: Wageningen Academic Publishers, pp. 927–33.

Blackmore, B. S. (2009) 'Futurefarm: the European farm of tomorrow', in E. J. van Henten, D. Goense and C. Lokhorst (eds) *Precision Agriculture '09*, Wageningen, The Netherlands: Wageningen Academic Publishers, pp. 887–92.

Bullock, D. S. and Bullock, D. G. (2000) 'From agronomic research to farm management guidelines: A primer on the economics of information and precision technology', *Precision Agriculture*, 2, 71–101.

Bullock, D. S. and Lowenberg-DeBoer, J. (2007) 'Using spatial analysis to study the values of variable rate technology and information', *Journal of Agricultural Economics*, 58, 517–35.

Dillon, C. R., Stombaugh, T. S., Kayrouz, B. M., Salim, J. and Koostra, B. K. (2007) 'An educational workshop on the use of precision agriculture as a risk management tool', in J. V. Stafford (ed.) *Precision Agriculture '07*, Wageningen, The Netherlands: Wageningen Academic Publishers, pp. 861–7.

Howard, W. H. (2010) 'Economics of Precision Agriculture', Module. Available online at <http://www.precisionag.org/html/ch14.html> (accessed April 2012).

Lambert, D. M., Lowenberg-DeBoer, J. and Bongiovanni, R. (2004) 'A comparison of four spatial regression models for yield monitor data: a case study from Argentina', *Precision Agriculture*, 5, 579–600.

Lambert, D. M., Lowenberg-DeBoer, J. and Malzer, G. L. (2006) 'Economic analysis of spatial-temporal patterns in corn and soybean response to nitrogen and phosphorus', *Agronomy Journal*, 98, 43–54.

Lark, R. M. and Cullis, B. R. (2004) 'Model-based analysis using REML for inference from systematically sampled data on soil', *European Journal of Soil Science*, 55, 799–813.

Lark, R. M. and Wheeler, H. C. (2003) 'A method to investigate within-field variation of the response of combinable crops to an input', *Agronomy Journal*, 95, 1093–104.

Liu, Y., Swinton, S. M. and Miller, N. R. (2006) 'Is site-specific yield response consistent over time? Does it pay?', *American Journal of Agricultural Economics*, 88, 471–83.

Lowenberg-DeBoer, J. (2000) 'Economic analysis of precision farming', in A. Borem, M. Gludice, D. Queiroz, E. Mantovanni, L. Ferreira, F. do Valle and R. Gomide (eds) *Agricultura de Precisao*, Vicosa, Brazil: Federal University of Vicosa, pp. 147–80.

Lowenberg-DeBoer, J. (2003) 'Precision farming or convenience agriculture', presented at 11th Australian Agronomy Conference, Geelong, Australia. Available online at <www.regional.org.au/au/asa/2003/i/6/lowenberg.htm> (accessed 22 May 2013).

Maine, N., Lowenberg-DeBoer, J., Nell, W. H. and Alemu, Z. G. (2010) 'Impact of variable-rate application of nitrogen on yield and profit: a case study from South Africa', *Precision Agriculture*, 11, 448–74.

Oliver, M. A. and Carroll, Z. L. C. (2004) *Description of Spatial Variation in Soil to Optimize Cereal Management*, Project Report 330, London: HGCA.

Pringle, M. J., Bishop, T. F. A., Lark, R. M., Whelan, B. M. and McBratney, A. B. (2010) 'The analysis of spatial experiments' in M. A. Oliver (ed.) *Geostatistical Applications for Precision Agriculture*, Dordrecht: Springer, pp. 243–67.

Robertson, M., Carberry, P. and Brennan, L. (2007) *The Economic Benefits of Precision Agriculture: Case Studies from Australian Grain Farms*, CSIRO Report, Campbell, ACT, Australia: CSIRO.

Ruffo, M., Bollero, G., Bullock, D. S. and Bullock, D. G. (2006) 'Site-specific production functions for variable rate corn nitrogen fertilization', *Precision Agriculture*, 7, 327–42.

Schirrmann, M., Gebbers, R., Kramer, E. and Seidel, J. (2011) 'Soil pH mapping with an on-the-go sensor', *Sensors*, 11, 573–98.

Stewart, C. M. and McBratney, A. B. (2001) 'Development of a methodology for the variable-rate application of fertiliser in irrigated cotton fields', in P. C. Robert, R. H. Rust and W. E. Larson (eds) *Proceedings of the 5th International Conference on Precision Agriculture, Bloomington, Minnesota, USA, 16–19 July, 2000*, Madison, MI: Agronomy Society of America, pp. 1–13.

Taylor, J. A., McBratney, A. B., and Whelan, B. M. (2007) 'Establishing management classes for broadacre agricultural production', *Agronomy Journal*, 99, 1366–76.

Tenkorang, F. and Lowenberg-DeBoer. J. (2008) 'On-farm profitability of remote sensing in agriculture', *Journal of Terrestrial Observation*, 1, 50–9.

12 Spatially distributed experimentation

Tools for the optimization of targeted management

Robert G. V. Bramley, Roger A. Lawes and Simon E. Cook

Background

Precision agriculture (PA) has been identified as one of several technologies with the potential to assist in enhancing agricultural productivity (Carberry *et al.*, 2011) and eco-efficiency (Keating *et al.*, 2010). In particular, it is seen as a means of maintaining agricultural output while reducing the risk associated with producing that output (Figure 12.1). This is an idea that is consistent with the suggestion that a basic principle of PA is to increase the likelihood of a beneficial outcome by better targeting of inputs to production potential (Cook and Bramley, 1998; Lawes and Robertson, 2011). Note that in Figure 12.1, 'risk' is synonymous with inputs (e.g. amount of fertilizer applied) or the cost of inputs. Of interest in the context of this chapter is the suggestion (Figure 12.1) that, whereas PA improves productivity through reduction of risk, the harnessing of genotype by environment by management (G × E × M) interactions is one of the principal avenues by which productivity may be increased for the same level of risk. Here, we contend that PA may also assist in raising productivity for the same level of risk by exploiting the G × E × M interactions that may exist at the sub-field scale.

Classical approaches to agronomic field experimentation are founded on the analysis of variance (ANOVA) and its variants in which the effects of site variation are assumed to be removed from the experimental results through the use of blocking and randomization. In the absence of methods to examine this variation, this is a reasonable approach (Cook and Bramley, 1998). It is consistent with both the 'null hypothesis of Precision Agriculture' (Whelan and McBratney, 2000) and the operating assumption employed by farmers prior to adopting PA, that uniform management is the optimal strategy. Thus, such trials, which are typically conducted in small plots, are assumed to represent the system that the farmer is faced with managing. However, as the various chapters and their illustrations in this book attest, the tools of PA (yield monitors, remote and proximal sensing, etc.) provide a means of indicating that both agricultural production and the land which supports it typically vary over short distances (a few metres or tens of metres); that is, agricultural production is influenced by

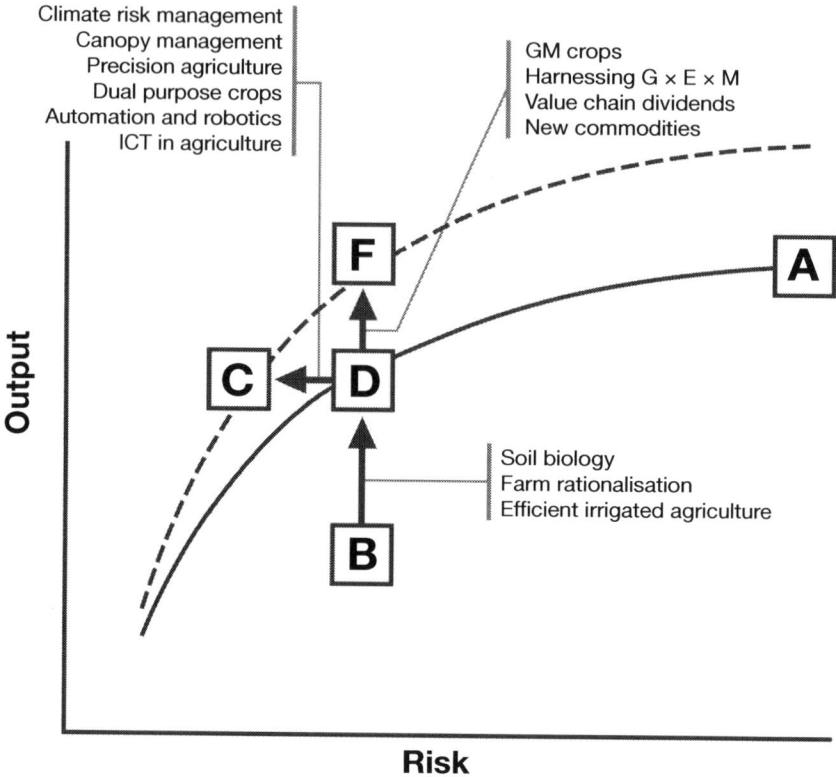

Figure 12.1 A return–risk framework and technologies that have the potential to affect Australian dryland agriculture over the next 20 years and beyond (Carberry *et al.*, 2011, adapted from Keating *et al.*, 2010).

Note
A and D are points on the efficiency frontier (solid line) for the best technologies used in 2010, and C and F are points on a new efficiency frontier (dashed line) which arises through the use of new technologies. B represents a position below the current efficiency frontier. GM = genetically modified; G × E × M = genotype by environment by management interactions; ICT = information and communication technologies. Note that Carberry *et al.* (2011) also identified technologies that affected Australian dryland agriculture in the 30 years prior to 2010. These included controlled traffic (B to D), conservation agriculture (D to F) and fertilizer management (D to C).

variations in edaphic, biotic and abiotic stresses. Recent experience in South Australia (R. Bramley, unpublished data) indicates that the patterns of yield variation commonly differ between crop types grown in rotation in the same fields (for example, legumes and pulses vs. cereals), so we can say that agricultural production is highly subject to G × E interactions. Many crop-yield maps also exhibit a 'stripiness' aligned with the row orientation that cannot be explained in terms of underlying variation in the land, and which may be attributed to artefacts of management. These might include a blocked spray nozzle or seed chute, or a missed or double pass when fertilizing. Thus, G × E interactions are also affected by M. Other elements of M such as depth of sowing, choice of rotation or, indeed, the intended product stream for a crop, which may impact on input expenditure, may also interact with G × E.

Against this background, Bramley *et al.* (2005) highlighted the potential importance of the choice of location for traditional plot-based agronomic field experiments (Figure 12.2) and the likelihood that this choice could impact markedly on both the results obtained and the merits or otherwise of extrapolating these results (or recommendations derived from them) to other locations. Thus, while the measured treatment response in a plot-based experiment may be affected by the underlying variation of the experimental site (Figure 12.2), the range of variation in a potentially useful covariate (e.g. clay content, plant available water, soil salinity, etc.) may be insufficient to promote robust extrapolation to other areas, if indeed such covariation is examined – which often it is not. An obvious solution to this problem, which reflects 'Krige's relation' (Webster and Oliver, 2007), is to conduct the experiment over a much larger area in which the range of variation in the useful covariate is also larger. Such a solution also offers the desirable benefit of enabling the experiment to be conducted at a scale that is meaningful to the farmer and using the same equipment that the farmer uses, rather than small-scale equipment or manual approaches. This solution therefore also lends itself to farmer-initiated experimentation, which is important, given that most farmers like to try new ideas out before implementing them across their entire businesses (Pannell *et al.*, 2006). In other words, the problem may be framed from the manager's perspective and experiments conducted at the scale at which management occurs. An additional problem with the traditional plot-scale approach is that, fundamentally, it seeks to address the question as to whether treatment 'A' is better than treatment 'B'. As will be illustrated below, for some experiments, this is less useful than an approach which recognizes that both 'A' and 'B' may be beneficial, albeit in different locations within the same field.

Using examples from the Australian grains and wine industries, this chapter illustrates a different approach to experimentation. Rather than small co-located plots (e.g. Figure 12.2) and ANOVA, we illustrate some spatially distributed approaches supported by geostatistical and spatial analysis. Our underlying premise in promoting these approaches is that in order to move from point D to point F in Figure 12.1, one has to progress via point C, with PA being the key to achieving the move to F. In other words, with appropriately crafted experimentation, PA may lead to increased agricultural productivity through reducing risk *and* increasing outputs.

It should be noted that in the experiments outlined in this chapter we focus on comparison of yield (or some index of yield) rather than profit as used to compare treatments in the Rosewood case study in Chapter 11. Other responses that could be used in experiments could be those related to crop quality, the incidence and severity of disease or an environmental goal, such as reduced amount of N leaching to groundwater.

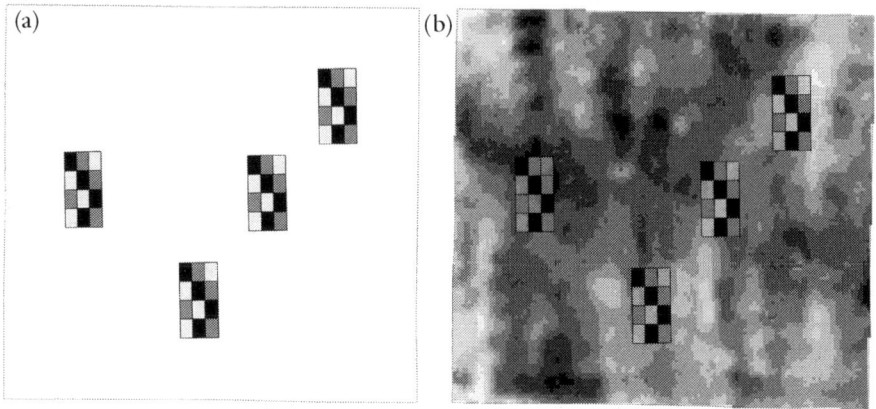

Figure 12.2 Possible locations for a plot-based experiment: (a) in the absence of any knowledge of variability and (b) when a yield map is available (Bramley *et al.*, 2005).

Note
It is easy to see that in (b) the results of the experiment may be compromised by whether the plots are located in a lower (light-grey) or higher (dark-grey) yielding area. Equally, the range of yield variation in the area under any group of plots is considerably less than for the block as a whole. If this variation was being driven by a covariate such as clay content, for example, extrapolation of the results from one of the possible experimental locations based on clay content would be problematic since it would only be possible to assess the effect of clay content on yield over a restricted range of clay contents.

Spatially distributed experimentation

Whole-of-block approaches

Early attempts at spatially distributed experimentation included the whole-of-block experiments of Adams and Cook (1997, 2000), Cook *et al.* (1999) and Pringle *et al.* (1999, 2004). These all used variable-rate fertilizer technology (VRT) to vary inputs using embedded designs (McBratney, 1985) of varying complexity. The simplest was the checkerboard (Adams and Cook, 1997; Cook and Bramley, 1998; Pringle *et al.*, 1999, 2004) in which a control plus three rates (i.e. four rates in total) of nitrogen (N) were applied in a highly replicated 'check' design (210 replications) over an entire field (~80 ha) of wheat (*Triticum aestivum*). The check design ensured that each treatment had all of the other treatments as its nearest neighbours. Therefore, it was possible to assess the response of any treatment by contrast with the neighbouring control using a moving window approach (Cook *et al.*, 1999). As an alternative, Pringle *et al.* (1999, 2004) used kriging to interpolate treatment-specific maps from which, for any given location, a simple linear or quadratic representation of yield response was derived using the interpolated location-specific response measure for each treatment. In such ways, the variable response to different treatments was explored. Simple map algebra (subtraction of the map for one treatment from that for another) could also be used to examine the differences between pairs of treatments over the entire field.

More complex designs were also used, taking advantage of the capacity of variable-rate (VRT) equipment to vary fertilizer rates continuously. These included the sine-wave design (Adams and Cook, 2000) and the 'stepped wave' (Cook *et al.*, 1999), both of which made use of wavelet analysis (Lark and Webster, 1999) for the assessment of treatment response.

While all of these experiments required the involvement of experts for their design, implementation and analysis, the checkerboard (Adams and Cook, 1997; Cook and Bramley, 1998; Pringle *et al.*, 1999, 2004) provided an important learning experience. As it turned out, unfavourable seasonal conditions meant that the checkerboard experiment could not be used by the collaborating farmer as a basis for decision-making in the following crop. However, because the scale of the experiment rendered it subject to spatially variable factors of practical significance (e.g. waterlogging, the effects of weeds, micronutrient deficiency), the farmer could see that the approach was a powerful learning aide for the improved management of fertilizer inputs. He therefore used a similar design over his entire farm the following year (Dr S. E. Cook, pers. comm.). This was arguably a more important result than the intended one, since it pointed towards the likelihood that PA could markedly add power to the natural inclination of farmers to experiment.

At much the same time, Doerge and Gardner (1999) used a 'split planter' to plant different varieties of maize in adjacent strips using a similarly highly replicated design. Following yield monitoring at harvest, separate maps for each variety could be interpolated for the entire field, and simple map algebra used to identify areas in the field that were particularly suited to the different varieties. As with the checkerboard, this approach provided a powerful learning tool for both farmers and agronomists seeking to optimize crop production – in this case by matching crop characteristics (i.e. G) to site characteristics (i.e. E).

Lanyon and Bramley (2004) and Bramley *et al.* (2005) tried a similar approach in a 6.8-ha vineyard in the Langhorne Creek region of South Australia which was subject to a number of soil constraints. Using a highly replicated strip design, they explored options for improving grapevine (*Vitis vinifera* cv. Shiraz) access to soil moisture through either ripping or the application of compost; these treatments were compared to the status quo (control). As with the earlier grains work (see above), their analysis depended on simple map algebra, albeit following the simple, but important, use of a common global variogram for the interpolation of each treatment (Lanyon and Bramley, 2004). This was achieved by offsetting the coordinates for each measure of treatment response (e.g. yield) in an easterly direction by a distance much larger than the dimensions of the vineyard, prior to variogram estimation. Thus, control data were not offset, treatment 1 data were offset by 1,000 m and treatment 2 by 2,000 m; the vineyard was approximately square, with sides of ~300 m. By using these offsets, the possible effects of treatment-specific differences in the variograms used for interpolation were removed from their analysis. All of the data for each treatment were used to characterize the spatial structure in the data, but each treatment was interpolated separately, albeit using the common variogram. Subtraction of the control from the two treatment maps allowed the spatially variable benefit of each to be assessed together with their relative merits.

A positive feature of the Lanyon and Bramley (2004) experiment was the value that the vineyard manager attributed to being able to see the spatially variable response to the treatments for himself – something that would not have been possible had small plots been used. Nevertheless, anecdotal evidence suggested that some viticultural researchers regarded this experiment with circumspection, given the lack of any measure of the statistical significance of treatment response. As has been pointed out before (e.g. Bramley *et al.*, 2005; Bramley, 2009), farmers do not generally make decisions on the basis of statistical significance, but instead use their own measures of 'significance'. These might be based on considerations such as the magnitude of the response (e.g. additional yield), the benefit:cost trade-off, or whether the benefit is large enough to justify the additional effort required in doing something new (i.e. 'Can I be bothered?'), among a raft of other possible considerations (e.g. Pannell *et al.*, 2006) set against the experience of the farmer in managing his or her land over many years. Indeed, whereas intuition derived from years of practical experience is a critical guide to decision making for many farmers (McCown 2002; Fountas *et al.*, 2005), researchers tend to rely on statistical significance alone.

A solution to this impasse has recently been provided by Bishop and Lark (2006, 2007), who proposed a geostatistically based method for the analysis of 'landscape-scale' experiments. This method is based on the assumption that the observed responses (e.g. yields) to a set of different treatments may be regarded as realizations of a set of spatially auto- and cross-correlated random variables, and enables estimation of these responses for any part of the experimental site. As a consequence, the estimation of contrasts between different treatments at different locations or over regions of different size and shape is also possible, as is estimation on a point-wise basis of the statistical significance of these contrasts. Of note with this methodology is the fact that it uses all of the data to map each treatment response, rather than just the data pertaining to each treatment alone. Through simple weight of numbers, this enhances the likelihood that robust treatment-specific maps are produced.

Application of the Bishop and Lark (2006) methodology has recently been demonstrated in whole-of-block experiments conducted in vineyards in the Clare Valley region of South Australia (Panten *et al.*, 2010; Panten and Bramley, 2011) and in Tasmania (Bramley *et al.*, 2011). In the latter case, a simple replicated strip design, like that used by Doerge and Gardner (1999), was used in two experiments which aimed to evaluate alternate spray management strategies for the control of either powdery mildew (*Erysiphe necator*) or botrytis bunch rot (*Botrytis cinerea*), both commercially important grapevine diseases. In the Clare Valley, a more complex strip design, similar to the one used by Lanyon and Bramley (2004) and Bramley *et al.* (2005), was used to explore different vineyard floor management options for increasing vine vigour (Figure 12.3). In this example, the vineyard manager was concerned that the organic management system being used may have led to an inadequate supply of N to the vines and or competition for water from the permanent ryegrass sward in the inter-rows. Accordingly, the treatments (application of compost or the planting of a cereal or legume cover crop) were chosen to counter these possible effects. The strip design used here involved splitting the row length into three so that each row had a third of its length allocated to each of the treatments. Subsequent analysis (Dr K. Panten, pers. comm.) suggested that this offered no benefit over the simpler strip design used by Bramley *et al.* (2011), which is therefore preferred for its greater ease of implementation.

As Figure 12.3 illustrates, the response to the treatments was spatially variable. Arguably of greater importance, was the finding that while the use of mulch (RM) did not give a response that was significantly different from the compost-based control (RC), the benefits delivered by the cereal or legume (CL) treatment were markedly spatially variable. Thus, only approximately one half of the

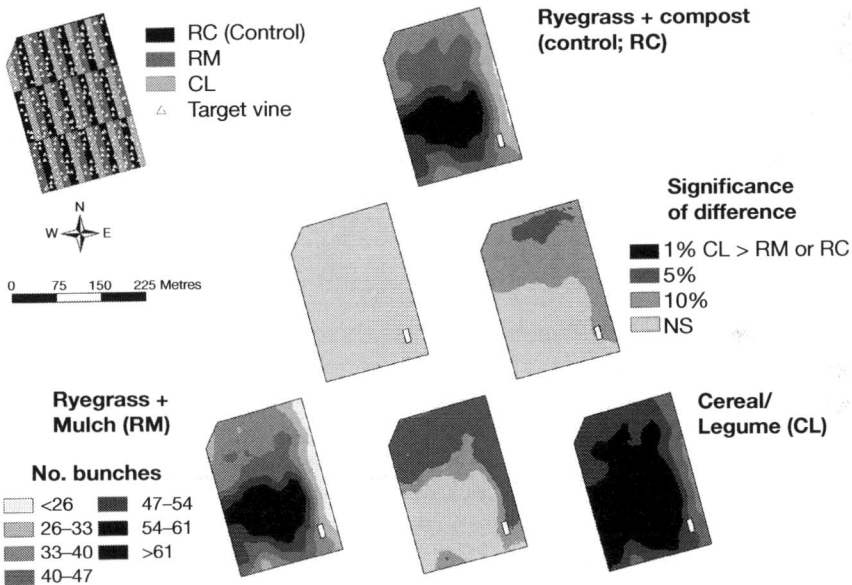

Figure 12.3 A whole-of-block experiment conducted in South Australia (Panten *et al.*, 2010; Panten and Bramley, 2011) in which a highly replicated design (top left) was applied in 2004 over an entire 4.8-ha vineyard to assess the merits of three mid-row management strategies.

Note
The method of Bishop and Lark (2006) was used to analyse treatment effects. Treatment-specific responses, in this case assessed by the number of bunches per m of row measured for 378 target vines at vintage in 2006, are shown in the maps in the corners of the triangle, with the significance of the difference between these shown in the maps positioned between them; the same legends apply to each bunch number or difference map. Although there is no significant difference between the RM and RC treatments, the benefits delivered by the CL treatment are markedly spatially variable.

vineyard would have benefitted from a change in vineyard floor management. As with the earlier examples, understanding of the spatial variation in treatment effect would not have been gained using a conventional plot-based design. It is also worth highlighting in relation to this experiment that neither the vineyard manager nor his staff considered implementation of the experimental design (Figure 12.3) prohibitively difficult. Further, the manager reported the benefit of being able to wander through the experiment and see for himself that the treatment effects differed in different parts of the vineyard. Likewise, the vineyard managers involved in the Tasmanian work (Bramley *et al.*, 2011) attached considerable value to the fact that the experiment was implemented at a commercial scale using their own equipment. Llewellyn (2007) has noted that the farmer who observes the results from a trial on their farm is likely to consider this information highly relevant; if they also consider the results to be true or favourable, they are likely to act on them.

Simple strips

In spite of the positive views of the collaborating farmers involved in the checkerboard and vineyard experiments, discussion of the whole-of-block approach with farmer members of SPAA-Precision Agriculture Australia (www.spaa.com. au) suggests a reluctance amongst grain growers to take on the whole-of-block approach. Aversion to risk is one reason for this (see Chapter 11), and lack of access to appropriate technical support (Cook and Bramley, 2000; Robertson *et al.*, 2011) is also of considerable importance. Another reason is that adopters of PA in the Australian grains industry have tended to pursue zone-based management rather than continuous variable rate. Mindful of the fact that the full range of yield variation in a field can often be encountered in a single row (Bramley, 2009), a possible acceptable alternative to a whole-of-block experiment is to use a single carefully located strip. This is the approach taken with so-called 'N-rich strips' (Raun *et al.*, 2002; Schepers and Raun, 2011).

The idea with N-rich strips is that a strip of crop, corresponding to a single pass of the application equipment, is given a luxury application of N either at sowing or early in the season. Prior to making a mid-season fertilizer decision, a proximal (or remote) crop sensor is then used to compare some index of crop performance (e.g. NDVI) between the N-rich strip and neighbouring areas that have received the normal dressing of N. Based on this comparison, the farmer can make a decision as to the need for and rate of any mid-season application. This decision may be aided by calculation of the response index (RI; Raun *et al.*, 2002) – the ratio of crop performance in the N-rich strip to that in a neighbouring part of the normally managed remainder. If the strip is placed in such a way as to traverse management zones, then a zone-based variable-rate strategy for mid-season application may be derived.

Somewhat surprisingly, we are unaware of attempts to analyse RI spatially. Typically, it is calculated on a whole-of-strip or zone basis. However, a simple method for doing this has recently been proposed by Lawes and Bramley (2012) based on a moving window *t*-test. Figure 12.4 shows the results of such an analysis for an 'N-rich strip' experiment implemented by a farmer in a 72-ha field of barley (*Hordeum vulgare* cv. Commander) in South Australia during the 2010 growing season. Yield maps obtained in 2005, 2006 and 2009, when cereals (either wheat or barley) had also been grown, were used together with a high resolution soil survey of the soil's apparent electrical conductivity, EC_a, (EM38 – see Chapter 6), to derive three management zones (Taylor *et al.*, 2007). These were characteristically relatively low, medium or high yielding. The barley was sown with a uniformly applied base rate of fertilizer in June 2010. Approximately one month later (i.e. at early tillering), an N-rich strip, oriented through the three zones, was applied through a boom spray (40-m wide) using a liquid N fertilizer. Further uniform applications of N were made to the entire field during the season such that overall, the strip received a total of 201 kg N ha^{-1}, with the remainder of the field receiving 100 kg N ha^{-1}. Of course, with timely analysis of RI, these mid-season applications could have been varied if the farmer thought it warranted.

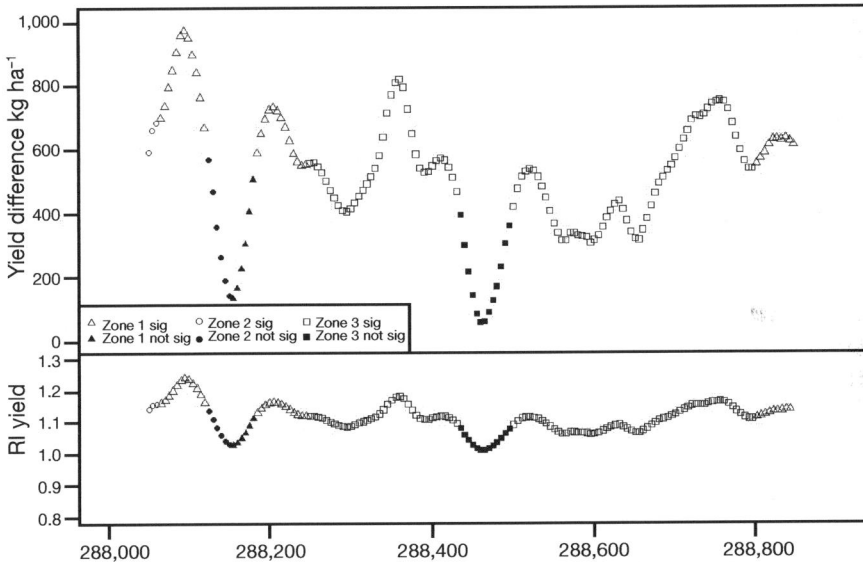

Figure 12.4 Variable response to luxury N application in an 'N-rich strip' experiment conducted with barley in South Australia in 2010 (Lawes and Bramley, 2012).

Note
Because the strip was oriented approximately east–west, it was convenient here to plot response as a function of its easting coordinate. Both yield differences (top) and RI (bottom) were calculated on a moving window basis (windows of 5 consecutive pixels in length) using paired 'strips' of data extracted from the yield map (pixels of 5 m). The pairs in this case comprised a strip of pixels extracted from the centre of the N-rich strip, and a control strip located 40 m to the north; the boom spray used to impose the strip was 40-m wide. Zones 1, 2 and 3 were characteristically relatively high, medium and low yielding. A moving window *t*-test was used to determine whether the yield difference was statistically significant (p < 0.05).

As can be seen in Figure 12.4, the magnitude of the yield difference between the N-rich strip and the adjacent control, whether this difference was statistically significant ($p < 0.05$), and the RI all varied along the strip; these parameters also varied within the zones. The latter result highlights an oft-forgotten fact that, while management zones are less variable than the field as a whole, they are *not* uniform. In fact, 2010 was an unusual year for cereal growing in South Australia. The expectation is for the crop to be water-limited in this environment, but heavy mid-season rains meant that it was N-limited (hence the high N application rates used) and unusually high yields for this region were achieved (the mean yield in this field was approximately 4.8 t ha^{-1}). Accordingly, RI values calculated from yield data were lower than expected, with the results collectively suggesting that a uniform application of the N-rich rate would have been a better strategy than either the lower (i.e. normal) rate, or a zone-based targeted strategy. However, given the fluctuating yield response to N, Figure 12.4 also suggests that if continuous variable rate had been used rather than zone-based management, it would probably have delivered a benefit.

Future directions

The suggestion that the tools of PA be used for experimentation 'under real farm conditions' was first made by Reetz (1996), who observed that 'such research will likely be more readily accepted and the results more easily adapted to other farms'. To this, one could add that just as PA promotes the ability to undertake the sort of spatially distributed experimentation described here, so too does its successful implementation depend on it, since the basis on which farmers make site-specific decisions does, itself, need to be site-specific. What then would facilitate adoption of this approach?

There are currently two major impediments to the easy implementation of the Bishop and Lark (2006) approach as used in the examples discussed above, one of which also applies to the moving window analysis of strips. First, farmers and or their advisers (who are more likely to be charged with experimental analysis) need easy-to-implement software which, within easy-to-define bounds, needs to do the experimental analysis for them. At the time of writing, software that performs the Bishop and Lark (2006) analysis is not available. The Lawes and Bramley (2012) moving window methodology is easily implemented in a spreadsheet program, but has yet to be conveniently packaged, while the moving window regression method used for analysis of the checkerboard (Cook *et al.*, 1999) also remains unpackaged (R. J. Corner, pers. comm.). However, the extraction of the paired strips of pixels from yield maps on which the Lawes and Bramley (2012) method depends is complex and needs the development of an additional software tool, and or access to high-level GIS skills, especially for fields in which the strip does not run either north–south or east–west. Both problems might be readily addressed by a software engineer working with a spatial analyst. Of more significant concern is the limit placed on the implementation of the Bishop and Lark (2006) method by the capacity of normal desk-top computers.

Figure 12.3 was derived from data collected by hand from 378 'target vines'. However, the obvious attraction of the whole-of-block approach lies in its potential for response measures (e.g. yield, NDVI) to be made using on-the-go technologies such as yield monitors or proximal crop sensors. Importantly, these will be likely to remove the requirement for the farmer to do anything additional to his normal practice; of course, they also dramatically increase the spatial resolution of the resulting maps. However, like more conventional kriging (e.g. Webster and Oliver, 2007), the Bishop and Lark (2006) method involves the inversion of large matrices during some of the processes (e.g. cross-validation), the size of which is governed by the size of the dataset. This matrix inversion may very quickly go beyond the computing capacity that is likely to be available to most potential users, especially if a global search window is used. Cressie and Kang (2010) identified a similar limitation in the analysis of the large datasets typical of PA when this is based on global kriging, as opposed to local kriging such as is used in yield map interpolation (Taylor *et al.*, 2007). Thus, to facilitate the use of the whole-of-block approach, there is a need for research aimed at either adapting the Bishop and Lark (2006) method to a local, rather than global basis, modifying it along the lines of the kriging method proposed by Cressie and Kang (2010) and or identifying an appropriate means of reducing the number of data points in large datasets without a significant loss of information. In the interim, the strip-based approach of Lawes and Bramley (2012) may offer a viable, albeit less sophisticated and spatially explicit alternative. Nevertheless, we see the development of such easy-to-use analytical tools for non-experts as essential because spatially distributed experimentation clearly offers a means by which underlying variability can be used as a useful experimental tool for understanding G × E interactions. With subsequent appropriate, spatially explicit management, they also offer the means by which farmers may move to a new efficiency frontier (Figure 12.1). The information provided by such experiments meets the requirements of salience, credibility and legitimacy (Cash *et al.*, 2003) and may therefore form the basis for knowledge systems that contribute to sustainable development.

Acknowledgements

Preparation of the chapter was funded by CSIRO under the auspices of the Sustainable Agriculture Flagship. It draws on research which in turn was funded by Australia's graingrowers, grapegrowers and winemakers through their investment bodies, the Grains Research and Development Corporation and the Grape and Wine Research and Development Corporation. The input and assistance of many colleagues and collaborators in that work are gratefully acknowledged, as is Dr Peter Carberry (CSIRO Sustainable Agriculture Flagship) for providing Figure 12.1. The comments of Dr Carberry and of Tony Webster (CSIRO Ecosystem Sciences) and of Dr Matthew Pringle (Queensland Department of Environment and Resource Management) on an earlier draft of this chapter are also gratefully acknowledged.

References

Adams, M. L. and Cook, S. E. (1997) 'Methods of on-farm experimentation using precision agriculture technology', presented at the 1997 ASAE Annual International Meeting (paper no. 97-3020), St. Joseph, MI: ASAE.

Adams, M. L. and Cook, S. E. (2000) 'On-farm experimentation: application of different analytical techniques for interpretation', in P. C. Robert, R. H. Rust and W. E. Larson (eds), *Proceedings of the 5th International Conference on Precision Agriculture*, Madison, MI: Agronomy Society of America, pp. 1–17.

Bishop, T. F. A. and Lark, R. M. (2006) 'The geostatistical analysis of experiments at the landscape-scale', *Geoderma*, 133, 87–106.

Bishop, T. F. A. and Lark, R. M. (2007) 'A landscape scale experiment on the changes in available potassium over a winter wheat cropping season', *Geoderma*, 141, 384–96.

Bramley, R. G. V. (2009) 'Lessons from nearly 20 years of precision agriculture research, development and adoption as a guide to its appropriate application', *Crop and Pasture Science*, 60, 197–217.

Bramley, R. G. V., Lanyon, D. M. and Panten, K. (2005) 'Whole-of-vineyard experimentation: an improved basis for knowledge generation and decision making', in J. V. Stafford (ed.) *Proceedings of the 5th European Conference on Precision Agriculture*, Wageningen, The Netherlands: Wageningen Academic Publishers, pp. 883–90.

Bramley, R. G. V., Evans, K. J., Dunne, K. J. and Gobbett, D. L. (2011) 'Spatial variation in response to "reduced input" spray programs for powdery mildew and botrytis identified through whole-of-block experimentation', *Australian Journal of Grape and Wine Research*, 17, 341–50.

Carberry, P. S., Bruce, S. E., Walcott, J. J. and Keating, B. A. (2011) 'Innovation and productivity in dryland agriculture: a return–risk analysis for Australia', *Journal of Agricultural Science*, 149, 77–89.

Cash, D. W., Clark, W. C., Alcock, F., Dickson, N. M., Eckley, N., Guston, D. H., Jager, J. and Mitchell, R. B. (2003) 'Knowledge systems for sustainable development', *Proceedings of the National Academy of Sciences (PNAS)*, 100, 8086–91.

Cook, S. E. and Bramley, R. G. V. (1998) 'Precision agriculture: opportunities, benefits and pitfalls of site-specific crop management in Australia', *Australian Journal of Experimental Agriculture*, 38, 753–63.

Cook, S. E. and Bramley, R. G. V. (2000) 'Precision agriculture: Using paddock information to make cropping systems internationally competitive', presented at Emerging Technologies in Agriculture: From Ideas to Adoption Conference, 25–26 July 2000, Bureau of Rural Sciences, Canberra, pp. 1–8. Available online at <http://adl.brs.gov.au/data/warehouse/brsShop/data/scook_paper.pdf> (accessed October 2011).

Cook, S. E., Adams, M. L. and Corner, R. J. (1999) 'On-farm experimentation to determine site-specific responses to variable inputs', in P. C. Robert, R. H. Rust and W. E. Larson (eds) *Proceedings of the 4th International Conference on Precision Agriculture*, Madison, WI: American Society of Agronomy, Crop Science Society of America and Soil Science Society of America, pp. 611–21.

Cressie, N. and Kang, E. L. (2010) 'High-resolution digital soil mapping: Kriging for very large datasets', in R. A. Viscarra-Rossel, A. B. McBratney and B. Minasny (eds) *Proximal Soil Sensing*, Dordrecht: Springer, pp. 49–63.

Doerge, T. A. and Gardner, D. L. (1999) 'On-farm testing using the adjacent strip comparison method', in P. C. Robert, R. H. Rust and W. E. Larson (eds),

Proceedings of the 4th International Conference on Precision Agriculture, Madison, WI: American Society of Agronomy, Crop Science Society of America and Soil Science Society of America, pp. 603–9.

Fountas, S., Blackmore, S., Ess, D., Hawkins, S., Blumhoff, G., Lowenberg-DeBoer, J. and Sorensen, C. G. (2005) 'Farmer experience with precision agriculture in Denmark and the US Eastern corn belt', *Precision Agriculture*, 6, 121–41.

Keating, B. A., Carberry, P. S., Bindraban, P. S., Asseng, S., Meinke, H. and Dixon, J. (2010) 'Eco-efficient agriculture: concepts, challenges and opportunities, *Crop Science*, 50, S109–S119.

Lanyon, D. M. and Bramley, R. G. V. (2004) *Ameliorating Soil Constraints to the Performance of Established Vineyards*, Final report to the Grape and Wine Research and Development Corporation on project CSL 01/01 Adelaide, Australia: GWRDC. Available online at <www.gwrdc.com.au/webdata/resources/project/CSL0101.pdf> (accessed October 2011).

Lark, R. M. and Webster, R. (1999) 'Analysis and elucidation of soil variation using wavelets', *European Journal of Soil Science*, 50, 185–206.

Lawes, R. A. and Bramley, R. G. V. (2012) 'A simple method for the analysis of on-farm strip trials', *Agronomy Journal*, 104, 371–7.

Lawes, R. A. and Robertson, M. J. (2011) 'Whole-farm implications on the application of variable rate technology to every cropped field', *Field Crops Research*, 124, 142–8.

Llewellyn, R. S. (2007) 'Information quality and effectiveness for more rapid adoption decisions by farmers', in N. Turner and T. Acuna (eds) *'Ground-Breaking Stuff': Proceedings of the 13th Australian Society of Agronomy Conference, Perth, Western Australia, 10–14 September 2006*, pp. 148–56.

McBratney, A. B. (1985) 'The role of geostatistics in the design and analysis of field experiments with reference to the effect of soil physical properties in the field', in D. R. Nielsen and J. Bouma (eds) *Soil Spatial Variability*, Wageningen, The Netherlands: Pudoc, pp. 3–8.

McCown, R. L. (2002) 'Changing systems for supporting farmers' decisions: problems, paradigms, and prospects' *Agricultural Systems*, 74, 179–220.

Pannell, D. J., Marshall, G. R., Barr, N., Curtis, A., Vanclay, F. and Wilkinson, R. (2006) 'Understanding and promoting adoption of conservation practices by rural landholders', *Australian Journal of Experimental Agriculture*, 46, 1407–24.

Panten, K. and Bramley, R. G. V. (2011) 'Viticultural experimentation using whole blocks: evaluation of three floor management options', *Australian Journal of Grape and Wine Research*, 17, 136–46.

Panten, K., Bramley, R. G. V., Lark, R. M. and Bishop, T. F. A. (2010) 'Enhancing the value of field experimentation through whole-of-block designs', *Precision Agriculture*, 11, 198–213.

Pringle, M. J., McBratney, A. B. and Cook, S. E. (1999) 'Some methods of estimating yield response to a spatially-varied input', in J. V. Stafford (ed.) *Precision Agriculture '99*, Sheffield, UK: Sheffield Academic Press, pp. 309–18.

Pringle, M. J., McBratney, A. B. and Cook, S. E. (2004) 'Field-scale experiments for site-specific crop management. Part II: A geostatistical analysis', *Precision Agriculture*, 5, 625–645.

Raun, W. R., Solie, J. B., Johnson, G. V., Stone, M. L., Mullen, R. W., Freeman, K. W., Thompson, W. E. and Lukina, E. V. (2002) 'Improving nitrogen use efficiency in cereal grain production with optical sensing and variable rate application', *Agronomy Journal*, 94, 815–20.

Reetz, H. F. (1996) 'On-farm research opportunities through site-specific management', in P. C. Robert, R. H. Rust and W. E. Larson (eds) *Proceedings of the 3rd International Conference on Precision Agriculture*, Madison, WI: American Society of Agronomy, Crop Science Society of America and Soil Science Society of America, pp. 1173–6.

Robertson, M. J., Llewellyn, R. S., Mandel, R., Lawes, R., Bramley, R. G. V., Swift, L., Metz, N. and O'Callaghan, C. (2011) 'Adoption of variable rate technology in the Australian grains industry: status, issues and prospects', *Precision Agriculture*, 13, 181–99.

Schepers, J. S. and Raun, W. R. (2011) 'Nitrogen sensors to fine tune the nutrient management decision making process', in J. A. Delgado and R. F. Follett (eds) *Advances in Nitrogen Management for Water Quality*, Ankeny, IA: SWCS, pp. 206–29.

Taylor, J. A., McBratney, A. B. and Whelan, B. M. (2007) 'Establishing management classes for broadacre agricultural production', *Agronomy Journal*, 99, 1366–76.

Webster, R. and Oliver, M. A. (2007) *Geostatistics for Environmental Scientists*, 2nd edn, Chichester, UK: John Wiley & Sons.

Whelan, B. M. and McBratney, A. B. (2000) 'The "null hypothesis" of precision agriculture management', *Precision Agriculture*, 2, 265–79.

Part 4
Case studies

CS1 Sampling and mapping in precision agriculture

Margaret A. Oliver

Introduction

In Chapter 1, I summarized the history and value of geostatistics in precision agriculture, and in Chapter 4 Marchant *et al.* considered the accuracy of geostatistical predictions for precision management decisions. Sampling plays an important role in the accuracy of such predictions, but this seldom receives sufficient attention. For precision farming surveys in the United Kingdom and many other countries, sampling is often at about one sample per hectare or even more sparse (Godwin and Miller, 2003) because of the costs involved in both obtaining and processing the samples. This approach to sampling, however, takes no account of either the spatial scale of variation or of how many sampling points might be needed for further analyses. As a result many of the maps created for precision agriculture are based on too few samples to provide accurate maps by interpolation, and are not suitable for site-specific management. Geostatistics can provide information about the scale of variation through the variogram and provide predictions at unsampled places by kriging.

If variograms of soil or crop properties are available from previous surveys for the area or another area with similar soil parent material or a similar crop, they can provide a guide to the spatial scale of variation. To ensure that the variogram (see below) can be estimated accurately the sampling interval should be between a third and a half the scale of variation; this kind of relation with the range is often used as a guide to sampling interval (see Kerry *et al.*, 2010). Variograms computed from ancillary data such as aerial photographs, remotely and proximally sensed images, yield maps and so on can also provide an indication of the approximate scale of spatial variation and guide sampling as described above. Variation over very short distances of a few metres can be smoothed by mixing together several small cores of soil or, for example, by taking several leaves from one or more plants over a given sample support to create a bulked or composite sample. In addition to issues of spatial scale in sampling there is the question of the number of samples to be taken. To compute an accurate variogram by the usual method of moments estimator (see below) requires at least 100 data (Webster and Oliver, 1992). Kerry and Oliver (2007) showed that a variogram estimated by the more advanced method of residual maximum likelihood (REML) can provide more accurate predictions with fewer data than one estimated conventionally (this method is beyond the scope of this book).

Geostatistics

Theory

Geostatistics is underpinned by the notion that values distributed in space tend to be spatially correlated at places near to one another. This spatial correlation can be described by the variogram, which is the central tool of geostatistics. It summarizes the way that properties vary from place to place. The usual method of moments estimator (Matheron, 1965) for computing the empirical semi-variances from data, $z(\mathbf{x}_1)$, $z(\mathbf{x}_2)$,... is:

$$\hat{\gamma}(\mathbf{h}) = \frac{1}{2m(\mathbf{h})} \sum_{i=1}^{m(\mathbf{h})} \{z(\mathbf{x}_i) - z(\mathbf{x}_i + \mathbf{h})\}^2, \tag{CS1.1}$$

where $z(\mathbf{x}_i)$ and $z(\mathbf{x}_i + \mathbf{h})$ are the actual values of the variable, Z, at places $\mathbf{x}_i$ and $\mathbf{x}_i + \mathbf{h}$ and $m(\mathbf{h})$ is the number of paired comparisons at lag $\mathbf{h}$, where $\mathbf{h}$ is a vector in both direction and distance. The experimental or sample variogram is obtained by changing $\mathbf{h}$.

The experimental variogram must be fitted by a model to describe the spatial variation (see Webster and Oliver, 2007). It is usual to fit several suitable models to determine which fits best in a least squares sense. The most commonly fitted models are the spherical and exponential ones. Figure CS1.1 shows an annotated spherical function with the model parameters indicated. The parameters from the best-fitting model can then be used for geostatistical interpolation or kriging.

Kriging is an optimal method of prediction or estimation in geographical space; it is optimal in the sense of unbiasedness and minimum variance. In addition, kriging provides not only predictions but also estimates of the kriging

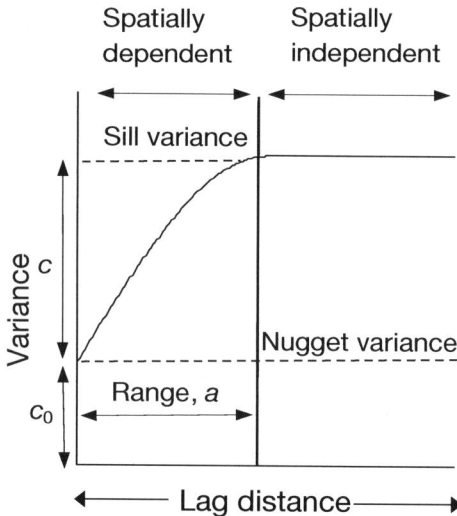

Figure CS1.1 Spherical variogram function with axes and model parameters labelled.

variances or errors at each prediction point. It is a method of local weighted moving averaging of the observed values of a random variable, Z, within a neighbourhood, V. It can be done for point supports (punctual kriging) or block supports of various size (block kriging). In precision agriculture, kriged predictions are often required over areas that are larger than the sample support of the data, therefore block kriging is more widely used. There are many types of kriging, but ordinary kriging is the method most widely used. If the variable, Z, has been measured at sampling points, $\mathbf{x}_i$, $i = 1,\dots N$, we use this information to estimate its value over the unknown block, B, by:

$$\hat{Z}(B) = \sum_{i=1}^{n} \lambda_i z(\mathbf{x}_i), \tag{CS1.2}$$

where n usually represents the data points within the local neighbourhood, V, and is much smaller than the total number in the sample, N, and λ_i are the weights. The weights are a function of the variogram and the location of the sampling points and target point. To ensure that the estimate is unbiased the weights are made to sum to one,

$$\sum_{i=1}^{n} \lambda_i = 1. \tag{CS1.3}$$

The estimation variance of $\hat{Z}(B)$ is:

$$\operatorname{var}[\hat{Z}(B)] = E[\{\hat{Z}(B) - Z(B)\}^2] = 2 \sum_{i=1}^{n} \lambda_i \bar{\gamma}(\mathbf{x}_i, B) - \sum_{i=1}^{n} \sum_{j=1}^{n} \lambda_i \lambda_j \gamma(\mathbf{x}_i, \mathbf{x}_j) - \bar{\gamma}(B,B), \tag{CS1.4}$$

where is $\bar{\gamma}(\mathbf{x}_i, B)$ the average semi-variance between data point $\mathbf{x}_i$ and the target block B, and $\bar{\gamma}(B,B)$ is the average semi-variance within B, the within-block variance.

Equation (CS1.4) for a block leads to the following kriging equations:

$$\sum_{i=1}^{n} \lambda_i \gamma(\mathbf{x}_i, \mathbf{x}_j) + \psi(B) = \bar{\gamma}(\mathbf{x}_j, B) \text{ for all } j$$

$$\sum_{i=1}^{n} \lambda_i = 1, \tag{CS1.5}$$

where the Lagrange multiplier, $\psi(B)$, is introduced to achieve minimization. The weights, λ_i, are inserted into Equation (CS1.2) to give the prediction of Z at B. The block kriging (prediction or estimation) variance is then obtained as

$$\sigma^2(B) = \sum_{i=1}^{n} \lambda_i \bar{\gamma}(\mathbf{x}_i, B) + \psi(B) - \bar{\gamma}(B,B). \tag{CS1.6}$$

The values of the property are usually estimated at the nodes of a fine grid for mapping, and the variation can then be displayed by isarithms or layer shading. The kriging variances or standard errors can be mapped similarly: they are a guide to the uncertainty of the estimates.

Soil and crop properties can vary at markedly different spatial scales both within and between fields. Therefore, the scales of variation in the properties of most importance for site-specific management should be used to guide sampling to obtain spatially dependent data. Surveys on a square grid have been favoured in PA because they are efficient for sample collection in the field, prediction and mapping. For geostatistical analysis it is advisable to supplement grid sampling with some samples at a shorter interval than that of the grid to ensure that the variogram represents the variation adequately. The following example shows how the variation can be quantified to produce accurate maps and how such maps can be used.

Mapping for precise management

Farmers can use many types of data to aid more efficient management of their land, which in turn helps to improve environmental protection. The study site is a field on the Yattendon Estate in Berkshire, south-central England (Oliver and Carroll, 2004). The field is 23 ha with a complex topography that comprises plateau areas in the north and east and a large dry valley in the centre and south of the field. The physiography is generally undulating, a characteristic of the chalk hills in England. Figure CS1.2 shows a map of elevation in this field. The soil, a Luvisol (IUSS, 2006), has developed on the Reading Beds, which comprise sediments with a range of particle sizes that overlie the Upper Chalk (Cretaceous age).

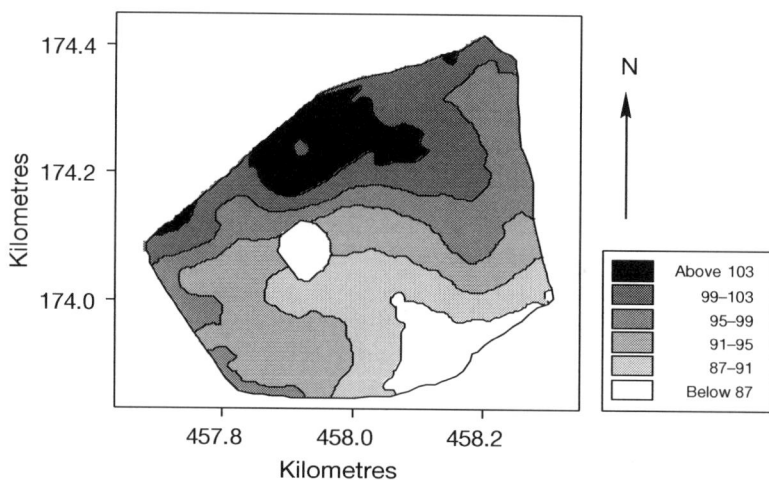

Figure CS1.2 Interpolated map of elevation (m) at the study site on the Yattendon Estate, Berkshire, England.

Before the survey, variograms were computed from previous records of yield to obtain an idea of the scale of the spatial variation within the field. Figure CS1.3b shows the experimental variogram (symbols) of yield for 1997 and the fitted nested or double spherical model, which were typical of the variograms for other years. The average range of variation of this variogram, taking the short- (28 m) and long-range (123 m) components into account, is about 75 m (Table

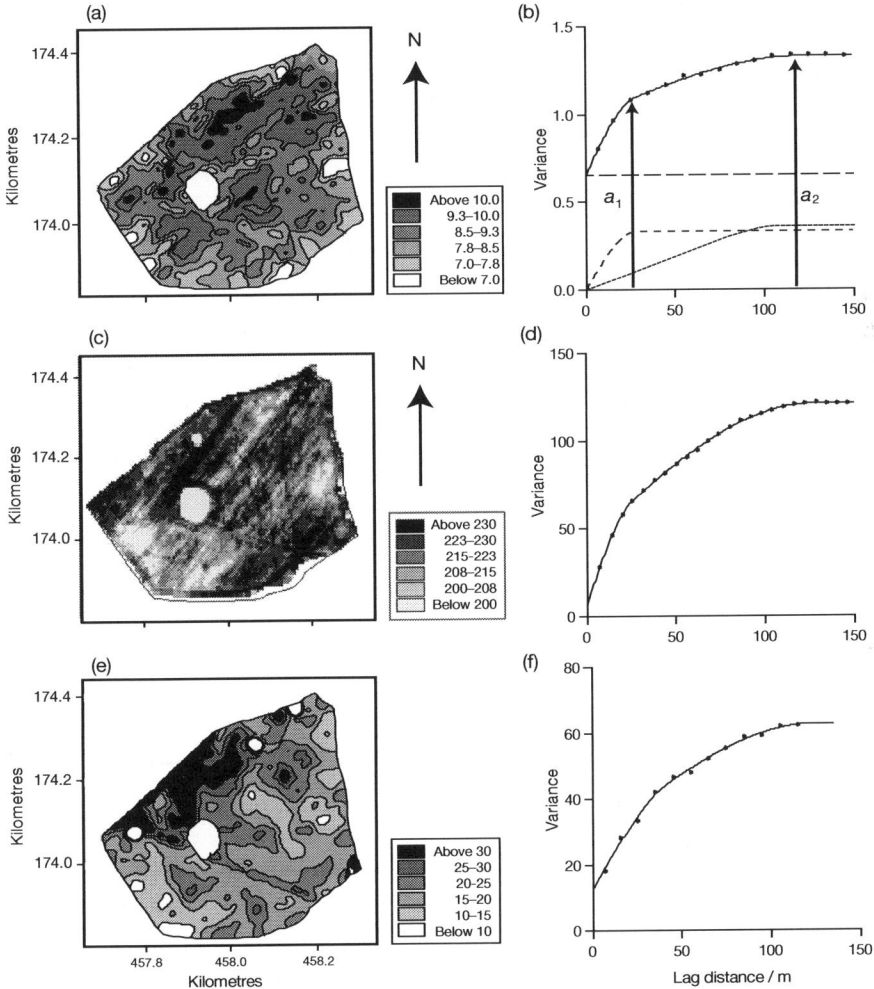

Figure CS1.3 (a) Kriged map of yield, (b) experimental variogram (symbols) of yield and the fitted double spherical model (solid line), the nugget variance (c_0 — — — —), short-range (a_1 - - - - -,) and long-range (a_2 ·········,) components of the model, (c) pixel map of red waveband from colour aerial photograph, (d) experimental variogram and fitted model of red waveband, (e) kriged map of soil's apparent electrical conductivity, EC_a and (f) experimental variogram and fitted model of EC_a.

CS1.2 gives the model parameters). The general pattern of variation in yield was similar for all years examined, but large- and small-yielding areas can reverse according to the weather conditions in some fields.

A colour aerial photograph of the field for 1991 (stubble was present) (Aerofilms Ltd., www.aerofilms.com) was scanned and Figure CS1.3c shows a pixel map of the red waveband. Figure CS1.3d shows the experimental variogram computed from the digital numbers of the red waveband, which was modelled with a nested spherical function as for yield. Table CS1.2 gives the model parameters. Electro-magnetic induction (EMI) surveys are very popular in PA (see Chapters 6 and 8). It measures the soil's apparent electrical conductivity (EC_a mS m^{-1}), which appears to be related to several soil physical properties such as moisture content (Sheets and Hendrickx, 1995), particle size distribution (Dalgaad *et al.*, 2001) and salinity (Triantifilis *et al.*, 2000). The EC_a was measured with an EM38 sensor in the vertical position, which provides a weighted depth reading to 1.5 m with a strong emphasis at the 0.3–0.5m depth (Dalgaad *et al.*, 2001). To reduce the effect of the soil's moisture status, measurements were made when the soil was close to field capacity. Observations were made along transects about 15 m apart, resulting in over 3,000 values. Figure CS1.3e shows the variogram computed from these data and the fitted double spherical model; the parameters are given in Table CS1.2. The variogram ranges show some similarity to those of yield and the aerial photograph data.

The map of yield in Figure CS1.3a shows considerable spatial variation with both small and larger features that reflect the two scales of variation present. The NE–SW alignment in the variation is also evident in the aerial photograph for 1991, Figure CS1.3c. The variation in yield also shows some relation to that of elevation, Figures CS1.3a and CS1.2, respectively, where the dry valley has moderate to large yields and the plateau area has the largest yields. The small digital numbers in the aerial photographs also appear to relate to the lower-yielding areas and to the smaller EC_a values (Figure CS1.3e and f). The pattern of variation in EC_a values shows a strong relation with yield in general. These results suggest that ancillary data such as yield, EC_a and aerial photographs probably relate to patterns of variation in the soil. Therefore, if a relation is suspected information about spatial scale from such data can be used to indicate an appropriate sampling interval. In this study, those for yield were used as these were available first.

The sampling interval selected, 30 m, was between a third and a half of the average range of the variogram for yield (see above). Sampling was on a square grid with additional samples along short transects at 15 m intervals from randomly selected grid nodes to identify any short-scale variation (Figure CS1.4a). At each sampling location 10 small cores of soil were taken from the topsoil (0–15 cm) in a 1 m^2 area (the sample support) and mixed together to form a bulked sample. The soil samples were air-dried and sieved; the < 2 mm fractions were analysed in the laboratory for extractable phosphorus (P), potassium (K) and magnesium (Mg) using the standard methods of DEFRA (2010). Organic matter content (OM) was estimated by loss on ignition (LOI). The particle size distribution was determined by laser diffraction grain sizing.

Results of soil spatial analysis

Table CS1.1 gives the summary statistics of the variables included in this case study. The skewness coefficients for the soil variables are within the limits usually used of ±1 to indicate near-normality, but those for yield and EC_a are outside these limits. However, variograms of the transformed data were little different from those of the raw data and so the latter were used. Experimental variograms were computed from all the topsoil data in Table CS1.1 and models were fitted. Table CS1.2 shows that the spherical function was the best-fitting model to all the selected properties, except for the variograms of the sub-sampled data for K which were fitted by circular functions. Both functions describe variation that is patchy, that is, areas with larger and smaller values with an average extent given by the variogram range. Figures CS1.4 and CS1.5 show the experimental variograms (symbols) of the soil variables and their fitted models. The variation in P has been least-well resolved by the sampling and this can be seen by the large nugget variance, c_0 (Figure CS1.5c), which mainly comprises variation that occurs over distances less than the sampling interval. Therefore, there is considerable variation in topsoil P over distances < 30 m. The variogram ranges for topsoil K, Mg and LOI and the long-range component of EC_a are similar at about 130 m (Table CS1.2). Sand and clay contents have similar ranges to one another (Figure CS1.5g and i) and also to P of about 266 m. The short-range component of EC_a (Table CS1.2) is similar to those of yield and the red waveband, which seems to relate to the lines of management (distance between tramlines) and lower-yielding areas where there has been traffic over the field. The long-range component of the red waveband is similar to the ranges of K and Mg, and to the long-range component of EC_a (Table CS1.2).

I use potassium to illustrate the effects of sample size and sampling intensity. As mentioned above, the recommended sample size to compute a reliable variogram by the usual method of moments estimator (Equation CS1.1) should be at least 100 (Webster and Oliver, 1992), and the sample spacing should be capable of resolving the variation within the field to enable site-specific management. The

Table CS1.1 Summary statistics of the soil properties, yield and EC_a

	No.	Mean	Minimum	Maximum	Standard deviation	Variance	Skewness
K (mg kg⁻¹)	230	143.0	48.1	254.0	37.0	1360.0	0.128
Mg (mg kg⁻¹)	230	67.3	12.8	136.0	22.9	524.4	0.290
P (mg kg⁻¹)	230	28.1	5.20	55.2	8.79	77.26	0.088
LOI (g 100g⁻¹)	230	3.38	1.88	6.29	0.914	0.8354	0.74
Sand (%)	230	50.8	14.0	83.0	14.4	207.4	0.02
Clay (%)	230	18.2	6.00	42.0	6.58	43.30	0.58
Yield₉₇ (t ha⁻¹)	4698	8.74	1.51	13.0	1.43	2.045	−1.80
EC_a (mS m⁻¹)	3275	21.1	6.50	82.5	9.35	87.42	2.25

variogram range of the full data, 127 m, suggested that the data could be sub-sampled to give an interval of 60 m (Figure CS1.4e). There were also samples at shorter distances (15 m and 30 m) along the original short transects to obtain a good estimate of the variogram near to the origin. Figure CS1.4f shows the vario-gram computed from this sub-sample; it was computed from 91 data, which is just less than the recommended number. The experimental variogram is quite similar to that computed on the full data, but it is more erratic at lags > 100 m, where the estimates become less reliable. However, the nugget variance, c_0, is almost three times as large as for the variogram of the full data, indicating that the variation is less well resolved. The experimental variogram of these data further sub-sampled to 56 sites is very erratic (Figure CS1.4j) and it was more difficult to model. The nugget variance is larger than that for the first sub-sample; it indicates that less of the variation has been resolved, as would be expected from this smaller sample size with a larger interval between sites. The variogram based on 56 data is not a reliable indication of the structure of the variation in these data.

Figure CS1.4c, g and k show the maps based on kriging with the above three variograms of topsoil K and the data. Figure CS1.4c, based on the full data, shows the detail in the variation, but as sampling becomes progressively sparse there is a considerable loss of detail. The map based on 91 data (Figure CS1.4g)

Table CS1.2 Model parameters of functions fitted to the experimental variograms of the topsoil variables

Variable	Model type	Model parameters		
		c_0	c	$a\,(m)$
K(230 sites – all data)	Spherical	158.5	1217.0	127.8
K(91 sites)	Circular	363.0	1057.0	140.9
K(56 sites)	Circular	456.0	978.0	151.2
Mg	Spherical	199.6	291.0	105.8
P	Spherical	51.45	30.81	254.5
LOI	Spherical	0	0.7245	168.7
Sand	Spherical	28.88	206.7	257.5
Clay	Spherical	11.80	38.86	286.7
Ancillary data Variable				
Yield 1997	Double spherical	0.6493	c_1 0.3309 c_2 0.3500	a_1 27.79 a_2 122.9
Red waveband 1991	Double spherical	6.85	c_1 35.94 c_2 78.50	a_1 25.8 a_2 127.5
EC_a	Double spherical	12.72	c_1 14.87 c_2 35.17	a_1 43.11 a_2 126.6

Note
The model parameters are: c_0 the nugget variance, c_1 and c_2 are the sills of the autocorrelated variance, and a_1 and a_2 the ranges of spatial dependence.

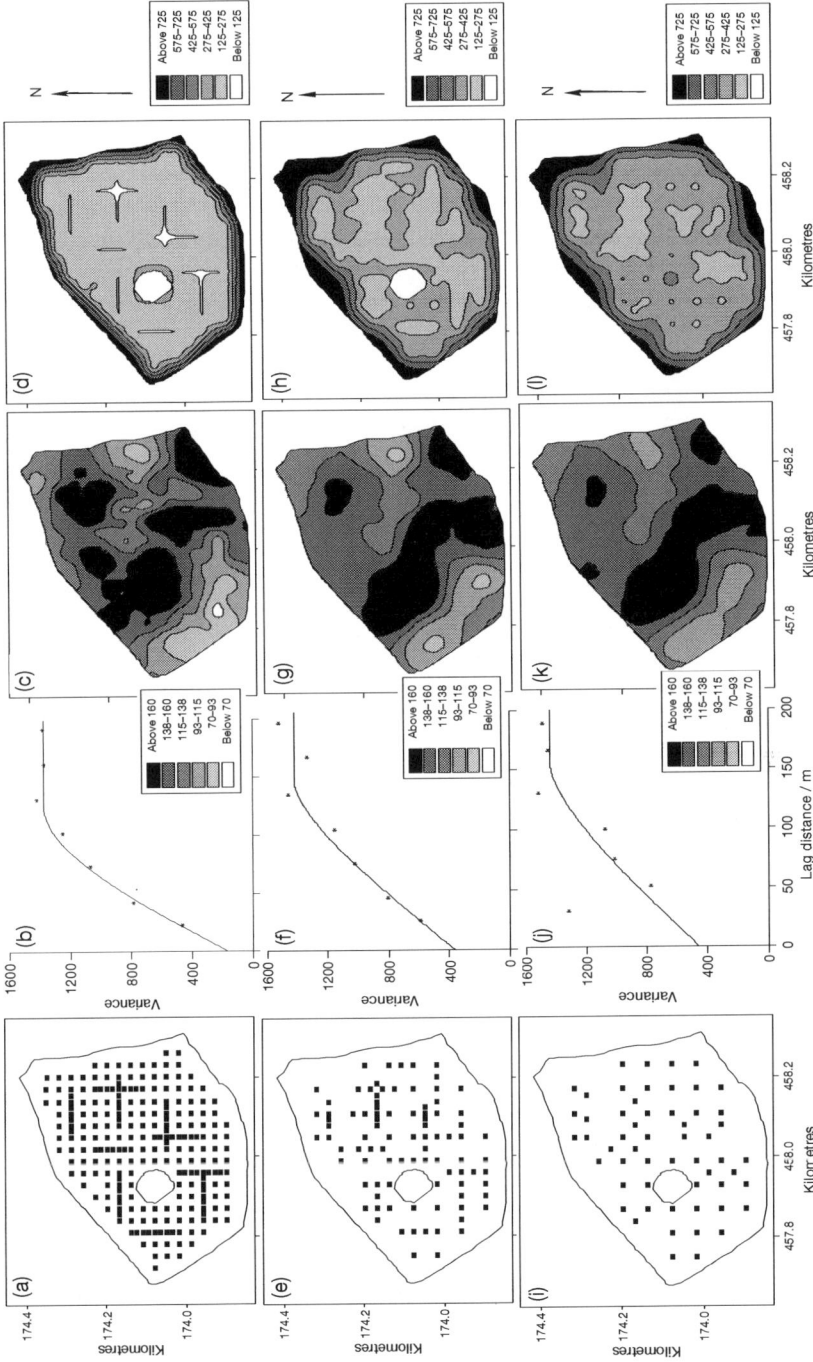

Figure CS1.4 Full sample (230 data) of field on the Yattendon Estate, Berkshire: (a) sampling scheme, (b) experimental variogram and fitted model, (c) kriged map and (d) map of kriging variances of potassium. First sub-sample (91 data): (e) Sampling scheme, (f) experimental variogram and fitted model, (g) kriged map and (h) map of kriging variances of potassium. Second sub-sample (91 data): (i) Sampling scheme, (j) experimental variogram and fitted model, (k) kriged map and (l) map of kriging variances of potassium.

would be reasonably acceptable for site-specific management, but that based on 56 data (Figure CS1.4k) would not be reliable because the areas where K needs to be added in the SW and NE of the field are no longer evident. These maps generally show that SSM would be beneficial in this field for K and would reduce the largest applications to only those areas where concentrations are < 115 mg l⁻¹. The kriging variances reflect the intensity of sampling, Figure CS1.4d, h and l. Figure CS1.4d shows the lines of the transects clearly because the kriging variances are smallest here. As sampling becomes sparser the kriging variances increase and show how unreliable the predictions become over much of the field with only 56 data.

The variogram model parameters were used with the data to produce kriged predictions for mapping. Figures CS1.4 and CS1.5 show kriged maps of the crop nutrients; those for K and Mg (Figures CS1.4c and CS1.5b, respectively) have similar patterns of variation. Both of these nutrients are in the very low and low range for crop management based on DEFRA's guidelines (DEFRA, 2010), index 1 and 2 for both K and Mg. Nevertheless, uniform applications would waste fertilizer as well as adding to the load of these elements in the environment. There are clearly defined areas where more K and Mg should be added than to the rest of the field. The variation in P (Figure CS1.5d) is somewhat different – in the western third of the field values are small (< index 2), whereas in the rest of the field (N and E) values are larger (> index 2). This map could be used to guide SSM of P so that P is added only to the western part of the field where it is < 26 mg l⁻¹. This would reduce the overall amount of P applied compared with uniform application. The map of topsoil LOI (Figure CS1.5f) shows some similarity with the variation in K and Mg. Soil organic matter is important in precise management because where its content in the soil is greater, the more resistance and tolerance the crop has to pests and diseases. There are also more microflora and organic matter which help to balance nutrient levels (Altieri and Nicholls, 2003). Furthermore, where there is more organic matter and clay in the soil, pesticides can be applied more freely because they are held in the soil for longer and degrade (Price *et al.*, 2009). Figure CS1.5h and j shows the maps of clay and sand contents, respectively. Patches of large and small values are quite extensive as was expressed by the variogram ranges. The patterns of variation in LOI, K and Mg are similar to those of clay even though there is more spatial variation present. The variation in sand content shows an inverse relation with these soil variables, and where sand content is large LOI is smaller because organic matter oxidizes faster on coarser-textured soil. The pattern of variation in particle size distribution corresponds closely with that for EC_a, Figure CS1.3f, which shows the effect of soil texture on this measurement. For this field, the map of EC_a would provide a good substitute for the maps of particle size content and K. It could also provide a basis for site-specific management – areas where values are large would need less K, less water and could take more pesticides (Price *et al.*, 2009), herbicides and greater planting density. These types of action epitomize PA and site-specific management, and they are both economically and environmentally sound.

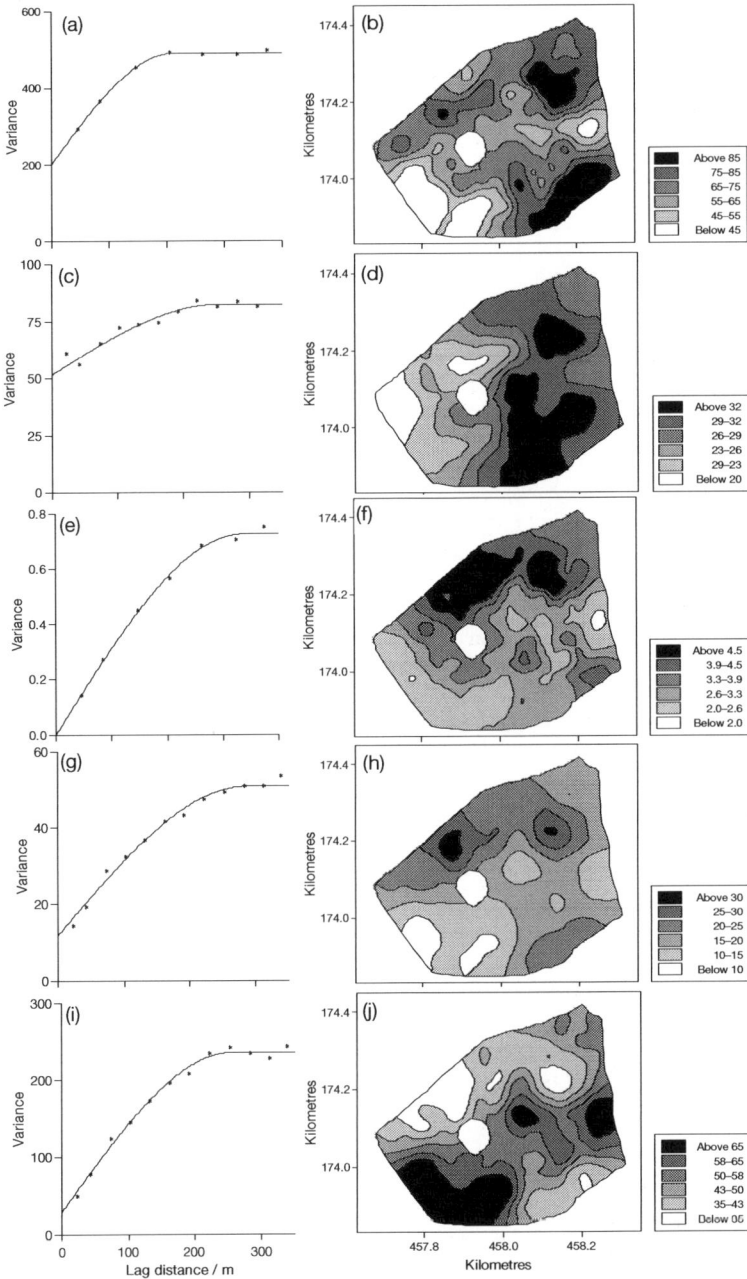

Figure CS1.5 Experimental variograms, fitted models and kriged maps for topsoil soil properties of the field on the Yattendon Estate, Berkshire: (a, b) magnesium, (c, d) phosphorus, (e, f) loss on ignition, (g, h) percentage clay content and (i, j) percentage sand content.

Conclusions

This case study has shown the importance of determining the spatial scale of variation before sampling the soil, crop or other environmental features. Sampling is crucial for obtaining data that are suitable for mapping and precise management. Too few data and or too large a sampling interval can result in too little detail about the variation to undertake appropriate management. This is a waste of money, time and effort that could lead to poor and damaging management decisions. Geostatistics with good sample data is an important tool for the modern farmer who wishes to optimize the use of inputs for both sound economic and environmental reasons.

References

Altieri, M. A. and Nicholls, C. I. (2003) 'Soil fertility management and insect pests: harmonizing soil and plant health in agroecosystems', *Soil & Tillage Research*, 72, 203–11

Dalgaad, M., Have, H. and Nehmdahl, H. (2001) 'Soil clay mapping by measurement of electromagnetic conductivity', in G. Grenier and S. Blackmore (eds) *Proceedings of the Third European Conference on Precision Agriculture*, Montpellier: Agro, vol. I, pp. 367–72.

DEFRA (2010) *Fertiliser Manual RB209*, 8th edn, Belfast: The Stationery Office.

Godwin, R. J. and Miller, P. C. H. (2003) 'A review of the technologies for mapping within-field variability', *Biosystems Engineering*, 84, 393–407.

IUSS Working Group WRB (2006) *World Reference Base for Soil Resources 2006*, 2nd edn, World Soil Resources Reports 103, Rome: FAO.

Kerry, R. and Oliver, M. A. (2007) 'Sampling requirements for variograms of soil properties computed by the method of moments and residual maximum likelihood', *Geoderma*, 140, 383–96.

Kerry, R., Oliver, M. A. and Frogbrook, Z. L. (2010) 'Sampling in precision agriculture', in M. A. Oliver (ed.) *Geostatistical Applications for Precision Agriculture*, Dordrecht: Springer, pp. 1–34.

Matheron, G. (1965) *Les variables régionalisées et leur estimation: une application de la théorie de fonctions aléatoires aux sciences de la nature*, Paris: Masson et Cie.

Oliver, M. A. and Carroll, Z. L. (2004) *Description of Spatial Variation in Soil to Optimize Cereal Management*, Project Report 330, London: HGCA.

Price, O. R., Oliver, M. A., Walker, A. and Wood, M. (2009) 'Estimating the spatial scale of herbicide and soil interactions by nested sampling, hierarchical analysis of variance and residual maximum likelihood', *Environmental Pollution*, 157, 1689–96.

Sheets, K. R. and Hendrickx, J. M. H. (1995) 'Non-invasive soil water content measurement using electromagnetic induction', *Water Resources Research*, 31, 2401–9.

Triantifilis, J., Laslett, G. M. and McBratney, A. B. (2000) 'Calibrating an electromagnetic induction instrument to measure salinity in soil under irrigated cotton', *Soil Science Society of America Journal*, 64, 1008–17.

Webster, R. and Oliver, M. A. (1992) 'Sample adequately to estimate variograms of soil properties', *Journal of Soil Science*, 43, 177–92.

Webster, R. and Oliver, M. A. (2007) *Geostatistics for Environmental Scientists*, 2nd edn, Chichester, UK: John Wiley & Sons.

CS2 Precision agriculture in sugarcane production

José P. Molin, Gustavo Portz and Lucas Rios do Amaral

Introduction

Sugarcane (*Saccharum* ssp.) is the main crop for the supply of sugar and ethanol production in tropical and subtropical areas. Annual sugar production in these areas has averaged almost 160 million tons over the last few years, which represents around 80 per cent of world production. Ethanol production in this area is approximately 85 billion litres, corresponding to 35 per cent of world production. Brazil is the main producer of sugarcane as a raw material. It accounts for more than one third of world production, with 670 million tons of cane, followed by India with 285 million tons (FAO, 2011). In 2009, the global sugarcane production was 1,661 million tons in total (FAO, 2011).

According to the Ministry of Agriculture, Livestock and Food Supply of Brazil, sugarcane production increased by 55 per cent over the last five years. The 2010/11 harvest was processed at 434 sugarcane mills and distilleries. Approximately 46 per cent of sugarcane production is allocated to sugar production and 54 per cent to ethanol, representing a production of 39 million tons of sugar and 27.7 billion litres of ethanol. Sugarcane is ranked third among Brazilian crops, with the largest cropping areas after soya bean and corn; it is grown on nearly 9 million hectares or 1 per cent of Brazilian territory.

The sugarcane industry has evolved to produce sugar and ethanol. More recently, with the increase in bioelectricity generation, the sector has been called *Sucroenergético* (from Portuguese, *sucro* = sugar; *energetic* = energy). Sugarcane is directly converted into energy that is available as sugar for consumption, ethanol for vehicles and bioelectricity for household and industrial use. Bioelectricity is produced from cane bagasse, the solid residue after sugar extraction, which is burned as fuel for boilers that generate electricity by moving turbines in the mills. The surplus bioenergy produced is sold to the national electrical grid.

Projections for growth in the sugarcane industry over the next few years are based on the economic value of each individual product-market segment. The sugar market, for instance, is well established in Brazil, with the lowest production cost around the world. Ethanol, in turn, has emerged in the internal market owing to the increased number of flexible fuel vehicles. Ethanol consumption exceeds that of gasoline and the export possibilities for this product are promising. Finally, the electrical energy produced from sugarcane bagasse accounts for 4.5 per cent of the Brazilian energy matrix and may grow with the exploitation of sugarcane residue, such as the leaves that are currently left to decompose in the field.

A survey performed by the University of São Paulo (USP) shows that in 2008 the sugarcane industry contributed US$28.1 billion to the gross domestic product (GDP) of Brazil, corresponding to nearly 2 per cent of the total, and generated tax revenues of US$9.86 billion. This sector also mobilizes a huge supply chain, as well as agricultural and industrial services. In 2008, the sugarcane industry was the third largest market for agricultural products in Brazil, representing 14 per cent of fertilizer sales (US$2.3 billion). In addition, it accounted for 9.5 per cent of agricultural pesticide sales.

Sugarcane farmers are responsible for approximately 22 per cent of cane production (nearly 139 million tons in the 2010/11 harvest), with the remaining provided by sugarcane mills (ORPLANA, 2011).

The sugarcane industry plays an important social role by creating jobs and generating income. In 2008, more than 1.2 million workers were involved in sugarcane and ethanol production in Brazil. Most workers have low levels of education, and if not employed in the sugarcane industry would have little chance of being included in the formal labour market. These workers, however, do not meet the technological requirements of the sector, which demands qualified labour.

Responding to international criticism of ethanol production in Brazil in 2009, the government launched the Agroecological Mapping of Sugarcane Cropping and the National Commitment for Improving Working Conditions in the Sugarcane Industry. Attempts to restrict the area available for the development of this sector by mapping it prohibited the use of the Amazon rainforest and Pantanal, the tropical wetland of more than 150,000 square kilometres in the west part of the country, and responded to doubts about the expansion of sugarcane crops in Amazonia. This commitment, in turn, was ratified by the government, the productive sector and the workers themselves, thereby guaranteeing that the growth of the sugarcane industry would be environmentally and socially sustainable (Dulci, 2008).

Sugarcane crop

Sugarcane is a large-sized grass (*Poaceae*) native to Southeast Asia. It has a large biomass production capacity because it uses the C4 photosynthetic pathway, characterized by a high internal concentration of CO_2, a low respiration rate and the capacity for photosynthesizing at high temperatures (> 35°C), exhibiting a higher net photosynthetic rate than C3 plants such as wheat and soya bean (Pimental, 1998).

The plant must be exposed to stress conditions to promote maturation and increase sugar (sucrose) production. Water stress caused by long dry periods favours sugar accumulation. Moreover, thermal stress, achieved at a temperature of 20°C, can increase sugar concentration, but if exposed to frost, the plant or its apical meristem may die. Sugarcane must therefore be grown only in areas with adequate water and temperature conditions. Thus, in addition to the illegality of commercial sugarcane cropping in the Amazon rainforest, the area is unsuitable for this activity because of the hot, wet climate throughout the year.

Sugarcane is a semi-perennial crop, that is, it is economically profitable for about five years after planting, allowing successive cuts and regrowth over the years, until replanting. A sugarcane crop is therefore less costly than annual crops. The annual production of an efficiently managed commercial sugarcane crop is currently around 80 Mg of stalks per hectare during an annual cycle, ranging from 140 Mg ha^{-1} in the first cut to 60 Mg ha^{-1} in the last cut, before replanting (Maule *et al.*, 2001).

The crop is replanted when sugarcane production is lower than expected. The stubble is then removed and the soil is tilled to a depth of 0.40 m to alleviate compaction, and lime and fertilizers are added. Many sugar mills also incorporate industrial residues as a fertilizer source in the pre-planting phase. Legumes such as peanut and soya bean can also be planted before the new sugarcane crop because this rotation eliminates weeds and diseases and promotes biological fixation of nitrogen for the initial growing period. Stalks with vegetative organs (buds) are used for planting. They are introduced into furrows spaced 1.1 to 1.5 m apart, with 15–25 viable buds per metre, depending on plant variety (Galvani *et al.*, 1997).

The first sugarcane harvest takes place from 12 to 18 months after planting, according to time of the year. Harvesting is either manual or by mechanized stalk-cutting. The crop can be burned to reduce leaf volume and facilitate harvesting, especially hand-harvesting. Producers are, however, increasingly adopting mechanized harvesting without burning to improve nutrient cycling, meet ecological standards and enhance labour quality in the sector. In fact, there are two major targets being managed by the community. One is for São Paulo State, where burning will be banned in 2017 and the other is the federal legislation that states the limit to 2031. Other states are discussing their own legislation. These changes do not pose a serious obstacle to the sector because mechanical harvesting is already less expensive than manual sugarcane cutting.

After annual harvesting, the crop is treated with herbicides and insecticides according to need. In addition, fertilizers are reapplied, especially nitrogen and potassium, which are widely exported by the crop.

Precision agriculture practices in Brazil

The sugarcane industry is increasingly seeking technologies that improve crop management, yield and product quality at lower costs and less environmental impact, as shown by research conducted in the state of São Paulo, where 67 per cent of Brazilian sugarcane is produced (Silva *et al.*, 2011). This study also indicated that 96 per cent of sugarcane mills intend to extend the use of precision agriculture practices.

These practices are used at all stages of sugarcane cropping (Bramley, 2009). In Brazil, two techniques predominate: geo-referenced soil sampling with site-specific management and the use of autopilots in tractors and harvesters. Soil sampling on a grid has been applied at densities of one sample per 2 ha to 4 ha, with composite soil samples from 8 to 15 cores in a radius of about 3 m to 5 m. Data are interpolated to produce maps that indicate the specific levels of lime and fertilizers that must be applied to each site. This practice is carried out only when crops are renewed, and complete soil tillage and fertility correction are carried out.

The use of autopilots significantly improves parallel alignment of planting rows (Figure CS2.1). This device also facilitates the work of machine operators for other crop management procedures such as the regular and non-overlapping application of fertilizers and other inputs with sprayers and other devices. In addition, the autopilot allows the harvester to follow exactly the rows mapped when planting, where the sugarcane has to be cut. Many sugarcane mills also use geographical information systems (GIS) to plan and allocate crop rows. This facilitates harvesting operations, especially at night or in sugarcane fields flattened by strong winds. The autopilot, usually integrated into the hydraulic system of vehicles, is guided by a global navigation satellite system (GNSS) receiver with signal correction using real-time kinematic (RTK) technology.

Some sugarcane mills and farmers that use controlled traffic prohibit vehicles from driving over sugarcane rows designed only for cropping. Some vehicles have adjustable wheel spacing arranged such that wheels pass between but not over crop rows, which is facilitated by the use of autopilots (Figure CS2.2). Controlled traffic limits the wheels to the location where there is no crop, increasing soil compaction in that area but eliminating it on the crop row. On

Figure CS2.1 Parallel alignment between planting rows performed with (right) or without (left) autopilot.

Source: F. Torres, Jacto S.A., with permission.

compacted soil, the poorer development of the root system results in a smaller soil volume being explored by roots and, consequently, less absorption of water and nutrients. Controlled traffic increases production and the productive life of the sugarcane crop, representing reduction in production costs. This practice is particularly beneficial during the harvesting period when machinery traffic is greater. Soil compaction is aggravated in sugarcane crops because most harvesters cut only one row at a time and are constantly followed in the field by sugarcane trailers.

Figure CS2.2 Mechanized crude sugarcane harvesting with tractors, sugarcane trailers and harvesters with wheel spacing arranged to pass between crop rows.

Sugarcane producers that use the autopilot technology have achieved larger crop yields and have mitigated a decline in production that occurs over time. In areas with controlled traffic under good conditions, the autopilot can prolong the productive life of crops from five to ten years, thereby allowing more cuts from the same crop (Braunack and McGarry, 2006). Another measure adopted by producers to avoid traffic-related damage is the use of low-pressure high-flotation tyres on tractors and trailers, and tracked rather than wheeled harvesters, thereby increasing soil surface contact and decreasing compaction (Raper *et al.*, 1995).

The development of yield mapping for sugarcane crops is still limited to research tests. Producers rarely understand and use this monitoring tool to support site-specific management, and those who do complain that yield maps are neither accurate nor robust. There is one yield monitor (Magalhães and Cerri, 2007) that measures the amount of material passing through the conveyor belt of the harvester before it is dumped into the trailer, thereby determining the weight of harvested stalks through load-cells installed in the harvester elevator. Alternative techniques to obtain yield maps from multispectral images, for instance, are still far from being established.

Based on the vast knowledge available for other crops, the study of the spatial variability of sugarcane yield might provide essential information for optimizing management. This includes precise determination of nutrient export capabilities and the delimitation of management zones where inputs should be applied according to information on local production capacity to enhance sugarcane crop profitability. It represents an area for research that requires more attention as it may improve the crop management practices of today and lead to environmental benefits in addition.

The by-products of the sugarcane industry have considerable nutritional value. The main by-products are vinasse, a residual liquid from distillation of raw sugarcane juice during ethanol production, filter cake, a residue from the filtration of sugarcane juice produced in the mills, and furnace ashes produced by bagasse or leaf burning as a heating source to the boilers. Many industrial plants transform filter cake and ashes into organic compounds that are applied to the crops. Vinasse can be sprayed directly on to the soil as a liquid fertilizer with a low concentration of nutrients. The distribution of these by-products has yet to be assessed by variable-rate technology, a promising approach for future use.

In general, the vinasse produced is applied after harvesting, providing K_2O and part of the nitrogen needed for sugarcane growth with average concentrations of 2 kg m^{-3} of K_2O and 0.36 kg m^{-3} of N (Rossetto *et al.*, 2010). The vinasse dose applied is determined by its potassium content, chemical features of the soil and environmental regulations regarding maximum allowable amounts. The proper application of vinasse provides a number of benefits for the physical, chemical and biological properties of soil. However, excess vinasse can delay plant maturation, decreasing sucrose levels and compromising sugarcane quality. In addition, continuous vinasse use can lead to excessive potassium in the soil and contaminated groundwater. Vinasse is usually applied to the crops by big irrigation sprinklers set on a cart or wound on to a self-propelled reel

(Figure CS2.3) and powered by a motor-pump that drains vinasse from an open channel or a tank. With this system, the equipment can be moved and positioned precisely in the cropped area, much like a vehicle autopilot. It avoids overlapping or incomplete vinasse distribution. Because sprinkle displacement speed as well as vinasse flow can be controlled, application dosage can be regulated in addition, and there is the potential for variable-rate application, not explored as yet.

Filter cake is composed of 1.2 to 1.8 per cent phosphorus, with nearly half of that amount readily available. It also contains high levels of calcium (2.1 per cent dry matter) and considerable amounts of micronutrients. Filter cake composting, with the addition of gypsum, furnace ashes, straw and other materials (Figure CS2.4), improves nutrient concentration and reduces water content, making the product more suitable for long-distance transport.

Figure CS2.3 Vinasse application to the crops by gun sprinklers wound on to a self-propelled reel. Vinasse is drawn from a tank by a motor-pump.

Figure CS2.4 Area for composting preparation from filter cake added to gypsum, furnace ashes and other materials.

With regard to chemical fertilizer application at variable rates, nitrogen-based fertilizer is particularly important because an efficient method for estimating soil nitrogen content is still not available for tropical sugarcane cropping areas. As a rule, based on many field tests, the standard dosage for fixed nitrogen application is 100 kg ha^{-1}. Thus, there is a search for remote sensing methods that estimate the spatial variation of nitrogen demands. These may involve orbital imaging, aerial photography or dedicated sensors coupled to land vehicles. There is still a great demand for research and development in this area as it represents an opportunity for reducing cost and environmental damage.

The use of ground-based active optical sensors, which are light emitting sources that capture crop reflectance next to the crop canopy, has been studied for nitrogen application at variable rates in sugarcane crops (Figure CS2.5). Although sugarcane response to nitrogen application is known to be uneven, the use of optical sensors to control nitrogen application has produced satisfactory results (Molin *et al.*, 2010; Portz *et al.*, 2012). This technology is likely to be adopted by producers because it can optimize the use of nitrogen fertilizer and has the capacity to assess nitrogen application throughout the crop cycle.

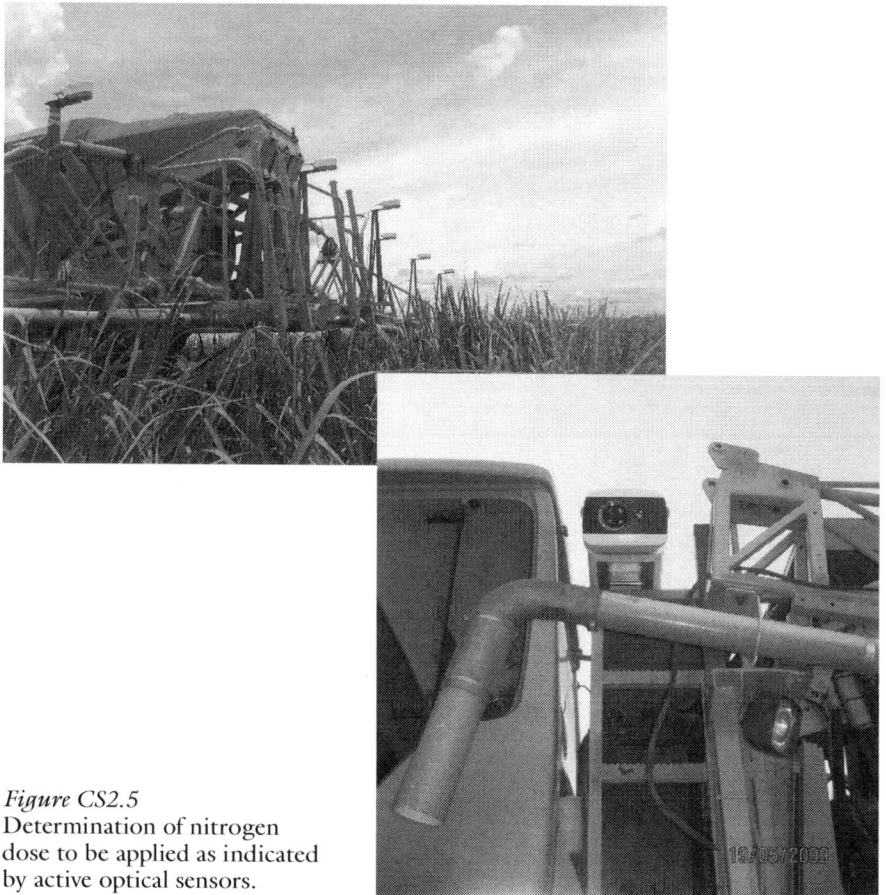

Figure CS2.5
Determination of nitrogen
dose to be applied as indicated
by active optical sensors.

Another method with a similar function is orbital remote sensing, which offers an interesting range of solutions for the sugarcane industry. Satellite imagery with different spatial and spectral resolutions can indicate nitrogen deficiency in the crops. However, the images are not available to producers early enough to allow interventions in the current crop year. The intensive growth stage of sugarcane and greatest demand for nitrogen coincide with the rainy season. At this time, orbital imagery acquisition becomes more difficult because the sky is frequently cloudy. An alternative procedure is the use of an image set from previous crop cycles to estimate crop biomass, define management zones and determine N doses to be applied in each zone based on the concepts of precision agriculture.

Sugarcane crops and their intrinsic characteristics offer a wide field for management studies, especially concerning the spatial variability of crops and the use of accurate techniques in precision agriculture, which are just beginning to be incorporated by sugarcane producers in Brazil.

References

Bramley, R. G. V. (2009) 'Lessons from nearly 20 years of precision agriculture research, development, and adoption as a guide to its appropriate application', *Crop and Pasture Science*, 60, 197–217.

Braunack, M. V. and McGarry, D. (2006) 'Traffic control and tillage strategies for harvesting and planting of sugarcane (*Saccharum officinarum*) in Australia', *Soil and Tillage Research*, 89, 86–102.

Dulci, L. S. (2008) 'Compromisso Nacional: aperfeiçoar as condições de trabalha na cana-de-açúcar' (National Compromise: improve labour conditions in sugarcane), General Secretariat of the Presidency, Brazil. Available online at <http://www.secretariageral.gov.br> (accessed 20 December 2011).

FAO (2011) FAOSTAT, Statistics Division of the Food and Agriculture Organization. Available online at <http://faostat.fao.org/> (accessed 15 August 2011).

Galvani, E., Barbieri, V., Pereira, A. B. and Villa Nova, N. A. (1997) 'Efeitos de diferentes espaçamentos entre sulcos na produtividade agrícola da cana-de-açúcar (*Saccharum spp.*)' (Effects of different furrow spacing on sugarcane yield), *Scientia Agrícola*, 54, 1–2.

Magalhães, P. S. G. and Cerri, D. G. P. (2007) 'Yield monitoring of sugar cane', *Biosystems Engineering*, 96, 1–6.

Maule, R. F., Mazza, J. A. and Martha, G. B. (2001) 'Produtividade agrícola de cultivares de cana-de-açúcar em diferentes solos e épocas de colheita' (Yield of sugarcane cultivars on different soils and harvesting time), *Scientia Agricola*, 58, 295–301.

Molin, J. P., Frasson, F. R., Amaral, L. R., Povh, F. P. and Salvi, J. V. (2010) 'Capacidade de um sensor ótico em quantificar a resposta da cana-de-açúcar a doses de nitrogênio' (Capacity of an optical sensor in verifying the sugarcane response to nitrogen rates), *Revista Brasileira de Engenharia Agrícola e Ambiental*, 14, 1345–9.

ORPLANA (2011) Organização dos Plantadores de Cana da Região Centro-Sul do Brasil (Sugarcane producers organization of South-Central Brazil). Available online at <http://www.orplana.com.br> (accessed 15 August 2011).

Pimentel, C. (1998) *Metabolismo de carbono na agricultura tropical* (Carbon metabolism in tropical agriculture), Seropédica, Brazil: Edur.

Portz, G., Molin, J. P. and Jasper, J. (2012) 'Active crop sensor to detect variability of nitrogen supply and biomass on sugarcane fields', *Precision Agriculture*, 13, 33–44.

Raper, R. L., Baley, A. C., Burt, E. C., Way, T. R. and Liberaty, P. (1995) 'The effects of reduced inflation pressure on soil–tire interface stresses and soil strength', *Journal of Terramechanics*, 32, 43–51

Rossetto, R., Dias, F. L. F., Vitti, A. C. and Tavares, S. (2010) 'Potássio' (Potash), in L. L. Dinardo-Miranda, A. C. M. Vasconcellos and M. G. A. Landell (eds) *Cana-de-açúcar* (Sugarcane), São Paulo, Brazil: Agronomic Institute.

Silva, C. B., de Moraes, M. A. F. D. and Molin, J. P. (2011) 'Adoption and use of precision agriculture technologies in the sugarcane industry of São Paulo state, Brazil', *Precision Agriculture*, 12, 67–81.

CS3 Precision rice farming for small-scale paddy fields in Asia

Byun-Woo Lee and Kyu-Jong Lee

Introduction

Paddy rice is a very intensive crop in terms of both material and labour inputs (Shibusawa *et al.*, 1999). Most traditional farming systems over-apply inputs such as seed, fertilizer and agro-chemicals to reduce the risk of crop failure. In particular, conventional uniform fertilizer application, which does not account for within-field variation in nutrient supply from the soil and the nutrient requirements of the crop, may result in some zones having an excess of nutrients and other zones having a nutrient deficit. Such inefficient fertilizer application has both economic and environmental impacts (Koutroubas and Ntanos, 2003). Precision farming manages field variability by matching resource application and agronomic practices with crop requirements and soil properties in order to improve crop performance and quality. Precision farming in rice production has focused on applying fertilizer more efficiently.

In most Asian rice-growing countries, farm sizes are smaller than 1 ha and rice fields are scattered within these farms (Yagi, 2012). To date little attention has been paid to the spatial variability in rice yield because the fields are so small. However, spatial variation in soil nutrients and rice yield has been observed even within small fields (Dobermann *et al.*, 1996). For example, in Figure CS3.1 we see substantial variation in rice yield and protein content within a field of less than 2/3 ha (Nguyen, 2005). This may be caused by variability in soil properties, nutrients, weeds or flood-irrigation.

Precision farming or site-specific management recognizes the spatial and temporal variability in crop production. The implementation of site-specific management requires (Batte and VanBuren, 1999): (1) the collection of data that measure the spatial and temporal variation of the farming system; (2) the analysis and interpretation of these data to support a range of management decisions and (3) application of a management response that is varied appropriately in both space and time. Precision farming has been conducted primarily in developed countries and has focused on the management of large fields using sophisticated technologies such as crop sensors, global positioning systems and remote sensing. This is unsuitable for most Asian rice farms where fields and the entire landholding are typically small. Thus, sophisticated sensors must be replaced by other rapid and cost-effective approaches of acquiring and processing

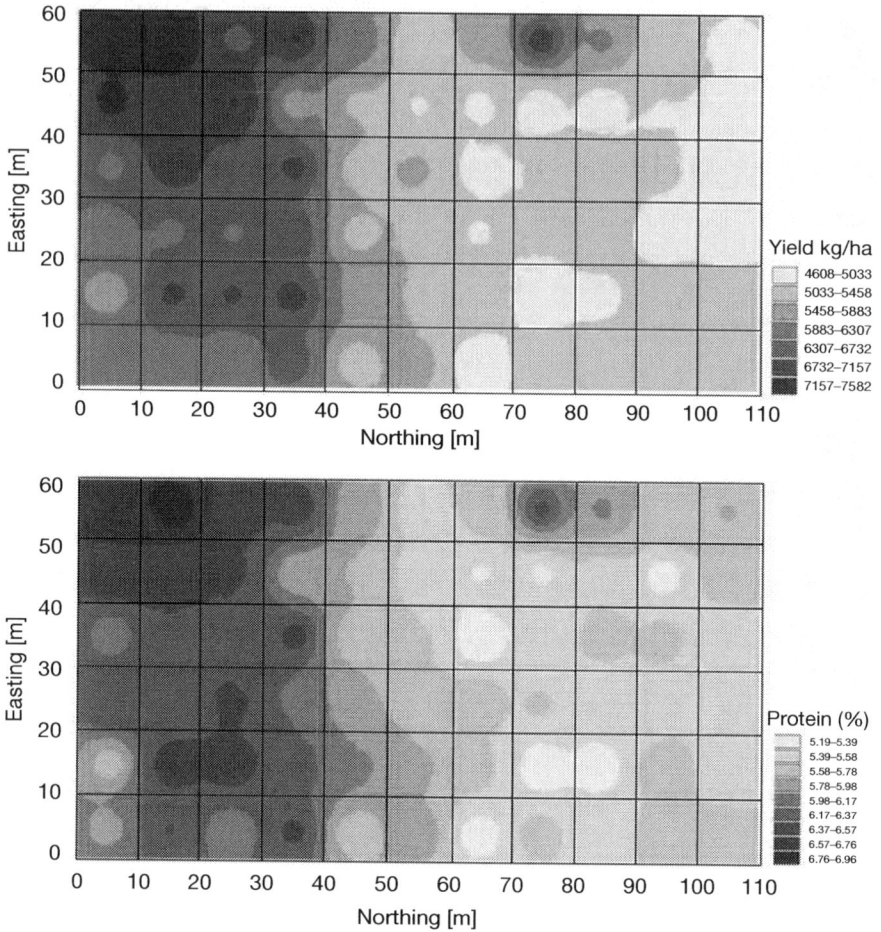

Figure CS3.1 Kriged map of spatial variation of rice yield and protein content in a small paddy field subjected to uniform N fertilizer top-dressing in Korea (Nguyen, 2005).

location-specific information within a field. For example, this case study describes a site-specific N management system where the crop growth is monitored by applying image analysis methods to digital photographs of the crop. Fertilizer inputs are then adapted according to the measured growth in order to optimize yield and quality. Digital photography is an inexpensive and non-destructive method of gathering data.

Site-specific recommendation of nitrogen fertilizer

The optimum timing and rate of N fertilizer applications depend on crop N demand, which in turn depends on N nutrition status of the crop at growth and development stages that are important for the formation of grain yield. It is recommended that N fertilizer is applied in several doses or 'splits' which correspond to critical stages in rice development. Fertilizer can be applied at transplanting to assist crop recovery, at the tillering stage to increase tiller and panicle numbers, at panicle initiation stage (PIS) to increase the number of spikelets per panicle and at full flowering to increase the 1,000-grain weight and quality measures such as the ripening percentage (De Datta, 1981). In Korea, N fertilizer is applied in three splits. The application at full-flowering stage is omitted as it increases the protein content of rice, which adversely affects its eating quality. Thus, top-dressing N fertilizer at panicle initiation stage is the last chance to fine-tune the rice growth for higher grain yield and quality.

The spatial variation in soil properties has been considered when prescribing the amount of N top-dressing in lowland rice crops (Miyama, 1988; Nguyen, 2005). However, some authors have indicated that even some stable soil properties, such as soil organic matter, texture, bulk density and cation exchange capacity, vary considerably from season to season and from year to year (Mohanty *et al.*, 2004; Dobermann, 1994; Cassman *et al.*, 1996). Also, the weather and crop management techniques can vary unavoidably from year to year and these factors combine to affect crop growth. Therefore, crop growth and nutrition status are ideal indicators of both spatial and temporal variation in soil, climate and management techniques. If these factors are monitored at the PIS and if the likely response of the crop to additional fertilizer is known, then it is possible to determine optimal recommendations for top-dressing. Therefore, inexpensive real-time methods of measuring crop growth and nutrition status are required together with mathematical models of crop growth in response to the fertilizer.

Models for determining rate of N top-dressing

Nguyen (2006) developed a multivariate regression model (Table CS3.1) to predict grain yield and milled-rice protein content of rice at harvest in response to different top-dressing rates. The model was based upon experiments conducted over three years with 46 different N rates and timing of treatments, and four different cultivars.

The yield and quality varied according to shoot N content at panicle initiation stage (PNup) and nitrogen uptake from panicle initiation stage to harvest (PHNup). The PHNup was determined from the N top-dressing rate at panicle initiation stage (Npi) and PNup.

Thus, if the N content at PIS is measured it is possible to use the model to determine the amount of N that should be applied to achieve a specified yield and quality. However, the model does not account for annual variation in weather or management technique which might alter the ideal N fertilizer rates (Lobell and Asner, 2004; Lory and Scharf, 2003). Therefore, Lee (2011) recalibrated the model by using not only N nutrition status of the crop at PIS, but also climatic factors during the reproductive stage of rice development as predictor variables (Table CS3.2). The recalibration was based on 334 datasets comprising nine years of experimental data from 2001 to 2009.

Table CS3.1 Models to predict grain yield, milled-rice protein and N uptake from panicle initiation stage to harvest

Equations for yield, protein and PHNup prediction	R^2
PHNup = 3.31 + 0.0692Npi + 0.678PNup – 0.105PNup^2	0.82
Yield = 754 + 825.7PNup – 29.6PNup^2 + 563.1PHNup – 19.2PHNup^2	0.87
Protein = 7.09 + 0.016PHNup^2 – 0.347PNup + 0.029PNup^2	0.73

Notes
Yield: grain yield (kg ha^{-1}); protein: milled-rice protein (%); PHNup: N uptake from PIS to harvest (gm^{-2}); Npi: N applied at PIS (kg N ha^{-1}); PNup: N content at PIS (g m^{-2}); R^2: coefficient of determination.

Table CS3.2 Stepwise multiple regression models to predict shoot N accumulation from PIS to harvest (PIINup), grain yield and milled rice protein content

Model equation (n = 334)	R^2	RMSEP	REP (%)
PHNup= –30.12 + 0.057Npi + 0.78PNup – 0.10 PNup^2 – 0.02Rd1 + 1.67T1	0.78	1.15	20.84
Yield = –35594.0 + 553.2PNup – 13.5PNup^2 + 399.7PHNup – 0.84PHNup^2 + 32447.3T1 – 610.8T1^2 – 33188.7T2 + 699.8T2^2	0.83	40.33	6.33
Protein = 273.71 + 0.013PNup^2 + 0.18PHNup + 0.001Rd2 – 28.09T1 + 0.53T1^2 + 8.42T2 – 0.18T2^2	0.73	0.55	8.33

Notes
Yield, protein, Npi, PNup and PHNup as in Table CS3.1; Rd1 and Rd2: sum of solar radiation (MJ m^{-2}) from 30 days before heading (DBH) to heading date and from heading to 40 days after heading (DAH), respectively; T1 and T2 (°C): average temperature from 30DBH to heading date and from heading date to 40DAH, respectively; R^2: coefficient of determination; RMSEP: root mean square error of prediction; REP: relative error of prediction.

Non-destructive monitoring of rice growth and nutrition status

Site-specific management systems require measurements of crop growth indicators and soil properties across the field to determine how fertilizer applications should vary. However, information about within-field variation is often limited because the measurement methods are time-consuming, laborious, destructive and expensive. Although many soil properties such as texture and horizon depths are relatively stable and therefore do not need to be measured regularly, others such as soil nutrient status are much more transient. Reliable measurements of soil nutrients from sampled soil cores are time-consuming and expensive. In recent years, various crop variables related to crop growth and biochemistry such as leaf-area index (LAI), plant N concentration, N uptake and chlorophyll content have been reliably predicted by remote sensing techniques (Diker and Bausch, 2003). These techniques provide a fast and non-destructive characterization of crop status and its variation in space (Nguyen *et al.*, 2008). Various remote sensing technologies are available. The spectral reflectance of a crop canopy is well correlated with crop growth and nutrition status (Haboudane *et al.*, 2004; Inoue *et al.*, 1998). Satellite measurements of spectral radiance have been used successfully to identify crop species and estimate crop area (MacDonald and Hall, 1980). However, these platforms have inherent limitations because of the high cost of images from aircraft, the infrequency of satellite overpasses and the need for atmospheric corrections and cloud screening of the data. These issues are particularly acute on Asian rice farms because of their small field sizes and the frequent presence of clouds in the monsoon rice-growing region.

Colour digital cameras are a much cheaper and more accessible alternative to remote sensing techniques. Image analysis of digital photographs has been proposed to evaluate the crop colour and nutrition status (Jia *et al.*, 2007; Li *et al.*, 2010), canopy cover (Laliberte *et al.*, 2007; Li *et al.*, 2010), growth characteristics (Behrens and Diepenbrock, 2006; Reiko and Hiroyuki, 2010) and weed densities (Vrindts *et al.*, 2002). The images can be acquired quickly and frequently to assess vegetation change over time and they can be easily archived (Pan *et al.*, 2007). Lee (2011) developed a non-destructive method for monitoring crop growth and N nutrition status with colour digital camera image analysis. Digital images of rice canopies grown under various N treatments were taken from 2 m above the canopy at nadir position periodically before the heading stage. The images covered a rice field of 0.45 m × 0.6 m and included six hills of rice. Immediately after the acquisition of digital camera images the rice plants were sampled to measure leaf-area index, shoot dry weight and shoot N accumulation. Canopy cover and 10 colour indices were calculated from digital camera images using standard image-analysis software. Nine of these colour indices and canopy cover showed significant correlations with the measurements of growth and N nutrition. All of these measurements were most strongly correlated with canopy cover. Stepwise multiple linear regression analysis was used to formulate models to estimate leaf-area index, shoot dry weight and shoot N accumulation from the parameters derived from the image analysis. Table CS3.3 illustrates that these models showed acceptable precision and accuracy, indicating that colour digital photographs can be used for characterizing the growth status of rice non-destructively.

Implementation of site-specific N recommendations

Lee (2011) integrated the image analysis of digital photographs with the mathematical models of rice development to form a system to recommend site-specific N treatments. The system was implemented in Visual Basic and consisted of the following stages (Figure CS3.2):

1   Estimation of the nitrogen content at PIS (PNup) from image analysis of digital photographs from each plot.

Table CS3.3 Statistical parameters for evaluating the performance of models for leaf-area index, shoot dry weight and shoot N accumulation using digital camera image analysis

Crop variable	Calibration			Validation		
	R^2	RMSEP	REP (%)	R^2	RMSEP	REP (%)
LAI	0.89	0.53	16.05	0.77	0.60	25.33
DW	0.83	63.54	29.98	0.76	63.03	28.59
Nup	0.85	0.74	18.72	0.80	0.74	19.21

Note
LAI, DW and Nup: leaf-area index, shoot dry weight and shoot nitrogen uptake, respectively.

2   Calculation of the required N uptake between PIS and harvest (PHNup) for target milled-rice protein content of 6.8 per cent based on the estimated PNup in each plot.
3   Calculation of the required N rates (Npi) for each plot that should be applied at PIS to achieve the PHNup values determined above. Each plot in the experiment was assumed to have a similar natural soil N supply of 3.3 kg ha^{-1} from panicle initiation stage to harvest and N uptake recovery of 69.2 per cent. These values were the averages of two years of experimental data (2003–4).

The system was used to achieve a target milled-rice protein content of 6.8 per cent. The application of the prescribed N rate treatment calculated from the model equations of Table CS3.1 increased grain yield by 421 kg ha^{-1} in comparison to uniform N treatments of 42.3 kg N ha^{-1} and zero N treatments. The variation in milled-rice protein content among plots under site-specific

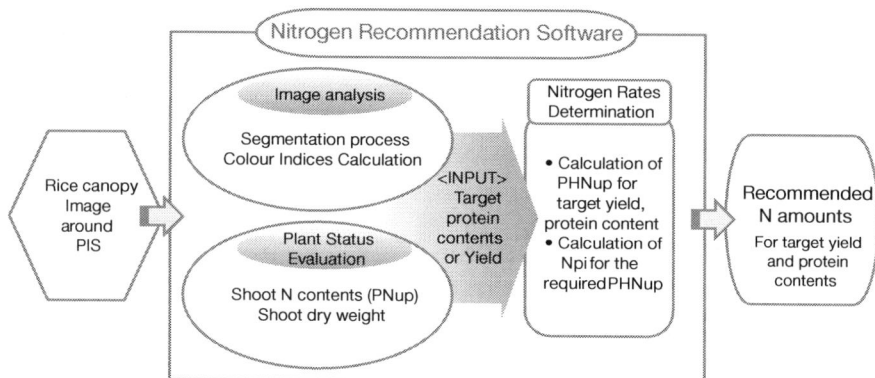

Figure CS3.2 Flow diagram of software to recommend panicle N fertilizer rate for target yield and milled-rice protein content.

Table CS3.4 Descriptive statistics of grain yield and milled-rice protein content in the prescribed N rate treatment (PRT) for the target milled-rice protein content of 6.8 per cent, fixed N rate treatment (FRT) and no N treatment at panicle-initiation stage of a rice cultivar

Treatment	Grain yield (kg ha⁻¹)				Milled-rice protein content (%)			
	Mean	CV(%)	Min.	Max.	Mean	CV(%)	Min.	Max.
No N	5,572	106	4,868	6,615	6.46	3.3	6.02	6.73
FRT	6,563	66	5,595	6,953	6.72	4.6	6.23	7.10
PRT	6,984	114	5,607	8,113	6.78	2.4	6.58	6.98

Notes
FRT: 36 kg N ha⁻¹; PRT: 42.3 kg N ha⁻¹ on average, ranging from 15.8 to 62.7 kg ha⁻¹ depending on plant N nutrition status at panicle initiation stage; CV: coefficient of variation.

management was significantly reduced from 3.3 per cent and 4.6 per cent with zero N and a fixed rate, respectively, to 2.4 per cent. The average of milled-rice protein content under site-specific management was 6.78 per cent; very close to the target milled-rice protein content of 6.8 per cent. These results suggest that an inexpensive system for precision farming with rice can be implemented for small-scale rice farming in Asia without the expensive sophisticated technologies adopted in Western large-scale precision agriculture.

References

Batte, M. T. and VanBuren, F. N. (1999) 'Precision farming: factors influencing productivity', paper presented at the Northern Ohio Crops Day meeting, Wood County, Ohio, 21 January 1999.

Behrens, T. and Diepenbrock, W. (2006) 'Using digital image analysis to describe canopies of winter oilseed rape (*Brassica napus* L.) during vegetative developmental stages', *Journal of Agronomy and Crop Science*, 192, 295–302.

Cassman, K. G., Dobermann, A., Sta Cruz, P. C., Gines, H. C., Samson, M. I., Descalsota, J. P., Alcantara, J. M., Dizon, M. A. and Olk, D. C. (1996) 'Soil organic matter and the indigenous nitrogen supply of intensive irrigated rice systems in the tropics', *Plant and Soil*, 182, 267–78.

De Datta, S. K. (1981) *Principles and Practices of Rice Production*, Singapore: John Wiley & Sons, Ltd.

Diker, K. and Bausch, W. C. (2003) 'Potential use of nitrogen reflectance index to estimate plant parameters and yield of maize', *Biosystems Engineering*, 85, 437–47.

Dobermann, A. (1994) 'Factors causing field variation of direct-seeded flooded rice', *Geoderma*, 62, 125–50.

Dobermann, A., Cassman, K. G., Sta Cruz, P. C., Adviento, M. A. A. and Pampolino, M. F. (1996) 'Fertilizer inputs, nutrients balance and soil nutrient supplying power in intensive, irrigated rice systems: III. Phosphorus', *Nutrient Cycling in Agroecosystems*, 46, 111–25.

Haboudane, D., Miller, J. R., Pattey, E., Zarco-Tejada, P. J. and Strachan, B. (2004) 'Hyperspectral vegetation indices and novel algorithms for predicting green LAI of crop canopies: modeling and validation in the context of precision agriculture', *Remote Sensing of the Environment*, 90, 337–52.

Inoue, Y., Moran, M. S. and Horie, T. (1998) 'Analysis of spectral measurements in paddy field for predicting rice growth and yield based on a simple crop simulation model', *Plant Production Science*, 1, 269–79.

Jia, L., Chen, X. and Zhang, F. (2007) 'Optimum nitrogen fertilization of winter wheat based on color digital camera image', *Communications in Soil Science and Plant Analysis*, 38, 1385–94.

Koutroubas, S. and Ntanos, D. (2003) 'Genotypic differences for grain yield and nitrogen utilization in Indica and Japonica rice under Mediterranean conditions', *Field Crops Research*, 83, 251–60.

Laliberte, A. S., Rango, A., Herrick, J. E., Fredrickson, L. E. and Burkett, L. (2007) 'An object-based image analysis approach for determining fractional cover of senescent and green vegetation with digital plot photography', *Journal of Arid Environments*, 69, 1–14.

Lee, K. J. (2011) *Diagnosis of Rice Growth Status and Recommendation of Panicle Nitrogen Topdressing Rate using Digital Camera Image Analysis*, PhD thesis, Seoul National University, Seoul, Korea.

Li, Y., Chen, D., Walker, C. N. and Angus, J. F. (2010) 'Estimating the nitrogen status of crops using a digital camera', *Field Crops Research*, 118, 221–7.

Lobell, D. B. and Asner, G. P. (2004) 'Cropland distributions from temporal unmixing of MODIS data', *Remote Sensing of the Environment*, 93, 412–22.

Lory, J. A. and Scharf, P. C. (2003) 'Yield goal versus delta yield for predicting fertilizer nitrogen need in corn', *Agronomy Journal*, 95, 994–9.

MacDonald, R. B. and Hall, F. G. (1980) 'Global crop forecasting', *Science*, 208, 670–9.

Miyama, M. (1998) 'A new method for standardization of fertilizer application for rice plant based on optimum nitrogen content', *Special Bulletin the Chiba Prefectual Agriculture Research Center*, 15, 1–92.

Mohanty, M., Painuli, D. K. and Mandal, K. G. (2004) 'Effect of puddling intensity on temporal variation in soil physical conditions and yield of rice (*Oryza sativa* L.) in a Vertisol of central India', *Soil Tillage Research*, 76, 83–94.

Nguyen, H. T. (2006) *Development of a Non-destructive Method for Assessing N-nutrition Status of Rice Plant and Prescribing N-fertilizer Rate at Panicle Initiation Stage for the Target Yield and Protein Content of Rice*, PhD thesis, Seoul National University, Seoul, Korea.

Nguyen, T. A. (2005) *Spatial Yield Variability and Site-Specific Nitrogen Prescription for the Improved Yield and Grain Quality of Rice*, PhD thesis, Seoul National University, Seoul, Korea.

Nguyen, T. H., Lee, K. J. and Lee, B. W. (2008) 'Recommendation of nitrogen topdressing rates at panicle initiation stage of rice using canopy reflectance', *Journal of Crop Science and Biotechnology*, 11, 141–50.

Pan, G., Li, F. and Sun, G. (2007) 'Digital camera based measurement of crop cover for wheat yield prediction', *Proceedings of IEEE International Geoscience and Remote Sensing Symposium, IGARSS, Barcelona, Spain*, pp. 797–800.

Reiko, I. and Hiroyuki, O. (2010) 'Use of digital cameras for phonological observations', *Ecological Informatics*, 5, 339–47.

Shibusawa, S., Sasao, A. and Sakai, K. (1999) 'Local variability of nitrate nitrogen in a small field', in J. V. Stafford (ed.) *Precision Agriculture '99*, Sheffield, UK: Sheffield Academic Press, pp. 377–86.

Vrindts, E., de Baerdemaeker, J. and Ramon, H. (2002) 'Weed detection using canopy reflection', *Precision Agriculture*, 3, 63–80.

Yagi, H. (2012) 'Farm size and distance to field in selected rice field areas with integration of plot and farm data', *Proceedings of the International Association of Agricultural Economists(IAAE) Triennial Conference, Foz do Iguacu, Brazil*, 18–24 August 2012.

CS4 Farmer perceptions of precision agriculture for fertilizer management of cotton

Roland K. Roberts, James A. Larson,
Burton C. English and J. Colby Torbett

Introduction

Cotton is an important crop in the USA with many opportunities for farmers to use precision agriculture (PA) methods profitably. The area farmed, crop value, cost of production and cost of fertilizer all lead one to expect that PA can be used effectively. In 2007 cotton producers, located primarily in the south and western USA (Figure CS4.1) (NASS, 2009), farmed 3.4% of the 129 million ha planted in principal crops. This made cotton the fifth largest crop by land area, following maize (27.6%), soya beans (24.7%), hay (18.7%) and wheat (17.0%) (NASS, 2011). Some counties in Texas had more than 50% of their cropland in cotton, while Alabama, Arkansas, Arizona, Georgia, Louisiana, North Carolina, Texas and Tennessee had counties with between 25% and 50% of their cropland in cotton (Figure CS4.2) (NASS, 2009). The average value of cotton production was US$1779 ha^{-1}, higher than for maize (US$1576 ha^{-1}), soya beans (US$1110 ha^{-1}) and wheat (US$531 ha^{-1}) (ERS, 2010, 2011). Higher value comes at greater cost; the average variable cost of producing cotton was US$1143 ha^{-1}, compared with US$638 ha^{-1} for maize, US$265 ha^{-1} for wheat and US$327 ha^{-1} for soya beans. Of these crops, cotton had the second largest average fertilizer cost of US$182 ha^{-1}, following maize (US$248 ha^{-1}), but greater than for soya beans (US$44 ha^{-1}) and wheat (US$102 ha^{-1}) (ERS, 2011). The high value and cost of cotton production, especially fertilizer cost, provide ample incentive for cotton farmers to adopt PA for fertility management (Larson *et al.*, 2005).

Although cotton is a perennial shrub adapted to survive in warm semi-arid environments, it is grown as an annual crop in the USA. Under the right growing conditions, cotton will first develop a root, stem and branch structure. Only after the plant has this basic structure will it begin to reproduce bolls that are harvested for cotton lint when mature. Limited stress management is required to encourage reproductive rather than vegetative growth. One management challenge is to maintain a vegetative–reproductive balance throughout the growing season that will optimize both lint yield and quality (Larson, 1992). Soil fertility management is important for maintaining this balance, given cotton's sensitivity to over application of fertilizer (especially nitrogen, N), which can lead to excess vegetative growth, affecting the value of harvested lint through reduced yield and quality (Cothren and Oosterhuis, 2009).

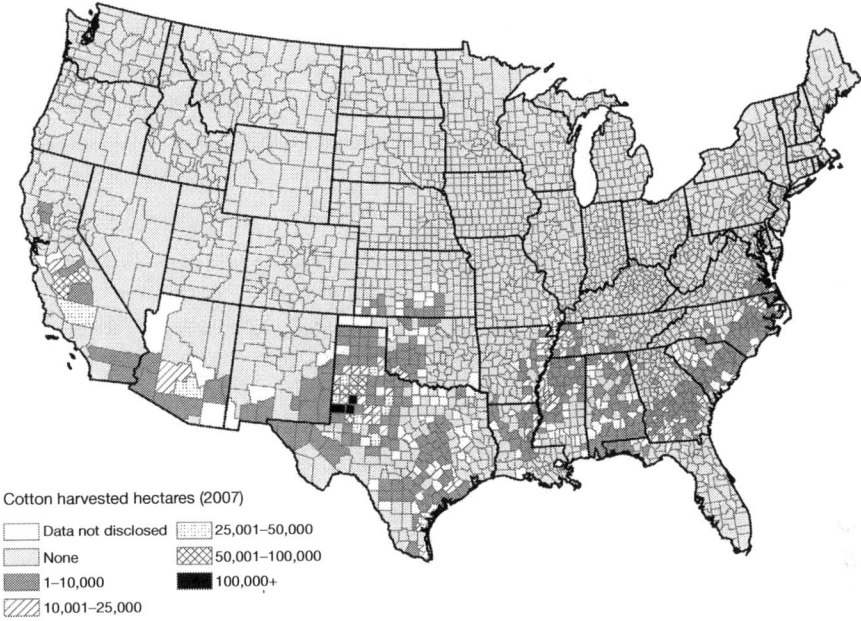

Figure CS4.1 Hectares of cotton harvested by county in 2007.

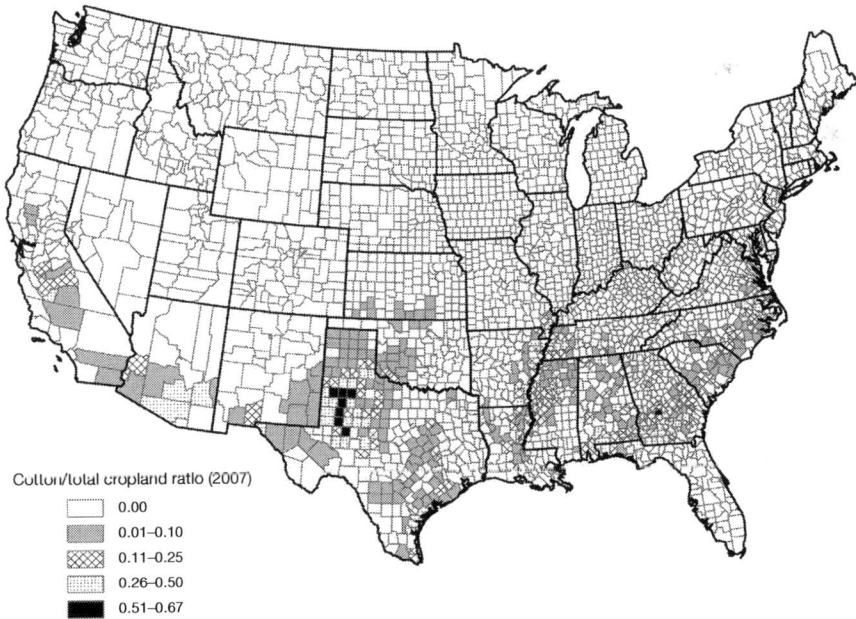

Figure CS4.2 Ratio of land in cotton to all cropland by county in 2007.

Cotton is a high-value, high-cost crop. Swinton and Lowenberg-DeBoer (1998) found that high-value crops, such as sugar beet, give the highest return with the adoption of PA. However, the adoption of PA for cotton occurred later than for grain or oil seed crops. The entry-level PA technology for grains and oil seeds was yield monitoring (Lowenberg-DeBoer, 1999). Yield monitoring first became commercially available in the USA in 1992 (Mangold, 1997; Ag Leader Technology, 2004) and by 2000 had been adopted by 29.6 per cent and 25.4 per cent of Midwestern maize and soya bean farmers, respectively (Daberkow *et al.*, 2002). In contrast, accurate yield monitors were not commercially available for cotton until 2000, so adoption of PA by cotton farmers lagged behind that of grain and oil seed farmers (Wolak *et al.*, 1999; Durrence *et al.*, 1999; Roades *et al.*, 2000). In their 2001 survey of cotton farmers in six states, Roberts *et al.* (2002) found that only 2.0 per cent and 0.7 per cent of respondents had adopted cotton yield monitors with and without GPS, respectively. In comparison, 11.5 per cent of respondents were using grid-based soil sampling to learn about the within-field variation of soil nutrient supplies and a further 8.8 per cent were conducting soil sampling within management zones. Hence, the entry-level PA technology for cotton farmers was geo-referenced soil sampling.

Yield monitoring for grains and oil seed crops and geo-referenced soil sampling for cotton were entry-level technologies because they were the first site-specific information technologies to become widely available. Such technologies provide information about variability in soil characteristics and crop yields within a field, and this information can be used to prescribe variable-rate input applications across the field. Uniform-rate fertilizer applications can lead to over-application in the portions of a field with relatively small fertilizer demands and under-application in areas with large demands. Variable-rate applications are intended to ensure each location receives the optimal amount of fertilizer that is required to maximize profits or to minimize waste and environmental impacts. Thus, compared with uniform-rate application methods practised in conventional farming, variable-rate application can prevent over-application in some portions of the field and under-application in others, potentially increasing field-average yield and or reducing field-average fertilizer application.

Although variable-rate fertilizer applications are intended to improve the efficiency of fertilizers compared with uniform-rate applications, the actual improvements in fertilizer efficiency achieved by farmers are difficult to quantify. The profit-maximizing farmer requires that the benefits exceed the costs associated with the three stages of variable-rate fertilizer management: (1) site-specific information gathering; (2) site-specific information analysis and input prescription mapping and (3) variable-rate application of fertilizers according to these maps. In a single field it might be possible to conduct field experiments to model the optimal variable-rate fertilizer applications and the associated benefits and costs, although as we see in Chapter 11 of this book there are many challenges associated with such experiments. In any case, it is not cost-effective for farmers to conduct such experiments in every field where cotton is grown. Therefore they must select their variable fertilizer application rates based on advice from soil-testing laboratories or Extension Services. This advice might be formulated from a number of experiments conducted across the region or state, but given the regional-scale variation in soil and growing conditions it is difficult to know how accurately the recommended fertilizer rates compare with the profit-maximizing fertilizer prescriptions for a particular field. The benefits of PA are likely to vary from field to field. In some fields uniform-rate applications will maximize profits but in others, where the growing conditions are more variable, the fertilizer applications must reflect these variations. Thus, given these potential prescription errors and differences in within-field variability across fields, farms and regions of the US cotton belt, it is not entirely clear whether farmers are benefitting from the adoption of variable-rate technology.

In January and February of 2001, the University of Tennessee and five other Land Grant Universities in the USA contracted with Cotton Incorporated to survey cotton farmers in six southern states (Alabama, Florida, Georgia, Mississippi, North Carolina and Tennessee) about their use of PA technologies (Roberts *et al.*, 2002). In 2000, 35.9 per cent of US cotton lint was produced in these six states (NASS, 2003). The survey results indicated that 23.0 per cent of respondents had used at least one PA technology. The PA adoption in these states grew to 45.8 per cent in 2005 (Cochran *et al.*, 2006) and 53.6 per cent in 2009 (Mooney *et al.*, 2010). Recognizing the difficulties in measuring improvements in efficiency from using PA compared with conventional farming methods, the 2001 Cotton Incorporated survey asked the farmers about the benefits they perceived. In particular, cotton precision farmers (farmers reporting the use of PA) were asked to rate from 1 (not important) to 5 (very important) the decision-making value of the PA technologies they had used in reducing (1) N use and (2) phosphorus and potassium (P&K) use (Roberts *et al.*, 2002). In comparison with uniform-rate application, efficient input use could mean increasing fertilizer application to increase yield on high yield-response portions of a field and substantially reducing fertilizer application without appreciably reducing yield on low yield-response portions of the field. The field-average result might be to reduce fertilizer application compared with uniform-rate application. If a cotton farmer indicates that PA contributes to reducing fertilizer use in cotton production, the farmer is likely to apply less fertilizer to produce the crop. Thus, cotton precision farmers' ratings of the importance of the decision-making value of PA technologies in reducing fertilizer use can be used to represent their perceptions about improved N and P&K efficiency from using PA relative to conventional farming methods. If the farmers' perceptions truly reflect the benefits of PA, then such surveys can be used to quantify, relatively cheaply, these benefits across a range of locations and growing conditions. The remainder of this chapter summarizes Torbett *et al.*'s (2007, 2008) statistical analyses of the survey results and discusses the conclusions that can be drawn regarding the improved efficiency of N and P&K applications from the use of PA technologies for fertilizer management.

Perceptions of improved fertilizer efficiency

Data

The analysis described in this chapter is based on data from the 2001 Cotton Incorporated survey (Roberts *et al.*, 2002). Information from later surveys (Cochran *et al.*, 2006; Mooney *et al.*, 2010) was not used because similar follow-up questions were not asked. Nevertheless, the 2001 data continue to provide insight into cotton precision farmers' perceptions about how PA improves N and P&K efficiency since geo-referenced soil sampling and variable-rate fertilizer application have remained the prevalent PA technologies used in cotton production (Roberts *et al.*, 2002; Cochran *et al.*, 2006; Mooney *et al.*, 2010).

Torbett *et al.* (2007, 2008) reported average importance ratings of 3.7 for improving both N and P&K efficiency, and 63 per cent of precision farmers responded to both questions with ratings of 4 or 5. Several farmers did not answer all the relevant questions for evaluating perceptions of improved fertilizer efficiency from using PA. In the final analysis, 135 observations with complete information were used to evaluate precision farmers' perceptions of improved N efficiency and 144 observations were used to evaluate perceived improvements in P&K efficiency.

Methods

The decision to adopt PA depends on the perceptions of the farmer about whether additional profit can be earned from using PA compared with conventional farming methods. Increased fertilizer efficiency could contribute to the additional profit, and the potential to increase efficiency will depend upon the particular characteristics of the farmer and farm in question. Torbett *et al.* (2007, 2008) proposed a number of farm and farmer characteristics which might influence the farmers' perception of PA. They used regression analyses to test the hypotheses that each factor influenced the farmers' perception. The dependent variable in these analyses was a categorical variable, namely the score assigned by the farmer which could take an integer value between 1 and 5. Therefore, the significance of the relationships was established using ordered logit regression methods which relate this categorical variable to an unobserved continuous variable that can have non-integer values (Torbett *et al.*, 2007, 2008). The survey asked a substantial number of questions about each farmer and farm. These answers can be treated as independent variables that might influence the farmers' perception of fertilizer management efficiency. We summarize the hypotheses considered by Torbett *et al.* (2007, 2008) below.

Hypotheses about site-specific information sources

Cotton precision farmers' perceptions of improved fertilizer efficiency can be affected by the site-specific information technologies they used to make fertility management decisions. These technologies are designed to help farmers prescribe variable-rate applications within a field. The information technologies hypothesized to affect perceptions of improved efficiency are yield monitoring with and without GPS, satellite imagery and or aerial photography, mapping of topography, slope, soil depth, etc. (henceforth, mapping), management-zone soil sampling, grid soil sampling, on-the-go sensing, plant tissue testing and soil survey maps. Each of these technologies is expected to have a positive effect on precision farmers' perceptions of improved N and P&K efficiency; however, some site-specific information technologies are expected to have greater influence on perceptions than others. For example, precision farmers using geo-referenced soil sampling are likely to perceive greater improvements in P&K efficiency than those using yield monitoring, because geo-referenced soil sampling provides more direct and easily interpretable information about site-specific crop fertilizer needs than yield monitoring.

Hypotheses about farm characteristics

The farm characteristics that might affect a cotton precision farmer's perceptions of improved N and P&K efficiency from using PA include farm size, land tenure and land quality. Farmers with more planted area (farm size) may have greater opportunity to observe spatial variation in soil N and P&K, and are more likely to benefit from greater efficiency through the use of PA. The ratio of area owned to total area farmed (land tenure) is expected to have a negative influence on perceptions of improved efficiency. There are limited incentives to determine the site-specific characteristics of rented land relative to owned land, especially if rental agreements are renewed annually. Also, farmers in the USA typically expand operations through land rental to avoid the fixed cost of land ownership; therefore, the farmer might have farmed the rented land for a relatively short time and have relatively little knowledge of its variation. Thus, site-specific knowledge of rented land may be less than site-specific knowledge of owned land, leading to greater perceived improvements in N and P&K efficiency from using PA on rented than owned land. Torbett *et al.* (2007, 2008) suggested that average yield (land quality) may be positively associated with yield variability within fields. Therefore greater land quality may indicate greater perceived opportunity to address this variability through variable-rate fertilizer management.

Hypotheses about farmer characteristics

The farmer characteristics that might affect perceptions of efficiency include educational attainment, age and computer use. The use of PA requires considerable analytical ability; thus, precision farmers who have attended college may perceive greater improvements in N and P&K efficiency if their educational attainment allows them to recognize and receive the benefits of PA better than less-educated precision farmers. Older, more experienced precision farmers may be better able to recognize improvements in the efficiency of N and P&K fertilization than their younger counterparts. Farmers who use computers for farm management might perceive greater benefits of PA technologies because they are more likely to be comfortable with PA computer-based systems and because they are more likely to keep accurate records that track improvements in N and P&K efficiency.

Hypotheses about general perceptions of precision agriculture

The survey asked each farmer about their perceptions of the current profitability of PA and its future importance. Positive answers to these questions are likely to indicate that the farmer has a generally favourable attitude to PA and is more likely to be positive about the potential for improved fertilizer efficiency.

Hypotheses about farm location

The geographic location of a farm can influence perceptions of improved efficiency because of differences in soil types, climate, prevalence of PA adoption and other factors specific to the farm's location (Larkin *et al.*, 2005; Roberts *et al.*, 2004). For example, Georgia cotton farmers may perceive greater improvements in N efficiency from PA relative to Tennessee farmers because of the poorer quality soil (NRCS, 2001) and greater average rainfall (Troeh and Thompson, 2005) in the major cotton-producing areas of Georgia (typically sandy soil) relative to Tennessee (upland silt loams). Both soils are susceptible to the leaching of N, but less so for the silt loam soil of western Tennessee (NRCS, 2001). With greater rainfall and more leachable soils, N deficiencies within cotton fields might be more variable and pronounced in Georgia and hence the perceived potential for precision fertilizer management might be greater. Similarly, perceptions of Mississippi and Tennessee precision farmers may differ because of the more variable soil (NRCS, 2001) and greater average rainfall (Troeh and Thompson, 2005) in the major cotton-producing areas of Mississippi (Southern Mississippi Valley alluvial soil with greater textural variation) than in Tennessee (NRCS, 2001). Alternatively, differences in farmer perceptions for P&K among states may be muted because of greater difficulty in identifying P&K deficiencies and variability in comparison to N (Phillips, 2009; Stevens *et al.*, 2002).

Results

The results for precision farmers' perceptions about improvements in N and P&K efficiency from using PA are summarized below. Further details can be obtained from Torbett *et al.* (2007, 2008).

Perceived improvements in N efficiency

According to the ordered logit regression models, four site-specific information technologies significantly increased the perception of improved efficiency in N use by PA methods. These were yield monitoring without GPS, management-zone soil sampling, grid soil sampling and on-the-go sensing. Unexpectedly, mapping had a significant negative relationship with the perceived importance of PA for improving the efficiency of N.

Precision farmers who rented a larger portion of the land they farmed perceived PA technologies as more important in improving N efficiency than those who owned more of the land they farmed. Consistent with the hypothesis above, this finding suggests that cotton precision farmers found PA technologies more useful for improving knowledge about the spatial N requirements of rented land than owned land. In addition, older precision farmers might have recognized the benefits of PA in improving N efficiency better than younger precision farmers.

Cotton precision farmers in Georgia and Mississippi were more likely to have positive perceptions of improved N efficiency from using PA than those in Tennessee. Precision farmers in the other states had perceptions similar to those in Tennessee.

Perceived improvements in P&K efficiency in contrast to N efficiency

Four site-specific information technologies were significant in explaining perceptions of PA in improving the efficiency of P&K in cotton production. The implications for mapping, management-zone soil sampling and grid soil sampling are similar to those for improving N efficiency. In the case of P&K, however, the use of a yield monitor without GPS did not affect perceptions, perhaps because P&K deficiencies are not so readily apparent through visual inspection as N deficiencies. Thus, yield monitoring without GPS and visual inspection throughout the crop season might provide farmers with better spatial knowledge of N deficiencies than of P&K deficiencies.

In contrast to the results for N efficiency, remote sensing (aerial and satellite imagery) was a significant factor affecting precision farmers' perceptions of P&K efficiency from using PA. Unexpectedly, precision farmers had a negative perception of this technology's importance in improving P&K efficiency. Nevertheless, the results for both N and P&K suggest that using high-altitude remote sensing to obtain site-specific field information did not increase precision farmers' perceptions of improved N or P&K efficiency. This finding may stem from remote sensing of cropped area in the USA being on the decline at the time of the survey in 2001 (Griffin *et al.*, 2004). Since that time, high-altitude remote sensing has become more popular among US cotton farmers (e.g. InTime, Inc., 2007; Brown and Wesch, 2006).

Many cotton precision farmers in the southeastern USA perform soil testing for P and K but not for N. Soil testing for N is usually unreliable because of extreme temporal and spatial variation in the N content of southeastern USA soil through volatilization, denitrification and leaching (NRCS, 2001). Nevertheless, perceptions of the value of management-zone soil sampling and grid soil sampling were significant for both N and P&K. The results suggest that farmers who used geo-referenced soil sampling to detect within-field variation in soil P&K may also have become more aware of, or sensitive to, the variation in N requirements.

On-the-go sensing affected precision farmers' perceptions of PA for improving N efficiency but not for improving P&K efficiency. This result was expected because on-the-go sensing technology typically measures N deficiencies through technologies such as GreenSeeker® (Trimble Navigation Limited, 2010). While not significant at the 5 per cent level, precision farmers who used on-the-go sensing perceived improvements in P&K fertilizer use efficiency at the 10 per cent level, suggesting that they may have become more aware of, or sensitive to, the variation in P&K when using this N-related technology in their fields.

The perceptions of cotton precision farmers in Georgia and Mississippi were different from those in Tennessee for N, but not for P&K. Differences for N between Georgia or Mississippi and Tennessee may have resulted from poorer quality and more variable soil and greater average rainfall relative to Tennessee, providing Georgia or Mississippi cotton precision farmers greater opportunity to perceive PA as more useful in improving N efficiency. The lack of significant differences in perceptions among states for P&K efficiency might result from the greater difficulty in identifying P&K deficiencies and variability relative to N, and the greater immobility of P&K in the soils relative to N (Comerford, 2005).

Discussion and conclusions

Substantial variation exists in the benefits and costs of PA. One source of variation enters through prescription errors in input recommendations from consultants, soil-testing laboratories and government agencies, and another comes from differences in within-field variability across fields, farms and regions. Consequently, many previous studies on the economic and environmental consequences of PA have been either hypothetical in nature (e.g. Batte and Ehsani, 2006; Roberts *et al.*, 2006; Mooney *et al.*, 2009; Shockley *et al.*, 2012) or experimental cases (e.g. Lambert and Lowenberg-DeBoer, 2000; Swinton and Lowenberg-DeBoer, 1998) often conducted at agricultural research facilities under better than average farming conditions. One way to obtain broader estimates of the economic and environmental consequences of PA has been to ask farmers through surveys about their experiences with PA (e.g. Larkin *et al.*, 2005). These types of surveys are inherently subjective, relying heavily on farmer perceptions. Nevertheless, if perceptions are correlated with reality, evaluation of farmer perceptions through survey data can provide relevant information about the broader economic and environmental consequences of widespread PA implementation. Farmer perceptions about PA under a wider range of climatic, soil and management conditions than those found at agricultural research facilities can help farmers in evaluating the utility of PA technologies. Their perceptions may also be useful to agribusinesses in developing PA technologies and to policymakers and government agencies interested in encouraging production practices that reduce the environmental impact of farming.

We have presented the results of two studies using data from a survey conducted in 2001 to evaluate the perceptions of US cotton farmers regarding the effects of PA on fertilizer efficiency. Similar follow-up questions were not asked in subsequent PA surveys of cotton producers. The usefulness of the data from the 2001 survey for improving knowledge about the economic and environmental consequences of practising PA suggests we might benefit from more current information by including similar questions in future surveys.

Notwithstanding the passage of time since the 2001 survey, the results presented in this case study provide several useful insights that address concerns about the negative environmental effects from excessive applications of fertilizer. First, results suggest that farmers who used PA perceived improvements in N and P&K fertilizer efficiency. If these perceptions correspond with reality, the increased efficiency would allow more of the N, P and K fertilizers to be harvested with the crop, with less remaining in the soil to run off into surrounding water bodies and leach into the groundwater. Second, greater perceptions of improved fertilizer efficiency may encourage cotton precision farmers who use yield monitors, geo-referenced soil sampling or on-the-go sensing to continue using these technologies to increase fertilizer efficiency relative to conventional farming methods. Third, if perceptions are close to reality, cotton precision farmers can achieve the greatest increase in fertilizer efficiency from the site-specific management of rented land, providing greater environmental benefits for society than if site-specific management of rented land were neglected. The results suggest that Extension personnel and others interested in reducing the negative environmental impacts of excessive use of crop fertilizers might want to develop programmes to stabilize rental agreements and promote PA on rented land where the greatest perceived benefit from improved efficiency can be achieved.

Acknowledgements

The authors thank Cotton Incorporated and the Agricultural Experiment Stations of the Land Grant Universities involved in the 2001 Cotton Incorporated survey – Auburn University, University of Florida, University of Georgia, Mississippi State University, North Carolina State University, and University of Tennessee – for funding this research.

References

Ag Leader Technology (2004) 'Product history', Ag Leader Technology, Ames, IA, USA. Available online at <www.agleader.com/about> (accessed 1 September 2011).

Batte, M. T. and Ehsani, M. R. (2006) 'The economics of precision guidance with auto-boom control for agricultural sprayers', *Computers and Electronics in Agriculture*, 53, 28–44.

Brown, T. and Wesch, R. (2006) 'Opti-grow aerial imagery', in D. Richter (ed.) *Proceedings of the Beltwide Cotton Conferences, 2006*, Memphis, TN: National Cotton Council of America, pp. 2293–4.

Cochran, R. L., Roberts, R. K., English, B. C., Larson, J. A., Goodman, W. R., Larkin, S. R., Marra, M. C., Martin, S. W., Paxton, K. W., Shurley, W. D. and Reeves, J. M. (2006) 'Precision farming by cotton producers in eleven states: results from the 2005 southern precision farming survey', Research Report 01-06, Department of Agricultural and Resource Economics, University of Tennessee, Knoxville, TN, USA.

Comerford, N. B. (2005) 'Soil factors affecting nutrient bioavailability', in H. BassiriRad (ed.), *Nutrient Acquisition by Plants: An Ecological Perspective*, New York: Springer, pp. 1–11.

Cothren, J. S. and Oosterhuis, D. M. (2009) 'Plant growth regulators in cotton', in J. M. Stewart, D. M. Oosterhuis, J. J. Heitholt and J. R. Mauney (eds), *Physiology of Cotton*, New York: Springer, pp. 289–303.

Daberkow, S. G., Fernandez-Cornejo, J. and Padgitt, M. (2002) 'Precision agriculture technology diffusion: Current status and future prospects', in P. C. Robert, R. H. Rust and W. E. Larson (eds) *Proceedings of the Sixth International Conference on Precision Agriculture and Other Precision Resources Management, Minneapolis, MN, 14–17 July*, Madison, WI: American Society of Agronomy, Crop Science Society of America and Soil Science Society of America (unpaginated CD-ROM).

Durrence, J. S., Thomas, D. L., Perry, C. D. and Vellidis, G. (1999) 'Preliminary evaluation of commercial yield monitors: the 1998 season in south Georgia', in C. P. Dugger and D. A. Richter (eds) *Proceedings of the Beltwide Cotton Conferences, 1999*, Memphis, TN: National Cotton Council America, pp. 366–72.

Economic Research Service (ERS) (2010) 'Commodity cost and returns: methods', US Department of Agriculture, 3 May. Available online at <www.ers.usda.gov/data/costsandreturns/methods.htm> (accessed 2 June 2011).

Economic Research Service (ERS) (2011) 'Commodity cost and returns: data', US Department of Agriculture, 25 May. Available online at <www.ers.usda.gov/Data/CostsAndReturns/testpick.htm#recent> (accessed 2 June 2011).

Griffin, T. W., Lowenberg-DeBoer, J., Lambert, D. M., Peone, J., Payne, T. and Daberkow, S. G. (2004) 'Adoption, profitability, and making better use of precision farming data', Staff Paper 04-06, Department of Agricultural Economics, Purdue University, West Lafayette, IN.

InTime, Inc. (2007) 'History of the company'. Available online at <www.gointime.com/about_history.jsp> (accessed 12 February 2007).

Lambert, D. M. and Lowenberg-DeBoer, J. (2000) 'Precision agriculture profitability review', Site Specific Management Center, School of Agriculture, Purdue University, West Lafayette, IN, 15 September. Available online at <http://agriculture.purdue.edu/SSMC/Frames/newsoilsX.pdf> (accessed 17 July 2012).

Larkin, S. L., Perruso, L., Marra, M. C., Roberts, R. K., English, B. C., Larson, J. A., Cochran, R. L. and Martin, S. W. (2005) 'Factors affecting perceived improvements in environmental quality from precision farming', *Journal of Agricultural and Applied Economics*, 37, 577–88.

Larson, J. A. (1992) *An Economic Analysis of the Sequential Decision Problem for Irrigated Cotton Production in Southwest Oklahoma*, unpublished PhD dissertation, Oklahoma State University.

Larson, J. A., Roberts, R. K., English, B. C., Cochran, R. L. and Wilson, B. S. (2005) 'A computer decision aid for the cotton yield monitor investment decision', *Computers and Electronics in Agriculture*, 48, 216–34.

Lowenberg-DeBoer, J. (1999) 'Risk management potential of precision farming technologies', *Journal of Agricultural and Applied Economics*, 31, 275–83.

Mangold, G. (1997) 'How many monitors?', *Ag Innovator News*, 5.3, 2.

Mooney, D. F., Larson, J. A., Roberts, R. K. and English, B. C. (2009) 'Evaluating investments in variable rate technology for agricultural sprayers in cotton production', *Journal of the American Society of Farm Managers and Rural Appraisers*, 31, 177–88.

Mooney, D. F., Roberts, R. K., English, B. C., Lambert, D. M., Larson, J. A., Velandia, M., Larkin, S. L., Marra, M. C., Martin, S. W., Mishra, A., Paxton, K. W., Rejesus, R., Segarra, E., Wang, C. and Reeves, J. M. (2010) 'Precision farming by cotton producers in twelve southern states: results from the 2009 summary', *Research Series*, 10–02, Department of Agricultural and Resource Economics, The University of Tennessee, Knoxville, TN, USA.

National Agricultural Statistics Service (NASS) (2003) 'Crop production 2002 summary', US Department of Agriculture, January. Available online at <http://usda.mannlib.cornell.edu/usda/nass/CropProdSu//2000s/2003/CropProd Su-01-10-2003.pdf> (accessed 2 June 2011).

National Agricultural Statistics Service (NASS) (2009) *2007 Census of Agriculture*, Washington DC: US Department of Agriculture.

National Agricultural Statistics Service (NASS) (2010) 'Acreage', US Department of Agriculture, June 30. Available online at <http://usda.mannlib.cornell.edu/usda/current/Acre/Acre–06–30–2011.pdf> (accessed 2 June 2011).

National Agricultural Statistics Service (NASS) (2011) 'Crop production 2010 summary', US Department of Agriculture, January. Available online at <http://usda.mannlib.cornell.edu/usda/nass/CropProdSu//2010s/2011/CropProdSu–01–12–2011_revision.pdf> (accessed August 2013).

National Resources Conservation Service (NRCS) (2001) *1997 National Resources Inventory (Revised December 2000)*, Washington, DC: US Department of Agriculture (CD-ROM).

Phillips, S. (2009) 'Nutrient deficiencies in cotton', International Plant Nutrition Institute, Presentation at the Louisiana Agricultural Technology and Management Conference, Alexandria, LA, February. Available online at <http://www.laca1.org/Presentations/2009/Cotton%20Deficiency%20Symptoms.pdf> (accessed 17 May 2012).

Roades, J. P., Beck, A. D. and Searcy, S. W. (2000) 'Cotton yield mapping: Texas experiences in 1999', in C. P. Dugger and D. A. Richter (eds) *Proceedings of the Beltwide Cotton Conferences, 2000*, Memphis, TN: National Cotton Council of America, pp. 404–7.

Roberts, R. K., English, B. C., Larson, J. A., Cochran, R. L., Goodman, W. R., Larkin, S. L., Marra, M. C., Martin, S. W., Shurley, W. D. and Reeves, J. M. (2002) 'Precision farming by cotton producers in six southern states: results from the 2001 southern precision farming survey', Research Report 0302, Department of Agricultural Economics, University of Tennessee, Knoxville, TN.

Roberts, R. K., English, B. C., Larson, J. A., Cochran, R. L., Goodman, W. R., Larkin, S. L., Marra, M. C., Martin, S. W., Shurley, W. D. and Reeves, J. M. (2004) 'Adoption of site-specific information and variable rate technologies in cotton precision farming', *Journal of Agricultural and Applied Economics*, 36, 143–58.

Roberts, R. K., English, B. C. and Larson, J. A. (2006) 'The variable-rate input application decision for multiple inputs with interactions', *Journal of Agricultural and Resource Economics*, 31, 391–413.

Shockley, J., Dillon, C. R., Stombaugh, T. and Shearer, S. (2012) 'Whole-farm analysis of automatic section control for agricultural machinery', *Precision Agriculture*, 13, 411–20.

Stevens, G., Motavalli, P., Scharf, P., Nathan, M. and Dunn, D. (2002) 'Crop nutrient deficiencies and toxicities', University of Missouri-Columbia Extension, College of Agriculture Food and Natural Resource Plant Protection Programs IPM1016. Available online at <http://ipm.missouri.edu/ipm_pubs/ipm1016.pdf> (accessed 23 May 2013).

Swinton, S. M. and Lowenberg-DeBoer, J. (1998) 'Evaluating the profitability of site-specific farming', *Journal of Production Agriculture*, 11, 439–46.

Torbett, J. C., Roberts, R. K., Larson, J. A. and English, B. C. (2007) 'Perceived importance of precision farming technologies in improving phosphorus and potassium efficiency in cotton production', *Precision Agriculture*, 8, 127–37.

Torbett, J. C., Roberts, R. K., Larson, J. A. and English, B. C. (2008) 'Perceived improvements in nitrogen fertilizer efficiency from cotton precision farming', *Computers and Electronics in Agriculture*, 64, 140–8.

Trimble Navigation Limited (2010) 'Nitrogen/defoliant/plant growth regulator: GreenSeeker RT200'. Available online at <http://www.greenseeker.com/green-seeker-RT200.html> (accessed 1 September 2011).

Troeh, F. M. and Thompson, L. M. (2005) *Soils and Soil Fertility*, 6th edn, Ames, IA: Blackwell Publishing.

Wolak, F. J., Khalilian, A., Dodd, R. B., Han, Y. J., Keshkin, M., Lippert, R. M. and Hair, W. (1999) 'Cotton yield monitor evaluation, South Carolina: year 2', in C. P. Dugger and D. A. Richter (eds) *Proceedings of the Beltwide Cotton Conferences, 1999*, Memphis, TN: National Cotton Council America, pp. 361–4.

Future prospects

Margaret A. Oliver, Thomas F. A. Bishop and Ben P. Marchant

The chapters in this book illustrate the range of possibilities that the approaches embodied in precision agriculture (PA) offer in terms of improving agricultural management, of reducing the effects of agriculture on the environment and of supporting the need for food security for an increasing world population. The adoption of PA has not been straightforward and this is probably because it was technology-driven at the outset. How could the 'new toys' that engineers produced be used? This applied very much to the yield monitor, but it is yield mapping that has driven PA forward. Farmers throughout time have recognized the variation present in their land and this was originally dealt with by managing small parcels (fields) in a similar way. As technology increased with new farm machinery, fields also increased in size and so did the variation within them. Farmers were intuitively aware of this, but the new yield mapping possibilities showed them the reality of the within-field variation in yield. Again farmers often know some of the causes of this variation at the broad scale, but not at the finer scale to gain greater insight that would help with better management. As the various scientific disciplines associated with agriculture such as soil science, hydrology, crop science, the study of weeds and diseases, nutrient management and so on became involved with PA it became clear that this was much more than a technological revolution but also an agricultural one. The demands of PA are now driving forward the technology and this can be seen clearly in weed management, where sophisticated sprayers are being developed that can switch on and off a herbicide or pesticide according to the distribution of weeds and pests.

As time has passed in PA many tools have come to its aid, in particular the proximal sensors which can provide the spatial resolution of information needed relatively cheaply. The variation that can be portrayed by these methods can be used by farmers and their managers to gain insight into the causes of the variation in crop yield and quality. This is crucial for achieving food security so that farmers can maximize yields without detriment to the environment. As the strictures placed on farmers in relation to environmental protection in all developed countries increase so their management must evolve to avoid unnecessary additions of nitrogen and phosphorus to the environment. One example of this is the nitrate vulnerable zones in the United Kingdom, within which farmers are constrained in their use of nitrogen fertilizers.

Site-specific management is the goal of precision farmers and requires detailed information on the variation present so that it can be managed effectively. Remote and proximal sensing provide much information on the variation in soil and the crop, but this requires effective interpretation. Apparent electrical conductivity (EC_a) of the soil provides excellent information on the degree of variation in the soil, but linking this with specific soil properties has been a stumbling block to its use in precise management. It does provide clues on soil texture and drainage, but these need to be verified by field information involving costly soil sampling and analysis. Digital maps of EC_a cannot act as a substitute for a map of soil phosphorus or any other crop nutrient required for optimal crop growth. The determination of soil pH in detail for site-specific management is one property where progress has been made with 'on-the-go' measurement and management. For site-specific management of all crop nutrients, seeding rates, irrigation and pesticides, rapid, inexpensive methods of measurement are required and this is likely to be where the next major breakthrough will come in PA.

Chapter 11 on the economics of PA shows that in many cases farmers have been reluctant to adopt PA because they are risk-averse and it is difficult to convince them of the financial benefits. However, they are often convinced by the environmental benefits first and the needs of best practice, especially where there are restrictions on management. Case study 4 concerns the adoption of PA by cotton farmers, and many were convinced by the more efficient use of N, P and K with the adoption of PA. It is likely that economics will encourage the greater adoption of PA over the next few decades as resources become increasingly scarce and expensive. This is true of the major crop nutrients and in particular water. The competition for water is already great in semi-arid regions of the world and it is there where more land is likely to be brought into use to increase agricultural output. Water has been a relatively cheap commodity, but competition will increase its cost and farmers will consider the need for variable-rate applications of water to those parts of fields that require it most. Optimal use of irrigation water should also help to avoid some of the problems associated with irrigation as described in Chapter 10.

What we have mentioned above is feasible at present, but is likely to become increasingly so as on-the-go soil sampling and analysis becomes established. However, we have yet to mention a currently intractable problem for many parts of the world, namely, the weather, which varies spatially and temporally in often unpredictable ways. Some fields show similar patterns of variation in yield regardless of the annual weather conditions, but many others show reversals in pattern between wet and dry years. Without being able to predict this in advance, the precise application of crop nutrients becomes less realistic. To understand how annual weather conditions affect yield requires several years of information to establish the temporal variation. We are a long way from annual weather forecasts, but within-season weather forecasting is improving constantly.

Experimentation has a role to play here in that performing the same experiment over several years under different weather conditions could lead to a better understanding of optimal application rates as they vary with weather. Then when farmers are given a seasonal forecast they can change the application rates for different parts of the field accordingly. As shown in Chapter 12, another benefit of experimentation in PA is that it is performed on a grower's own field, which makes the results more relevant and localized rather than the adoption of district or regional application rates, which is the norm. However, much research is still needed to develop experimental designs that are efficient both statistically and economically, for example, how much of an area should have zero or low application rates? Then, as outlined in Chapter 12, readily available and easy-to-use software is required to analyse the experimental results.

Precision agriculture has been associated with large-scale modern farming in the developed world, but it is clear from Chapter 3 in this book and also the broader literature that the principles of PA are being adopted at many levels of farming. Rice cultivation in paddy fields is generally small scale and Case study 3 shows what progress can be made with PA. There are also examples of date cultivation where site-specific management is done manually to reduce inputs and maintain quality (e.g. Mazloumzadeh *et al.*, 2010). The outlook for the small-scale farmer is good because expensive equipment is not a requirement for PA, but the same benefits can accrue with a reduction in inputs to make their use more efficient and economical and limit damage to the environment.

The need to use resources more economically and for preservation of the environment for sustainability contribute to the goal of food security. Economic use of seeds, crop nutrients, water and pesticides will mean that overall costs should reduce, that the available resources can be spread more widely and that fewer resources need to be used to reverse the adverse effects of agriculture on the environment. Precision agriculture should also play a major role in reducing the yield gap, which is the difference between actual productivity and potential yield, the best that is possible on a given area of land. To obtain the best yields possible depends on how farmers might use 'seeds, water, nutrients, pest management, soils, biodiversity and knowledge' (Godfray *et al.*, 2010). Lobell *et al.* (2009) showed that based on maize, rice and wheat grown in several countries the gap between actual and potential yield varied between 20 per cent and 80 per cent. Precision irrigation could play a major role in closing the yield gap in semi-arid environments and also in rainfed agriculture, where crops may be periodically under stress in some parts of fields as a result of water deficiency.

The future for PA looks positive in a world where land will be under pressure and where efficient and sustainable agriculture will be valued highly. This book should provide undergraduates in agriculture and the environmental sciences with a springboard to look further into the benefits of PA. It should also provide land managers, farmers, administrators and politicians with some background to what PA involves and what can be achieved.

References

Godfray, H. C. J., Beddington, J. R., Crute, I. R., Haddad, L., Lawrence, D., Muir, J. F., Pretty, J., Robinson, S., Thomas, S. M. and Toulmin, C. (2010) 'Food security: the challenge of feeding 9 billion people', *Science*, 327, 812–18.

Lobell, D. B., Cassman, K. G. and Field, C. B. (2009) 'Crop yield gaps: their importance, magnitudes, and causes', *Annual Review of Environment and Resources*, 34, 179–204.

Mazloumzadeh, S. M., Shamsi, M. and Nezamabadi-pour, H. (2010) 'Fuzzy logic to classify date palm trees based on some physical properties related to precision agriculture', *Precision Agriculture*, 11, 258–73.

Index of authors

Subject index

Printed in Great Britain
by Amazon